Communications in Computer and Information Science 542

Commenced Publication in 2007
Founding and Former Series Editors:
Alfredo Cuzzocrea, Dominik Ślęzak, and Xiaokang Yang

More information about this series at http://www.springer.com/series/7899

Mikhail Yu. Khachay · Natalia Konstantinova
Alexander Panchenko · Dmitry I. Ignatov
Valeri G. Labunets (Eds.)

Analysis of Images, Social Networks and Texts

4th International Conference, AIST 2015
Yekaterinburg, Russia, April 9–11, 2015
Revised Selected Papers

 Springer

Editors

Mikhail Yu. Khachay
Krasovsky Institute of Mathematics
  and Mechanics
Yekaterinburg
Russia

Natalia Konstantinova
University of Wolverhampton
Wolverhampton
UK

Alexander Panchenko
Technische Universität Darmstadt
Darmstadt
Germany

Dmitry I. Ignatov
National Research University Higher School
  of Economics
Moscow
Russia

Valeri G. Labunets
Ural Federal University
Yekaterinbug
Russia

ISSN 1865-0929   ISSN 1865-0937 (electronic)
Communications in Computer and Information Science
ISBN 978-3-319-26122-5   ISBN 978-3-319-26123-2 (eBook)
DOI 10.1007/978-3-319-26123-2

Library of Congress Control Number: 2015953242

Springer International Publishing AG Switzerland is part of Springer Science+Business Media
(www.springer.com)

Preface

This volume contains the proceedings of the 4th Conference on Analysis of Images, Social Networks, and Texts (AIST 2015)[1]. The first three conferences during 2012–2014 attracted a significant number of students, researchers, academics, and engineers working on interdisciplinary data analysis of images, texts, and social networks.

The broad scope of AIST makes it an event where researchers from different domains, such as image and text processing, exploiting various data analysis techniques, can meet and exchange ideas. We strongly believe that this may lead to cross-fertilisation of ideas between researchers relying on modern data analysis machinery. Therefore, AIST brings together all kinds of applications of data mining and machine learning techniques. The conference allows specialists from different fields to meet each other, present their work, and discuss both theoretical and practical aspects of their data analysis problems. Another important aim of the conference is to stimulate scientists and people from industry to benefit from the knowledge exchange and identify possible grounds for fruitful collaboration.

The conference was held during April 9–11, 2015. Following an already established tradition, the conference was organised in Yekaterinburg, a cross roads between European and Asian parts of Russia, the capital of the Urals region. The key topics of AIST are analysis of images and videos; natural language processing and computational linguistics; social network analysis; pattern recognition, machine learning, and data mining; recommender systems and collaborative technologies; Semantic Web, ontologies, and their applications.

The Program Committee and the reviewers of the conference included well-known experts in data mining and machine learning, natural language processing, image processing, social network analysis, and related areas from leading institutions of 22 countries including Australia, Bangladesh, Belgium, Brazil, Cyprus, Egypt, Finland, France, Germany, Greece, India, Ireland, Italy, Luxembourg, Poland, Qatar, Russia, Spain, The Netherlands, UK, USA, and Ukraine.

This year the number of submissions doubled and we received 140 submissions mostly from Russia but also from Algeria, Bangladesh, Belgium, India, Kazakhstan, Mexico, Norway, Tunisia, Ukraine, and USA. Out of 140, only 32 papers were accepted as regular oral papers (24 long and eight short). Thus, the acceptance rate was around 23 %. In order to encourage young practitioners and researchers we included five industry papers in the main volume and 26 papers in the supplementary proceedings. Each submission was reviewed by at least three reviewers, experts in their fields, who supplied detailed and helpful comments.

The conference also featured several invited talks and tutorials, as well as an industry session dedicated to current trends and challenges.

---

[1] http://aistconf.org/

The conference featured the following invited talks:

- Pavel Braslavski (Ural Federal University, Yekaterinburg, Russia), Questions Online: What, Where, and Why Should We Care?
- Mikhail Yu. Khachay (Krasovsky Institute of Mathematics and Mechanics UB RAS & Ural Federal University, Yekaterinburg, Russia), Machine Learning in Combinatorial Optimisation: Boosting of Polynomial Time Approximation Algorithms
- Valeri G. Labunets (Ural Federal University, Yekaterinburg, Russia), Is the Human Brain a Quantum Computer?
- Sergey Nikolenko (National Research University Higher School of Economics and Steklov Mathematical Institute, St. Petersburg, Russia), Probabilistic Rating Systems
- Andrey Savchenko (National Research University Higher School of Economics, Nizhny Novgorod, Russia), Sequential Hierarchical Image Recognition Based on the Pyramid Histograms of Oriented Gradients with Small Samples
- Alexander Semenov (International Laboratory for Applied Network Research at HSE, Moscow, Russia), Attributive and Network Features of the Users of Suicide and Depression Groups of Vk.com

The were also two tutorials:

- Alexander Panchenko (Technische Universität Darmstadt, Germany), Computational Lexical Semantics: Methods and Applications
- Artem Lukanin (South Ural State University, Chelyabinsk, Russia), Text Processing with Finite State Transducers in Unitex

The industry speakers also covered a wide variety of topics:

- Dmitry Bugaichenko (OK.ru), Does Size Matter? Smart Data at OK.ru
- Mikhail Dubov (National Research University Higher School of Economics, Moscow, Russia), Text Analysis with Enhanced Annotated Suffix Trees: Algorithmic Base and Industrial Usage
- Nikita Kazeev (Yandex Data Factory), Role of Machine Learning in High-Energy Physics Research at LHC
- Artem Kuznetsov (SKB Kontur), Family Businesses: Relation Extraction Between Companies by Means of Wikipedia
- Alexey Natekin (Data Mining Labs), ATM Maintenance Cost Optimisation with Machine Learning Techniques
- Konstantin Obukhov (Clever Data), Customer Experience Technologies: Problems of Feedback Modeling and Client Churn Control
- Alexandra Shilova (Centre IT), Centre of Information Technologies: Data Analysis and Processing for Large-Scale Information Systems

We would also like to mention the best conference paper selected by the Program Committee. It was written by Oleg Ivanov and Sergey Bartunov and is entitled "Learning Representations in Directed Networks."

We would like to thank the authors for submitting their papers and the members of the Program Committee for their efforts in providing exhaustive reviews. We would also like to express special gratitude to all the invited speakers and industry representatives.

We deeply thank all the partners and sponsors, and owe our gratitude to the Ural Federal University for substantial financial support of the whole conference, namely, the Center of Excellence in Quantum and Video Information Technologies. We would like to acknowledge the Scientific Fund of Higher School of Economics for providing AIST participants with travel grants. Our special thanks goes to Springer's editorial team, who helped us, starting from the first conference call to the final version of the proceedings. Last but not least, we are grateful to all organisers, especially Eugeniya Vlasova and Dmitry Ustalov, and the volunteers, whose endless energy saved us at the most critical stages of the conference preparation.

We would like to mention the Russian word "aist" is more than just a simple abbreviation (in Cyrillic), it means a "stork." Since it is a wonderful free bird, a symbol of happiness and peace, this stork inspired us to organise the AIST conference. So we believe that this young and rapidly growing conference will be bringing inspiration to data scientists around the world!

April 2015

Mikhail Yu. Khachay
Natalia Konstantinova
Alexander Panchenko
Dmitry I. Ignatov
Valeri G. Labunets

Organisation

The conference was organised by a joint team from Ural Federal University (Yekaterinburg, Russia), Krasovsky Institute of Mathematics and Mechanics, Ural Branch of Russian Academy of Sciences (Yekaterinburg, Russia), and the National Research University Higher School of Economics (Moscow, Russia). It was supported by a special grant from Ural Federal University for the Center of Excellence in Quantum and Video Information Technologies: from Computer Vision to Video Analystics (QVIT: CV → VA).

Program Committee Chairs

Mikhail Yu. Khachay — Krasovsky Institute of Mathematics and Mechanics of UB RAS, Russia
Natalia Konstantinova — University of Wolverhampton, UK
Alexander Panchenko — Technische Universität Darmstadt, Germany & Université catholique de Louvain, Belgium

General Chair

Valeri G. Labunets — Ural Federal University, Russia

Organising Chair

Eugeniya Vlasova — National Research University Higher School of Economics, Russia

Proceedings Chair

Dmitry I. Ignatov — National Research University Higher School of Economics, Russia

Poster Chairs

Nikita Spirin — University of Illinois at Urbana-Champaign, USA
Dmitry Ustalov — Krasovsky Institute of Mathematics and Mechanics and Ural Federal University, Russia

International Liaison Chair

Radhakrishnan Delhibabu — Kazan Federal University, Russia

Organising Committee and Volunteers

Alexandra Barysheva	National Research University Higher School of Economics, Russia
Liliya Galimzyanova	Ural Federal University, Russia
Anna Golubtsova	National Research University Higher School of Economics, Russia
Vyacheslav Novikov	National Research University Higher School of Economics, Russia
Natalia Papulovskaya	Ural Federal University, Yekaterinburg, Russia
Yuri Pekov	Moscow State University, Russia
Evgeniy Tsymbalov	National Research University Higher School of Economics, Russia
Andrey Savchenko	National Research University Higher School of Economics, Russia
Dmitry Ustalov	Krasovsky Institute of Mathematics and Mechanics and Ural Federal University, Russia
Rostislav Yavorsky	National Research University Higher School of Economics, Russia

Industry Session Organisers

Ekaterina Chernyak	National Research University Higher School of Economics, Russia
Alexander Semenov	National Research University Higher School of Economics, Russia

Program Committee

Mikhail Ageev	Lomonosov Moscow State University, Russia
Atiqur Rahman Ahad	University of Dhaka, Bangladesh
Igor Andreev	Mail.Ru, Russia
Nikolay Arefiev	Moscow State University and Digital Society Lab, Russia
Jaume Baixeries	Polytechnic University of Catalonia, Spain
Pedro Paulo Balage	Universidade de São Paulo, Brazil
Sergey Bartunov	Lomonosov Moscow State University and National Research University Higher School of Economics, Russia
Malay Bhattacharyya	Indian Institute of Engineering Science and Technology, India
Vladimir Bobrikov	Imhonet.ru, Russia
Victor Bocharov	OpenCorpora and Yandex, Russia
Daria Bogdanova	Dublin City University, Ireland
Elena Bolshakova	Lomonosov Moscow State University, Russia

Aurélien Bossard   Orange Labs, France
Pavel Botov   Moscow Institute of Physics and Technology, Russia
Jean-Leon Bouraoui   Université Catholique de Louvain, Belgium
Leonid Boytsov   Carnegie Mellon University, USA
Pavel Braslavski   Ural Federal University and Kontur Labs, Russia
Andrey Bronevich   National Research University Higher School
                              of Economics, Russia
Aleksey Buzmakov   LORIA (CNRS-Inria-Université de Lorraine), France
Artem Chernodub   Institute of Mathematical Machines and Systems
                              of NASU, Ukraine
Vladimir Chernov   Image Processing Systems Institute of RAS, Russia
Ekaterina Chernyak   National Research University Higher School of
                              Economics, Russia
Marina Chicheva   Image Processing Systems Institute of RAS, Russia
Miranda Chong   University of Wolverhampton, UK
Hernani Costa   University of Malaga, Spain
Florent Domenach   University of Nicosia, Cyprus
Alexey Drutsa   Lomonosov Moscow State University and Yandex,
                              Russia
Maxim Dubinin   NextGIS, Russia
Julia Efremova   Eindhoven University of Technology, The Netherlands
Shervin Emami   NVIDIA, Australia
Maria Eskevich   Dublin City University, Ireland
Victor Fedoseev   Samara State Aerospace University, Russia
Mark Fishel   University of Zurich, Germany
Thomas Francois   Université catholique de Louvain, Belgium
Oleksandr Frei   Schlumberger, Norway
Binyam Gebrekidan Gebre   Max Planck Computing and Data Facility, Germany
Dmitry Granovsky   Yandex, Russia
Mena Habib   University of Twente, The Netherlands
Dmitry Ilvovsky   National Research University Higher School
                              of Economics, Russia
Dmitry Ilvovsky   National Research University Higher School
                              of Economics, Russia
Vladimir Ivanov   Kazan Federal University, Russia
Sujay Jauhar   Carnegie Mellon University, USA
Dmitry Kan   AlphaSense Inc., USA
Nikolay Karpov   National Research University Higher School
                              of Economics, Russia
Yury Katkov   Blue Brain Project, Switzerland
Mehdi Kaytoue   INSA de Lyon, France
Laurent Kevers   DbiT, Luxembourg
Mikhail Yu. Khachay   Krasovsky Institute of Mathematics and Mechanics UB
                              RAS, Russia
Evgeny Kharitonov   Moscow Institute of Physics and Technology, Russia

Vitaly Khudobakhshov	Saint Petersburg University and National Research University of Information Technologies, Mechanics and Optics, Russia
Ilya Kitaev	iBinom, Russia and Voronezh State University, Russia
Ekaterina Kochmar	University of Cambridge, UK
Sergei Koltcov	National Research University Higher School of Economics, Russia
Olessia Koltsova	National Research University Higher School of Economics, Russia
Natalia Konstantinova	University of Wolverhampton, UK
Anton Konushin	Lomonosov Moscow State University, Russia
Andrey Kopylov	Tula State University, Russia
Kirill Kornyakov	Itseez, Russia and University of Nizhny Novgorod, Russia
Maxim Korolev	Ural State University, Russia
Anton Korshunov	Institute of System Programming of RAS, Russia
Yuri Kudryavcev	PM Square, Australia
Valentina Kuskova	National Research University Higher School of Economics, Russia
Sergei O. Kuznetsov	National Research University Higher School of Economics, Russia
Valeri G. Labunets	Ural Federal University, Russia
Alexander Lepskiy	National Research University Higher School of Economics, Russia
Benjamin Lind	National Research University Higher School of Economics, Russia
Natalia Loukachevitch	Research Computing Center of Moscow State University, Russia
Ilya Markov	University of Amsterdam, The Netherlands
Luis Marujo	Carnegie Mellon University, USA and Universidade de Lisboa, Portugal
Sérgio Matos	University of Aveiro, Portugal
Julian Mcauley	The University of California, San Diego, USA
Yelena Mejova	Qatar Computing Research Institute, Qatar
Vlado Menkovski	Eindhoven University of Technology, The Netherlands
Christian M. Meyer	Technische Universität Darmstadt, Germany
Olga Mitrofanova	St. Petersburg State University, Russia
Nenad Mladenovic	Brunel University, UK
Vladimir Mokeyev	South Ural State university, Russia
Gyorgy Mora	Prezi Inc., USA
Andrea Moro	Universitá di Roma, Italy
Sergey Nikolenko	Steklov Mathematical Institute and National Research University Higher School of Economics, Russia
Vasilina Nikoulina	Xerox Research Center Europe, France

Damien Nouvel	National Institute for Oriental Languages and Civilizations, France
Dmitry Novitski	Institute of Cybernetics of NASU
Georgios Paltoglou	University of Wolverhampton, UK
Alexander Panchenko	Université catholique de Louvain, Belgium
Denis Perevalov	Krasovsky Institute of Mathematics and Mechanics, Russia
Georgios Petasis	National Centre of Scientific Research Demokritos, Greece
Andrey Philippovich	Bauman Moscow State Technical University, Russia
Leonidas Pitsoulis	Aristotle University of Thessaloniki, Greece
Lidia Pivovarova	University of Helsinki, Finland
Vladimir Pleshko	RCO, Russia
Jonas Poelmans	Alumni of Katholieke Universiteit Leuven, Belgium
Alexander Porshnev	National Research University Higher School of Economics, Russia
Surya Prasath	University of Missouri-Columbia, USA
Delhibabu Radhakrishnan	Kazan Federal University and Innopolis, Russia
Carlos Ramisch	Aix Marseille University, France
Alexandra Roshchina	Institute of Technology Tallaght Dublin, Ireland
Eugen Ruppert	TU Darmstadt, Germany
Mohammed Abdel-Mgeed M. Salem	Ain Shams University, Egypt
Grigory Sapunov	Stepic, Russia
Sheikh Muhammad Sarwar	University of Dhaka, Bangladesh
Andrey Savchenko	National Research University Higher School of Economics, Russia
Marijn Schraagen	Utrecht University, The Netherlands
Vladimir Selegey	ABBYY, Russia
Alexander Semenov	National Research University Higher School of Economics, Russia
Oleg Seredin	Tula State University, Russia
Vladislav Sergeev	Image Processing Systems Institute of the RAS, Russia
Andrey Shcherbakov	Intel, Russia
Dominik Ślęzak	University of Warsaw, Poland and Infobright Inc
Gleb Solobub	Agent.ru, Russia
Andrei Sosnovskii	Ural Federal University, Russia
Nikita Spirin	University of Illinois at Urbana-Champaign, USA
Sanja Stajner	University of Lisbon, Portugal
Rustam Tagiew	Alumni of TU Freiberg, Germany
Irina Temnikova	Qatar Computing Research Institute, Qatar
Christos Tryfonopoulos	University of Peloponnisos, Greece
Alexander Ulanov	HP Labs, Russia
Dmitry Ustalov	Krasovsky Institute of Mathematics and Mechanics and Ural Federal University, Russia
Natalia Vassilieva	HP Labs, Russia

Yannick Versley	Heidelberg University, Germany
Evgeniya Vlasova	Higher School of Economics
Svitlana Volkova	Johns Hopkins University, USA
Konstantin Vorontsov	Forecsys and Dorodnicyn Computing Center of RAS, Russia
Ekaterina Vylomova	Bauman Moscow State Technical University, Moscow
Patrick Watrin	Université catholique de Louvain, Belgium
Rostislav Yavorsky	National Research University Higher School of Economics, Russia
Roman Zakharov	Université catholique de Louvain, Belgium
Marcos Zampieri	Saarland University, Germany
Sergei M. Zraenko	Ural Federal University, Russia
Olga Zvereva	Ural Federal University, Russia

Invited Reviewers

Sujoy Chatterjee	University of Kalyani, India
Alexander Goncharov	CVisionLab, Russia
Vasiliy Kopenkov	Image Processing Systems Institute of RAS, Russia
Alexis Moinet	University of Mons, Belgium
Ekaterina Ostheimer	Capricat LLC, USA
Sergey V. Porshnev	Ural Federal University, Russia
Paraskevi Raftopoulou	Technical University of Crete, Greece
Ali Tayari	Technical University of Crete, Greece

Sponsors and Partners

Ural Federal University
Krasovsky Institute of Mathematics and Mechanics
GraphiCon
Exactpro
IT Centre
SKB Kontur
JetBrains
Yandex
Ural IT Cluster
NLPub
Digital Society Laboratory
CLAIM

Contents

Pattern Recognition and Machine Learning

Social Network Analysis

Text Mining and Natural Language Processing

Industry Talk

Industry Papers

Invited Papers

A Probabilistic Rating System for Team Competitions with Individual Contributions

Sergey Nikolenko[1,2,3]([✉])

[1] National Research University Higher School of Economics, St. Petersburg, Russia
sergey@logic.pdmi.ras.ru
[2] Steklov Mathematical Institute at St. Petersburg, St. Petersburg, Russia
[3] Kazan (Volga Region) Federal University, Kazan, Russia

Abstract. We study the problem of constructing a probabilistic rating system for team competitions. Unlike previous studies, we consider a setting where the competition can be broken down into relatively small individual tasks, and it is reasonable to assume that each task is done by a single team member. We begin with a simplistic naïve Bayes approach which is this case reduces to logistic regression and then develop it into a more complex model with latent variables trained by expectation–maximization. We show experimental results that validate our approach.

Keywords: Probabilistic rating systems · EM algorithm · Presence-only data

1 Introduction

The motivation for probabilistic rating systems is best understood in the context of various competitions. In many sports and other competitions, it is important to be able to compare players and/or teams that have never actually played in the same tournament or have done so only a few times. For instance, probably the first Bayesian rating system was the Elo rating for chess developed in the 1950s [1]; it is designed to provide a unified estimate of a chessplayer's skill. This is important not only to hail the best players but also for matchmaking: the most interesting match for a player is another player with a close skill. Many modern computer games that allow users to play online matches also face a similar problem: the system has to find a good match for a user willing to play.

However, actual ratings and matchmaking are not the only uses for probabilistic rating systems. Mathematically speaking, a rating system is simply a way to compare several elements with respect to a certain characteristic with access only to small-scale noisy comparisons, without a way to compare all elements precisely and at once. Thus, for instance, Bradley–Terry models can be used for multiclass classification based on binary classification results between two classes [2,3]. Similar problems, where one has to recover a single ordering (ranking) by noisy partial comparisons, arise in image processing [4], comparing patterns for UCT algorithms [5], and other models. One important application

M.Y. Khachay et al. (Eds.): AIST 2015, CCIS 542, pp. 3–13, 2015.
DOI: 10.1007/978-3-319-26123-2_1

was CTR prediction for context ads on the Web [6]; this application was based on the TrueSkill Bayesian rating system [7].

Since the Elo rating, Bayesian rating systems have been generalized to handle more complex situations. In chess, there are only two players, but other competitions may have a more complex structure. In particular, the problem becomes much more complicated when it deals with team competitions: we now have to infer individual player skills from the results of noisy comparisons of team performances.

One important class of rating models are Bradley–Terry models [8–10]. They introduce a set of parameters that reflect the skills of individual players with a simple assumption on how the probability of winning depends on the parameters; for example, the simplest Bradley–Terry model assigns ratings γ_i to players and assumes that $p(i \text{ beats } j) = \gamma_i/(\gamma_i + \gamma_j)$. The ratings are trained with a minorization–maximization algorithm [10]. Bradley–Terry models perform well for individual players but are hard to generalize directly to team competitions and even to multiple players in a single tournament [11,12]; recent attempts to do so involve a parametrization modeled in the form of a neural network [13].

An important step in probabilistic ratings for team competitions was the TrueSkill model developed in [7] for the purposes of online gaming on Xbox Live servers and later used for web advertising [6]. TrueSkill can be viewed as a direct generalization of the Elo rating:

- every player's "true" skill s has a Gaussian prior distribution $\mathcal{N}(s; \mu, \sigma)$, where μ and σ are the player's rating parameters;
- on every tournament, a player is assumed to have a certain performance p distributed normally around the skill as $\mathcal{N}(p; s, \beta^2)$ with the same model parameter β for every player;
- team performance is assumed to be the sum of individual performances, $t = \sum_i p_i$;
- on the bottom level of the model, teams are ordered according to their relative standings, and special factors are introduced that represent the evidence: if team 1 scored higher than team 2, it means that its performance t_1 was significantly higher than that of t_2, $t_1 > t_2 + \epsilon$, and so on.

Since the factor graph of the TrueSkill model does not contain cycles, the training reduces to message passing from top to bottom and back. However, the factors at the bottom introduce complicated factor that do not propagate back easily; to approximate these factors, one has to use the Expectation Propagation algorithm [14]. The TrueSkill model has a flaw in that it does not handle ties between several teams correctly; an improved model with an additional layer of "place performances" was introduced in [15] with a training algorithm also based on Expectation Propagation. Later work experimented with additional information on tournament results, taking into account for the score differences between different places [16]. In this work, we take this line of research a step further, using instead of resulting scores the actual tasks successfully performed by each team; this leads to a completely different approach which is not based on TrueSkill.

In this work, we develop a novel probabilistic rating system for a specific sort of team competitions. Namely, we consider competitions where each tournament consists of individual tasks, and results of the teams on each individual task are known. Thus, each tournament is broken down into a large number of small "tournaments" with binary results. Our sample dataset comes from the results of a game where teams answer questions, and it is known which questions each team has answered. Other motivation may include, for instance, CTR prediction for context ads in situations when each user action is logged, and CTRs can be broken down into individual user decisions. For this problem, we propose two models, a simple model based on logistic regression and a model with latent variables; our evaluation results indicate that the latter significantly outperforms the former.

The paper is organized as follows. In Sect. 2, we present the problem setting and introduce the specific dataset we have tested our approach on. Section 3.1 shows the first natural idea for tackling this problem, a baseline model based on classical logistic regression. In Sect. 3, we present the model with latent variables that account for unknown individual contributions. Section 4 shows experimental results, and Sect. 5 concludes the paper.

2   Problem Setting

Consider a set of tournaments $\mathcal{D} = \{d_1, \ldots, d_{|\mathcal{D}|}\}$. Each tournament contains a certain number n_d of individual tasks $Q^{(d)}$, $d \in \mathcal{D}$, and each task has binary results: it can be either performed correctly or not. Note that this means that there are only $n_d + 1$ different possible places in a tournament, from performing all tasks correctly to none at all.

Teams $\mathcal{T}^{(d)} = \{t_1^{(d)}, \ldots, t_{M^{(d)}}^{(d)}\}$ participate in tournament $d \in \mathcal{D}$, and we denote their number by $M^{(d)} = |\mathcal{T}^{(d)}|$. Each team $t \in \mathcal{T}^{(d)}$ consists of players $P_t = \{p_1, \ldots, p_{m_t}\}$, a total of m_t players in the team. Note that players can change teams freely between the tournaments, a team is simply a collection of players for a specific tournament.

For each task $q \in Q^{(d)}$ we know for every participating team $t \in \mathcal{T}^{(d)}$ whether this task has been fulfilled; we denote these bits by x_{tq}, and they comprise the input dataset.

3   Modeling with Unknown Individual Contributions

3.1   Baseline: Logistic Regression

In this problem, we need to model the performance of a team as a function of individual player performances on individual tasks. We begin with the basic naive Bayes assumption: we assume that players perform tasks individually and the probability of a player to fulfill a task is independent of the other players on the team. Naturally, in real life this assumption does not hold, as interactions between players are often important. However our situation is conceptually similar to text mining with regard to naive Bayes: in a rating system, we do not need to say how many individual tasks will a team fulfill but only need to be able to predict final standings well, just like in text categorization we do not have to model the probabilities of categories but simply choose which is the best one or rank a number of alternatives; see [17,18] for an explanation of why naive Bayes works well in these situations.

Our baseline model is a simple logistic regression: we model each player i with his or her skill s_i, each task q with its complexity score c_q, add the global average μ, and train the logistic model

$$p(x_{tq} \mid s_i, c_q) \sim \sigma(\mu + s_i + c_q)$$

for each player $i \in t$ of a participating team $t \in \mathcal{T}^{(d)}$ and each task $q \in Q^{(d)}$, where $\sigma(x) = 1/(1+e^x)$ is the logistic sigmoid. Logistic regression can be trained with, for instance, the IRLS algorithm (iterative reweighted least squares); see, e.g., [14] for details.

3.2   Model with Latent Variables and the EM Algorithm

The logistic model basically assumes that each player successfully performed every task that the team has performed. While this is not a completely unreasonable simplifying assumption, in fact we do not know which player or players have done it. However, we do know for sure that if the team has not fulfilled a task successfully then no one from this team has. This situation is similar in spirit to the presence-only data models found in ecology, when, for instance, a species can be observed in a certain region but absence of such observations does not prove conclusively that the species does not live there [19–21]; other ideas for similar problems include the SVM approach [22].

This leads us to consider a model with latent variables. For each player-task pair, we add an unobserved variable z_{iq} with the semantics "player i has successfully performed task q". For these variables, we have the following constraints:

- if $x_{tq} = 0$ then $z_{iq} = 0$ for every player $i \in t$;
- if $x_{tq} = 1$ then $z_{iq} = 1$ for at least one player $i \in t$.

We still model each player with his or her skill and each task with its complexity, modeling z_{iq} as

$$p(z_{iq} \mid s_i, c_q) \sim \sigma(\mu + s_i + c_q).$$

To train this model, we use the classical expectation–maximization (EM) approach:

- on the E step, fix s_i and c_q for every player i and every task q and compute expected values of latent variables z_{iq} as follows: denoting by t the team of player i in tournament with task q,

$$\mathbb{E}\left[z_{iq}\right] = \begin{cases} 0 & \text{if } x_{tq} = 0, \\ p(z_{iq} = 1 \mid \exists j \in t \; z_{jq} = 1) = \frac{\sigma(\mu+s_i+c_q)}{1-\prod_{j \in t}(1-\sigma(\mu+s_j+c_q))}, & \text{if } x_{tq} = 1; \end{cases}$$

– on the M step, fix $\mathbb{E}\left[z_{iq}\right]$ and train the logistic model

$$\mathbb{E}\left[z_{iq}\right] \sim \sigma(\mu + s_i + c_q)$$

(note that this is not exactly classical logistic regression since responses are not 0–1 variables, but still a common generalized linear model with well known inference algorithms).

4   Experimental Evaluation

4.1   Dataset

As the dataset, we have used the results collected from a Russian intellectual game "What? Where? When?" which provides a good example of this rating problem. In the game, teams of players answer a certain set of questions, and whichever team answers most questions correctly wins. An early version of this dataset contained only tournament standings (team positions); it was used in [15] that introduced a modification of the TrueSkill model. In this work, we use a recent version that does contain data on specific questions a team has answered correctly. The dataset contains 2538 tournaments from early 2000s to the end of 2014, among them 794 tournaments with known results of individual questions, each tournament ranging from 30–40 to 200 questions, 44681 questions in total. We computed the ratings monthly from January 2012 (118 tournaments with question-wise data, 6443 questions) to January 2015 (794 tournaments, 44681 questions). The dataset contains 98207 players in total but many of them played in very few tournaments; we cut this long tail at the 200 question mark, using 30884 players with ≥ 200 played questions in the model and replacing all others with a single dummy "novice" player.

4.2   Evaluation Metrics

We have been modeling the events of fulfilling a specific task. However, as we have already noted, in the problem we do not have to predict these events specifically but rather aim to predict the final standings of the teams participating in a given tournament. Hence, for each team in a tournament we have predicted its final standings based on the probability of fulfilling an average task

$$p(\exists i \in t \; z_{iq} = 1) = 1 - \prod_{i \in t}(1 - \sigma(\mu + s_i))$$

and then rank the teams according to this probability; this is equivalent to ranking according to the expectation of the number of fulfilled tasks with $c_q = 0$. To evaluate our models, we have applied standard metrics for ranking quality:

– AUC – Area Under (ROC) Curve [23,24], a metric that measures ranking quality across the entire table of results; AUC measures the probability that the relative standings of a randomly selected pair of teams have been predicted correctly;

- WTA (winner takes all) – this metric equals 1 if the top rated team has won the tournament, and 0 otherwise;
- Top3 – share of correctly predicted teams among the top 3 scoring teams in the tournament;
- Top5 – share of correctly predicted teams among the top 5 scoring teams in the tournament;
- MAP (mean average precision) – share of correctly predicted teams among the top 10 scoring teams in the tournament.

We break ties in favor of the models: if, for instance, a model predicted the team to be in third place, and it tied for places 3–5, we counted it as a success in the Top3 metric.

4.3   Results

Results of our experiments are shown on Figs. 1, 2 and 3. We have evaluated both models, logistic regression (denoted LR) and model with latent variables and EM training (denoted EM), on the training dataset starting from January 2012 to September 2014 (so that we have at least three months more for testing). There is one open question in the model: it is still unclear how to incorporate time in such a way that later tournaments are more important for the rating than earlier ones; without this property, a player would be forever held back by the early tournament when he or she was still learning to play. To model this, at present we simply used a hard cutoff, evaluating five models: LR trained on all data, last 2 years, and last 1 year of tournaments, and EM trained on all data and last 2 years of tournaments.

Figures 1 and 2 show the main results: quality metrics evaluated on the test dataset consisting of 1 month of tournaments starting from the training date. It is clear that EM (black solid and dashed lines) outperforms LR (grey lines and black dotted line), and its advantage is more prominent at the bottom of the tournament results: LR predicts winners almost as well as EM but loses very significantly on AUC. This is an important property since it is relatively easy to predict who will be on top (there are not so many top players) so it is important for a rating system to try and stay meaningful as deep as possible down the player/team standings.

Figure 3 shows how prediction quality deteriorates as we look further ahead into the future. It shows the best model (EM trained on the entire dataset) quality on four different test sets: one, two, three, and six months ahead. Results indicate that while it becomes harder to predict top teams in significant time, AUC does not suffer all that much.

5   Conclusion

In this work, we have considered the problem of predicting the rankings in a team competition based on a sequence of tasks fulfilled by a single player from

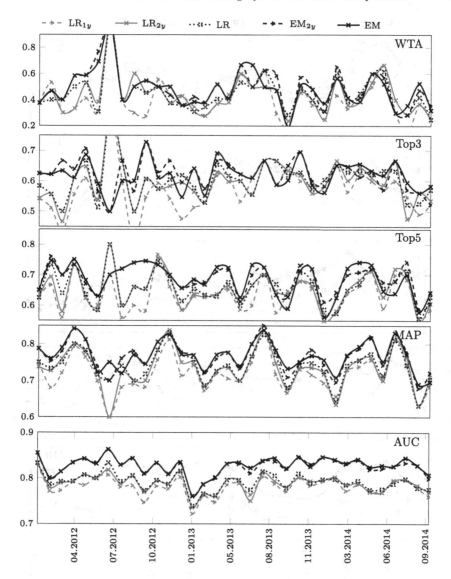

Fig. 1. Evaluation quality for three versions of LR and 2 versions of EM, 1 month ahead. Top to bottom: WTA, Top3, Top5, MAP, AUC

the team. We have developed a new probabilistic rating model that operates with presence-only data and showed with extensive evaluation that it performs better than the baseline logistic regression model.

Further work on this problem can follow several different paths. First, both proposed models can be extended by assigning several features to each player and each task, performing an SVD-like decomposition instead of logistic regression. This will probably improve predictions if task properties are known, although

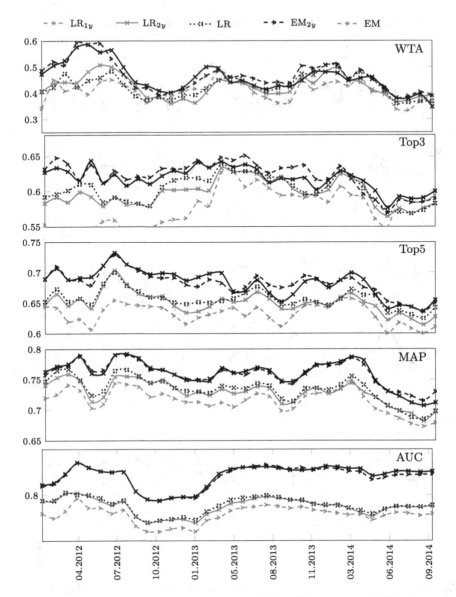

Fig. 2. Evaluation quality for three versions of LR and 2 versions of EM, 6 months ahead. Top to bottom: WTA, Top3, Top5, MAP, AUC

they are not known in practice, and it will become less clear how to rank the players when they are represented by vectors.

Second, an important problem is how to incorporate time into this model. Recent tournaments should have a larger weight than old ones, and a player should not be held back by the fact that he or she played badly years ago but improved since then. We have tried to assign weights to the logistic regression

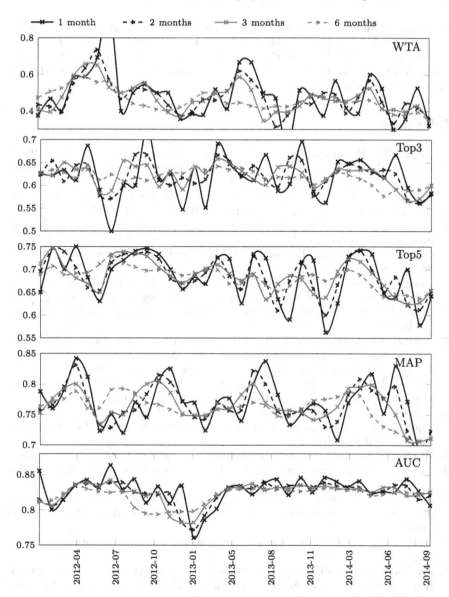

Fig. 3. Evaluation quality for different look-ahead times, EM trained on the entire dataset. Top to bottom: WTA, Top3, Top5, MAP, AUC

error function, allowing for a larger margin of error on older tournaments, but predictions only deteriorated. In practice, we have to resort to artificial means: we train the model on the last two years of tournaments, disregarding the rest; a conceptually sound solution to this problem is yet to be found.

Acknowledgements. This research has been partially supported by the Russian Foundation for Basic Research grant no. 15-29-01173, Government of the Russian Federation grant 14.Z50.31.0030, and the Presidential Grant for Leading Scientific Schools, NSh-3856.2014.1. I also thank Alexey Tugarev for providing access to the database of "What? Where? When?" tournament results.

References

1. Elo, A.: The Ratings of Chess Players: Past and Present. Arco, New York (1978)
2. Wu, T.F., Lin, C.J., Weng, R.C.: Probability estimates for multi-class classification by pairwise coupling. J. Mach. Learn. Res. **5**, 975–1005 (2004)
3. Huang, T.K., Weng, R.C., Lin, C.J.: Generalized bradley-terry models and multi-class probability estimates. J. Mach. Learn. Res. **7**, 85–115 (2006)
4. Stein, A., Aryal, J., Gort, G.: Generalized bradley-terry models and multi-class probability estimates. IEEE Trans. Geosci. Remote Sens. **43**, 852–856 (2005)
5. Coulom, R.: Computing Elo ratings of move patterns in the game of Go. ICGA J. **30**(4), 198–208 (2007)
6. Graepel, T., Candela, J.Q., Borchert, T., Herbrich, R.: Web-scale bayesian click-through rate prediction for sponsored search advertising in microsoft's bing search engine. In: Proceedings of the 27th International Conference on Machine Learning, pp. 13–20 (2010)
7. Graepel, T., Minka, T., Herbrich, R.: TrueSkill(tm): a bayesian skill rating system. In: Schölkopf, B., Platt, J., Hoffman, T. (eds.) Advances in Neural Information Processing Systems 19, pp. 569–576. MIT Press, Cambridge (2007)
8. Bradley, R.A., Terry, M.E.: Rank analysis of incomplete block designs. I. The method of paired comparisons. Biometrika **39**, 324–245 (1952)
9. Agresti, A.: Categorical Data Analysis. Wiley, New York (1990)
10. Hunter, D.R.: MM algorithms for generalized bradley-terry models. Ann. Stat. **32**(1), 384–406 (2004)
11. Plackett, R.L.: The analysis of permutations. J. Appl. Stat. **24**, 193–202 (1975)
12. Marden, J.I.: Analyzing and Modeling Rank Data. Chapman and Hall, London (1995)
13. Menke, J.E., Martinez, T.R.: A bradley-terry artificial neural network model for individual ratings in group competitions. Neural Comput. Appl. **17**(2), 175–186 (2008)
14. Bishop, C.M.: Pattern Recognition and Machine Learning. Springer, New York (2006)
15. Nikolenko, S.I., Sirotkin, A.V.: A new bayesian rating system for team competitions. In: Proceedings of the 28th International Conference on Machine Learning, pp. 601–608 (2011)
16. Nikolenko, S.I., Serdyuk, D.V., Sirotkin, A.V.: Bayesian rating systems with additional information on tournament results. SPIIRAS Proc. **22**, 189–204 (2012)
17. Zhang, H.: The optimality of naive bayes. In: Barr, V., Markov, Z. (eds.) Proceedings of the Seventeenth International Florida Artificial Intelligence Research Society Conference (FLAIRS 2004). AAAI Press (2004)
18. Zhang, H., Su, J.: Naive bayesian classifiers for ranking. In: Boulicaut, J.-F., Esposito, F., Giannotti, F., Pedreschi, D. (eds.) ECML 2004. LNCS (LNAI), vol. 3201, pp. 501–512. Springer, Heidelberg (2004)
19. Ward, G., Hastie, T., Barry, S., Elith, J., Leathwick, J.R.: Presence-only data and the EM algorithm. Biometrics **65**(2), 554–563 (2009)

20. Royle, J.A., Chandler, R.B., Yackulic, C., Nichols, J.D.: Likelihood analysis of species occurrence probability from presence-only data for modelling species distributions. Methods Ecol. Evol. **3**(3), 545–554 (2012)
21. Divino, F., Golini, N., Lasinio, G.J., Penttinen, A.: Bayesian modeling and MCMC computation in linear logistic regression for presence-only data (2013). arXiv:1305.1232 [stat.CO]
22. Elkan, C., Noto, K.: Learning classifiers from only positive and unlabeled data. In: Proceedings of the 14th ACM SIGKDD International Conference on Knowledge Discovery and Data Mining, KDD 2008, pp. 213–220. ACM, New York (2008)
23. Fawcett, T.: An introduction to ROC analysis. Pattern Recogn. Lett. **27**(8), 861–874 (2006)
24. Ling, C.X., Huang, J., Zhang, H.: AUC: a statistically consistent and more discriminating measure than accuracy. Proc. Int. Joint Conf. Artif. Intel. **2003**, 519–526 (2003)

Sequential Hierarchical Image Recognition Based on the Pyramid Histograms of Oriented Gradients with Small Samples

Andrey V. Savchenko[1(✉)], Vladimir R. Milov[2],
and Natalya S. Belova[3]

[1] Laboratory of Algorithms and Technologies for Network Analysis,
National Research University Higher School of Economics,
Nizhny Novgorod, Russia
avsavchenko@hse.ru
[2] Nizhny Novgorod State Technical University n.a. R.E. Alekseev,
Nizhny Novgorod, Russia
vladimir.milov@gmail.com
[3] National Research University Higher School of Economics, Moscow, Russia
nbelova@hse.ru

Abstract. In this paper we explore an application of the pyramid HOG (Histograms of Oriented Gradients) features in image recognition problem with small samples. A sequential analysis is used to improve the performance of hierarchical methods. We propose to process the next, more detailed level of pyramid only if the decision at the current level is unreliable. The Chow's reject option of comparison of the posterior probability with a fixed threshold is used to verify recognition reliability. The posterior probability is estimated for the homogeneity-testing probabilistic neural network classifier on the basis of its relation with the Bayesian decision. Experimental results in face recognition are presented. It is shown that the proposed approach allows to increase the recognition performance in 2–4 times in comparison with conventional classification of pyramid HOGs.

Keywords: Image recognition · Hierarchical recognition · Sequential analysis · Chow's reject option · Probabilistic neural network · HOG (Histograms of oriented Gradients) · PHOG (Pyramid HOG)

1 Introduction

Nowadays the recognition of complex images (faces, gestures, medical objects) [1] becomes all the more acute. Several sufficiently reliable descriptors has been recently proposed, e.g., SIFT (Scale-Invariant Feature Transform) [2], HOG (Histograms of Oriented Gradients) [3], SURF (Speeded-Up Robust Features) [4], etc. However, though image recognition technology has reached a certain level of maturity [5], many researchers try to improve the recognition quality by exploiting the hierarchical processing of analyzed objects. Such processing is known to be one of the known characteristic of human intellect [6, 7]. As a result, several hierarchical image recognition

© Springer International Publishing Switzerland 2015
M.Y. Khachay et al. (Eds.): AIST 2015, CCIS 542, pp. 14–23, 2015.
DOI: 10.1007/978-3-319-26123-2_2

methods have been presented, e.g., pyramids of features (in particular, PHOG (Pyramid HOG) descriptor [8]), hierarchical temporal memory [6], wavelet-analysis [9], deep convolution neural networks [10], etc.

The majority of these methods include the comparison of features of query image and models from the given database at *each* level of hierarchy (pyramid). Hence, their recognition performance is much worse in comparison with conventional non-hierarchical approach. For instance, the matching of the PHOG descriptors [7] is usually 1.5–3 times slower than the matching of the state-of-the-art HOGs [3]. Thus, most popular hierarchical methods usually cannot be implemented in real-time applications [11], e.g., in video-based face recognition [12]. In this paper we propose to perform sequential analysis of query image to overcome this drawback. Namely, more detailed representation of the query image is analyzed at the next level of pyramid *only if* it is impossible to obtain a reliable solution at the current level [13, 14]. To reject unreliable solution, we use the Chow's rule and compare the maximal posterior probability with the fixed threshold [13]. This probability is estimated on the basis of the asymptotic properties of the homogeneity-testing probabilistic neural network (HT-PNN) which was proved to be a good classifier in several pattern recognition tasks [15].

The rest of the paper is organized as follows: Sect. 2 briefly presents the classification of conventional HOG and PHOG features with the HT-PNN. In Sect. 3, the sequential hierarchical approach is introduced. In Sect. 4, we present the experimental results of the proposed method in the face recognition task [12]. Finally, concluding comments are given in Sect. 5.

2   Classification of the PHOG with the HT-PNN

Let a training set of $R > 1$ model images $\{X_r\}, r \in \{1, \ldots, R\}$ with height U_r and width V_r be specified. It is assumed that the class $c(r) \in \{1, \ldots, C\}$ of the rth model image is known. Here C is the total number of distinct classes. The task is to assign a query image X with height U and width V to one of C classes [16]. We assume the practically important case of small training sample $C \approx R$ [17]. Let's the objects of interest (say, faces in face recognition) are preliminary detected in either query or model images, hence, each of them contains only one object.

Let every image to be associated with a set of PHOG features [8]. At first, the whole image is divided into a regular squared grid of $K^{(1)} \times K^{(1)}$ blocks where $K^{(1)}$ is the number of rows and columns in the lowest level of pyramid with the most rough image approximation. The histogram $H_r^{(1)}(k_1, k_2) = [h_{r;1}^{(1)}(k_1, k_2), \ldots, h_{r;N}^{(1)}(k_1, k_2)]$ of gradient orientation is evaluated for each block $(k_1, k_2), k_1, k_2 \in \{1, \ldots, K^{(1)}\}$ of the rth model image. Here N is the number of bins in the histogram. Next, according to the PHOG method [8], the image is divided into a squared grid $K^{(2)} \times K^{(2)}$ blocks where $K^{(2)} > K^{(1)}$ and its histogram $H_r^{(2)}(k_1, k_2), k_1, k_2 \in \{1, \ldots, K^{(2)}\}$ is evaluated. This procedure is repeated for $L = const$ pyramid levels. Similarly, the histograms $H^{(l)}(k_1, k_2) = [h_1^{(l)}(k_1, k_2), \ldots, h_1^{(l)}(k_1, k_2)], l = \overline{1, L}, k_1, k_2 \in \{1, \ldots, K^{(l)}\}$ for the query image are estimated.

The second part is classifier design [16]. If $C \approx R$, the nearest neighbor rule is applied as it usually outperforms other more complex state-of-the-art machine learning techniques (MLP, SVM, etc.) in this particular case [17]. Though several conventional classifiers (e.g., CNN) can be applied in this task to select the features [18], the final decision is usually done with simple nearest neighbor rule. In view of the small spatial deviations due to misalignment after object detection, the following similarity measure with mutual alignment of blocks and comparison of the histograms in Δ-neighborhood of each block is used [19]

$$\rho^{(l)}(X, X_r) = \sum_{k_1=1}^{K^{(l)}} \sum_{k_2=1}^{K^{(l)}} \min_{\substack{|\Delta_1| \leq \Delta, \\ |\Delta_2| \leq \Delta}} \rho_H\Big(H_r^{(l)}(k_1 + \Delta_1, k_2 + \Delta_2), H^{(l)}(k_1, k_2)\Big). \quad (1)$$

Here ρ_H is any distance between HOGs. In this paper, we explore the square of Euclidean distance and our HT-PNN [15]:

$$\rho\Big(H_r^{(l)}, H^{(l)}\Big) = \sum_{i=1}^{N} \left(h_i^{(l)} \ln \frac{2\tilde{h}_i^{(l)}}{\tilde{h}_i^{(l)} + \tilde{h}_{r;i}^{(l)}} + h_{r;i}^{(l)} \ln \frac{2\tilde{h}_{r;i}^{(l)}}{\tilde{h}_i^{(l)} + \tilde{h}_{r;i}^{(l)}} \right), \quad (2)$$

where

$$\begin{aligned} \tilde{h}_{r;i}^{(l)} &= \sum_{j=1}^{N} K_{ij} h_{r;i}^{(l)} \\ \tilde{h}_i^{(l)} &= \sum_{j=1}^{N} K_{ij} h_i^{(l)} \end{aligned}, \quad (3)$$

is the convolution of the HOGs with the Gaussian Parzen kernel K_{ij} [20] and indexes (k_1, k_2) are missed for clarity.

In general case ($L > 1$) the outputs (1) at each level are aggregated [8] and X is assigned to the class of the closest model in terms of the following similarity measure

$$\rho_{PHOG}(X, X_r) = \sum_{l=1}^{L} w^{(l)} \cdot \rho^{(l)}(X, X_r), \quad (4)$$

Here weights of every pyramid level $w^{(l)} > 0$ are usually chosen experimentally. Unfortunately, though the accuracy of the PHOG is usually higher than the accuracy of the HOG [8], performance of the nearest neighbor rule with similarity measure (4) is

quite low as it requires $\sum_{l=1}^{L} (K^{(l)} \cdot K^{(l)})/K^{(1)} \cdot K^{(1)}$-times more calculations in comparison with conventional HOG [3] ($L = 1$). Hence, in the next section we present s way to improve the PHOG's performance by sequential analysis at each level.

3   Sequential Classification of the PHOG

The key idea of our paper is to perform recognition at each level independently and terminate the classification procedure if the decision was assumed to be reliable [21]. Otherwise, the next level of the hierarchy is analyzed in the same way. To verify the decision's reliability, let's use the statistical approach [16]. It is known [15, 22] that if the hypothesis $W_r^{(l)}$ for homogeneity of query X and model objects X_r at lth level are tested and it is assumed that the prior probabilities of each class are equal, then the nearest neighbor rule with the HT-PNN (1)–(3)

$$v(l) = \underset{r \in \{1,\dots,R\}}{\arg\min} \, \rho^{(l)}(X, X_r) \tag{5}$$

is equivalent to the maximal likelihood decision. In such case, an optimal Bayesian criterion for rejection of unreliable decision is achieved with the Chow's rule [13]

$$P\left(W_{v(l)}^{(l)} \middle| X\right) \leq p_0 = const, \tag{6}$$

where $P\left(W_{v(l)}^{(l)} \middle| X\right)$ is the posterior probability of hypothesis $W_{v(l)}^{(l)}$ and threshold

$$p_0 = \frac{\Pi_{10} - \Pi_{ro}}{\Pi_{01} + \Pi_{10} - \Pi_{ro}}. \tag{7}$$

Here Π_{10} is the losses of incorrect decision (5), which has not been rejected (6), Π_{01} is the losses of rejection (6) of correct decision (5) and Π_{ro} is the cost of reject option (6) (obviously, $\Pi_{ro} \leq \Pi_{10}$).

Hence, the task is to estimate the posterior probability $P(W_{v(l)}^{(l)} | X)$. If the classes are equiprobable, the Bayes theorem can be used

$$P\left(W_{v(l)}^{(l)} \middle| X\right) = \frac{P\left(X | W_{v(l)}^{(l)}\right)}{\sum_{r=1}^{R} P\left(X | W_r^{(l)}\right)}. \tag{8}$$

Here $P\left(X | W_r^{(l)}\right)$ is the conditional probability (likelihood) of the rth class. Fortunately, given the relationship of the HT-PNN and maximal likelihood [15], we get the final estimation of the posterior probability:

$$\hat{P}\left(W^{(l)}_{v(l)}\Big|X\right) = \frac{\exp\left(-UV \cdot \rho^{(l)}(X, X_{v(l)})\right)}{\sum\limits_{r=1}^{R} \exp(-UV \cdot \rho^{(l)}(X, X_r))}. \tag{9}$$

Our sequential recognition process is represented in Fig. 1. At first, the most rough approximations of images with small number of blocks $K^{(1)}$ are analyzed. If in this case it is possible to obtain a reliable solution with high posterior probability (9), the process is terminated and $c\left(v^{(1)}\right)$ (5) becomes the resulted class. Otherwise the description of the query object is detailed and the process is repeated until obtaining the reliable solution $c\left(v^{(l)}\right)$ at the lth level.

The last open question here is the final processing if decisions at all L levels are unreliable. In this case it is necessary to obtain the best solution from the set of candidates $\left\{c\left(v^{(l)}\right)\right\}, l \in \{1,\ldots,L\}$. The most evident way here is to perform fusion of classifiers (1) for each level. The most complex fusion methods (bagging, boosting, etc.) [16] require the large training set and cannot be used if $C \approx R$. Thus, in this paper we propose to use conventional principle of maximum posterior probability [16] - the final decision is taken in favor of class $c\left(v^{(l^*)}\right)$, where

$$l^* = \arg\max_{l \in \{1,\ldots,L\}} \hat{P}\left(W^{(l)}_{v(l)}\Big|X\right). \tag{10}$$

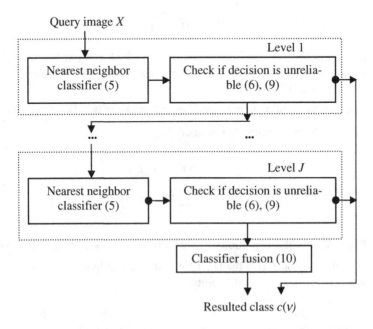

Fig. 1. Proposed sequential hierarchical procedure of image recognition

Thus, in the proposed method (5), (6), (9), (10) we implement an approach widely used in hierarchical image recognition [6, 16], namely, the build of the pyramid of features for various image resolution [8]. However, unlike the known similar PHOG-based methods (4) with simultaneous comparison of the union of features for all levels, we perform sequential analysis (compare with [14]) of detailed approximations of the query object and move to the next pyramid level *only if* the solution at the current level is not reliable (6). As a result, the performance of our approach should be much better in comparison with conventional nearest neighbor rule with similarity measure (4). The next section experimentally supports this claim.

4   Experimental Results

Our experimental study deals with the face recognition task. We combined 3 datasets, namely, AT&T [23], Yale [24] and JAFFE [25], into one set to demonstrate the flexibility of hierarchical recognition system (Fig. 1). AT&T dataset contains 400 images of 40 different people varied in pose. Yale database (165 photos of 15 persons) is used to test recognition with various light conditions. JAFFE dataset contains 213 images of 10 Japanese female persons varied in facial expressions. In total, our dataset consists of 778 photos of $C = 65$ persons. The faces were detected with the LBP cascade classifier [26] from OpenCV library [27] which was proved to be one of the best known face detection algorithms [28]. The median filter with window size (3×3) was applied to remove noise in detected faces. The number of bins in the HOG $N = 8$. Threshold p_0 for posterior probability in the Chow's rule (6) is equal to 0.85. The neighborhood size in (1) is equal to $\Delta = 1$ [19].

We compared the performance of the proposed sequential recognition with the PHOG (4) and conventional HOG (number of levels $L = 1$). We use hierarchies with 2 and 3 levels. In the first case, $K^{(1)} = 10$ and $K^{(2)} = 20$. In the second case, $K^{(1)} = 10$, $K^{(2)} = 15$ and $K^{(3)} = 20$. Weights in (4) were found experimentally to obtain the higher accuracy.

We evaluate the error rate (in %) and the average time (in ms) to recognize one test image with a modern laptop (4 core i7, 6 Gb RAM) and Visual C++ 2013 compiler and optimization by speed. We use multithreading to make brute-force search (1), (5) faster. Each thread is implemented with Windows ThreadPool API and operates only on a subset of the database. The whole training sample is divided into 8 distinct parts, i.e., we look for the nearest neighbor (5) in 8 parallel threads. The recognition performance was estimated by the following cross-validation procedure. At first, the number of photos per one person $n_p = const$ is fixed. For each person, we randomly choose n_p photos and put them into the model database $\{X_r\}$. Other photos are put into the test set. Then we estimate the error rate of test set recognition. This experiment is repeated 20 times. Finally, we estimate the mean of the error rate and recognition time for all experiments. The error rate and average recognition time for the HT-PNN (1)–(3) distance in dependence on the size of the training set R are shown in Figs. 2 and 3, respectively. Here "10 × 10" bar stands for the HOG with 10 × 10 grid (i.e., $L = 1$, $K^{(1)} = 10$), "20 × 20" bar represents the results of the HOG with 20 × 20 grid ($L = 1$, $K^{(1)} = 20$), "PHOG (10 × 10 + 20 × 20)" and "PHOG

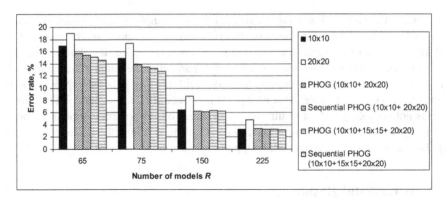

Fig. 2. Dependence of the error rate on the size of the training set R, HT-PNN (2), (3).

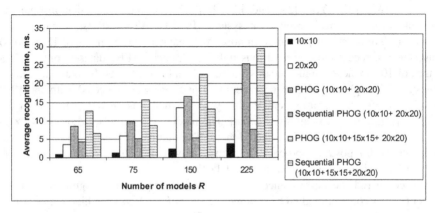

Fig. 3. Dependence of the average recognition time (ms.) on the size of the training set R, HT-PNN (2), (3).

$(10 \times 10 + 15 \times 15 + 20 \times 20)$" stand for conventional PHOG (4) with $L = 2$ (grids $K^{(1)} = 10$ and $K^{(2)} = 20$) and $L = 3$ (grids $K^{(1)} = 10$, $K^{(2)} = 15$ and $K^{(3)} = 20$) levels, respectively. The results of proposed sequential analysis (Fig. 1) are represented by the bars "Sequential PHOG $(10 \times 10 + 20 \times 20)$" and "Sequential PHOG $(10 \times 10 + 15 \times 15 + 20 \times 20)$" for $L = 2$ and $L = 3$ hierarchical levels with the same grid size, respectively.

Here one can notice that the accuracy of hierarchical approach is higher than the accuracy of the state-of-the-art HOG. It is especially true for small size R of the training sample. The difference in error rates of the PHOG (4) and our approach (Fig. 1) is not statistically meaningful (Fig. 2). However, the average recognition time (Fig. 3) for the proposed method is 1.5–3.5 times lower in comparison with conventional aggregation (5) as in most cases (especially for large R) the reliable solution (6) was found at the first level ($K^{(1)} = 10$).

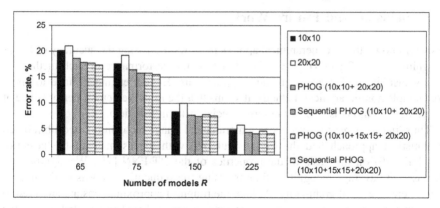

Fig. 4. Dependence of the error rate on the size of the training set R, Euclidean metric.

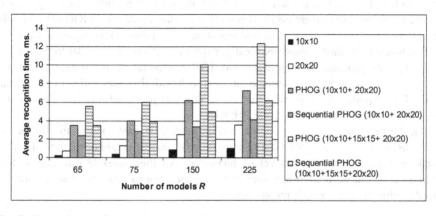

Fig. 5. Dependence of the average recognition time (ms.) on the size of the training set R, Euclidean metric.

In the second experiment we used conventional Euclidean distance to compare HOGs in (1) instead of the HT-PNN (2), (3). Error rates and average recognition time are shown in Figs. 4 and 5.

Based on these results, it is possible to draw the following conclusions. First, the error rate of Euclidean distance is 1–3.5 % higher than the HT-PNN's error rate. Second, comparing Figs. 2 and 4, the losses in the accuracy of conventional non-hierarchical approach (HOG [3]) in the experiment with Euclidean distance are much more noticeable than for the HT-PNN. Third, as the Euclidean distance is much more simple than the HT-PNN in computational sense, in some cases (with low R) the 1-ms gain in performance of our approach over aggregation (4) is not practically noticeable. Finally, though it is reasonable to use $L = 3$ levels for the HT-PNN with low values of the size R of the training set (Fig. 2), there is no reason to prefer it to the simple case of $L = 2$ levels of the hierarchy for Euclidean metric.

5 Conclusion and Future Work

It is well-known that the hierarchical approach allows increasing the accuracy of image recognition (Figs. 2 and 4) [7, 8]. Unfortunately, the performance of such methods is usually insufficient for many practical applications. Hence, engineers have to apply more simple nonhierarchical methods. It seems that the improvement of performance of hierarchical methods is one of the most crucial tasks in this field.

Thus, in this paper we introduced a hierarchical image recognition algorithm based on statistical approach and the Chow's rule (6) with the estimate of the posterior probability (9) on the basis of the properties of the HT-PNN [15]. We significantly decreased the average recognition time by using the sequential analysis of the images with different level of granularity. At first, the roughest approximations are analyzed to speed-up the recognition procedure. The images are analyzed in detailed way if the current decision is not reliable (7). The proposed approach (Fig. 1) showed its efficiency in the practically important face recognition problem. In some cases (Fig. 4) our approach is even able to increase the recognition accuracy over the PHOG (4). Though the posterior probability (9) is estimated based on the HT-PNN, this expression can be used with other similarity measures. Really, expression (9) is very similar to the output of widely used probabilistic neural network [20]. For instance, in our experiment we have shown the possibility to combine our approach with the state-of-the-art Euclidean metric. It is important to emphasize that the accuracy of proposed solution is even 0.4–0.8 % higher in comparison with the PHOG.

The further research of our approach (Fig. 1) can be continued in the following directions. First, it is an application of approximate nearest neighbor methods [11] to speed-up recognition at each level of hierarchy. Another possible direction is the application of our sequential hierarchical method in various tasks of classification of complex objects, e.g., speech recognition.

Acknowledgements. Andrey V. Savchenko is supported by RSF (Russian Science Foundation) grant 14-41-00039 in the National Research University Higher School of Economics.

References

1. Sonka, M., Hlavac, V., Boyle, R.: Image Processing, Analysis, and Machine Vision, 4th edn. Cengage Learning, Boston (2014)
2. Lowe, D.: Distinctive image features from scale-invariant keypoints. Int. J. Comput. Vis. **60** (2), 91–110 (2004)
3. Dalal N., Triggs B.: Histograms of oriented gradients for human detection. In: International Conference on Computer Vision and Pattern Recognition, pp. 886–893 (2005)
4. Bay, H., Ess, A., Tuytelaars, T., Van Gool, L.: SURF: speeded up robust features. Comput. Vis. Image Underst. **110**(3), 346–359 (2008)
5. He, K., Zhang, X., Ren, S., Sun, J.: Delving Deep into Rectifiers: Surpassing Human-Level Performance on ImageNet Classification. arXiv:1502.01852 [cs], http://arxiv.org/abs/1502.01852 (2015)
6. Hawkins, J., Blakeslee, S.: On Intelligence. Times Books, New York (2004)

7. Munoz, D., Bagnell, J.A., Hebert, M.: Stacked hierarchical labeling. In: Daniilidis, K., Maragos, P., Paragios, N. (eds.) ECCV 2010, Part VI. LNCS, vol. 6316, pp. 57–70. Springer, Heidelberg (2010)
8. Bosch, A., Zisserman, A., Munoz, X.: Representing shape with a spatial pyramid kernel. In: 6th ACM International Conference on Image and Video Retrieval CIVR 2007, pp. 401–408 (2007)
9. Zhai, J.-H., Zhang, S.-F., Liu, L.-J.: Image recognition based on wavelet transform and artificial neural networks. In: IEEE International Conference on Machine Learning and Cybernetics, pp. 789–793 (2008)
10. Cireşan, D., Meier, U., Masci, J., Schmidhuber, J.: Multi-column deep neural network for traffic sign classification. Neural Netw. **32**, 333–338 (2012)
11. Savchenko, A.V.: Directed enumeration method in image recognition. Pattern Recogn. **45** (8), 2952–2961 (2012)
12. Chellappa, R., Du, M., Turaga, P., Zhou, S.K.: Face tracking and recognition in video. In: Handbook of Face Recognition, pp. 323–351 (2011)
13. Chow, C.K.: On optimum recognition error and reject trade–off. IEEE Trans. Inf. Theory **16**, 41–46 (1970)
14. Wald, A.: Sequential Analysis. Dover Publications, New York (2013)
15. Savchenko, A.V.: Probabilistic neural network with homogeneity testing in recognition of discrete patterns set. Neural Netw. **46**, 227–241 (2013)
16. Theodoridis, S., Koutroumbas, K.: Pattern Recognition, 4th edn. Elsevier Inc., Amsterdam (2009)
17. Tan, X., Chen, S., Zhou, Z.H., Zhang, F.: Face recognition from a single image per person: a survey. Pattern Recogn. **39**(9), 1725–1745 (2006)
18. Taigman, Y., Yang, M., Ranzato, M., Wolf, L.: DeepFace: closing the gap to human-level performance in face verification. In: 2014 IEEE Conference on Computer Vision and Pattern Recognition (CVPR 2014), pp. 1701–1708 (2014)
19. Savchenko, A.V.: Nonlinear transformation of the distance function in the nearest neighbor image recognition. In: Zhang, Y.J., Tavares, J.M.R.S. (eds.) CompIMAGE 2014. LNCS, vol. 8641, pp. 261–266. Springer, Heidelberg (2014)
20. Specht, D.F.: Probabilistic neural networks. Neural Netw. **3**(1), 109–118 (1990)
21. Yao, Y.: Granular computing and sequential three-way decisions. In: Lingras, P., Wolski, M., Cornelis, C., Mitra, S., Wasilewski, P. (eds.) RSKT 2013. LNCS, vol. 8171, pp. 16–27. Springer, Heidelberg (2013)
22. Kullback, S.: Information Theory and Statistics. Dover Publications, New York (1997)
23. AT&T database of faces. http://www.cl.cam.ac.uk/research/dtg/attarchive/facedatabase.html
24. Yale face database. http://vision.ucsd.edu/content/yale-face-database
25. Japanese Female Facial Expression (JAFFE) database. http://www.kasrl.org/jaffe.html
26. Liao, S., Zhu, X., Lei, Z., Zhang, L., Li, S.Z.: Learning multi-scale block local binary patterns for face recognition. In: Lee, S.-W., Li, S.Z. (eds.) ICB 2007. LNCS, vol. 4642, pp. 828–837. Springer, Heidelberg (2007)
27. OpenCV library. http://opencv.org/
28. Degtyarev, N., Seredin, O.: Comparative testing of face detection algorithms. In: Elmoataz, A., Lezoray, O., Nouboud, F., Mammass, D., Meunier, J. (eds.) ICISP 2010. LNCS, vol. 6134, pp. 200–209. Springer, Heidelberg (2010)

Discerning Depression Propensity Among Participants of Suicide and Depression-Related Groups of Vk.com

Aleksandr Semenov[1]([⊠]), Alexey Natekin[2,3,4], Sergey Nikolenko[5,6], Philipp Upravitelev[7], Mikhail Trofimov[8], and Maxim Kharchenko[9]

[1] International Laboratory for Applied Network Research,
National Research University Higher School of Economics, Moscow, Russia
avsemenov@hse.ru
[2] Data Mining Labs, St. Petersburg, Russia
natekin@dmlabs.org
[3] Deloitte Analytics Institute, Moscow, Russia
[4] Technical University Munich, Garching, Germany
[5] Laboratory for Internet Studies,
National Research University Higher School of Economics, St. Petersburg, Russia
sergey@logic.pdmi.ras.ru
[6] Steklov Mathematical Institute at St. Petersburg, St. Petersburg, Russia
[7] Consultant Plus LLC, Moscow, Russia
upravitelev@gmail.com
[8] Moscow Institute of Physics and Technology, Moscow, Russia
mikhail.trofimov@phystech.edu
[9] Undev LLC, Moscow, Russia
ma.kharchenko@gmail.com

Abstract. In online social networks, high level features of user behavior such as character traits can be predicted with data from user profiles and their connections. Recent publications use data from online social networks to detect people with depression propensity and diagnosis. In this study, we investigate the capabilities of previously published methods and metrics applied to the Russian online social network VKontakte. We gathered user profile data from most popular communities about suicide and depression on VK.com and performed comparative analysis between them and randomly sampled users. We have used not only standard user attributes like age, gender, or number of friends but also structural properties of their egocentric networks, with results similar to the study of suicide propensity in the Japanese social network Mixi.com. Our goal is to test the approach and models in this new setting and propose enhancements to the research design and analysis. We investigate the resulting classifiers to identify profile features that can indicate depression propensity of the users in order to provide tools for early depression detection. Finally, we discuss further work that might improve our analysis and transfer the results to practical applications.

Keywords: Depression · Social network analysis · Online social networks · Egocentric networks · Data mining

© Springer International Publishing Switzerland 2015
M.Y. Khachay et al. (Eds.): AIST 2015, CCIS 542, pp. 24–35, 2015.
DOI: 10.1007/978-3-319-26123-2_3

1   Introduction

Over the recent years, social media data has been increasingly used to predict or identify important events, features and phenomena. Researchers tried to predict stock [3] and precious metalls markets [10], political elections [14], spread of epidemics [1] and viral messages [11] to name just a few areas of application.

However, the idea that social media data can help to detect users with depression propensity is a relatively recent one. One of the first serious work in this direction that we know of came from Microsoft Research in 2013 [5]. They recruited volunteers via Amazon's Mechanical Turk who had been diagnosed with MDD using a standard clinical test on depression called CES-D[1] (Center of Epidemiological Studies-Depression Scale).

Then the authors asked verified depressive users to provide their Twitter accounts for data collection and gathered data for a year prior to their diagnosis to measure their engagement, linguistic style, depressive language use, mentions of antidepressants, and their egocentric social network structure. They compared these features for people with and without MDD (Major Depressive Disorder) and built a classifier that was able to successfully predict depression with 70 % accuracy and 0.74 precision. In this study, it turned out that egocentric networks of users who were diagnosed with depression had higher clustering coefficients, which means that their friends in general were connected too. Moreover, models with network features worked much better than those with engagement and demographic features, suggesting that structural properties of egocentric networks do have a meaningful connection with depression.

Later the same team used a similar approach to detect post-partum depression (PPD) among women on Facebook [4]. In this case, the data included self-reports on postpartum experience from new mothers along with the results of their tests on PPD evaluation. After that, they analyzed public Facebook data of these women and built predictive models from the resulting features, including the number, type, and frequency of their postings on the wall, variation and mean power of the number of posts per week. Network characteristics were present again, this time via the "likes" and "comments" features of Facebook: each arc in the directed graph based on likes/comments represented a like or comment that one user posted to another. These measures represented social support, and the resulting models showed that they were the best predictors of PPD.

However, it is not always possible to find people who are certifiably depressive and acquire their permission to fill in questionnaires and provide access to their data in social media. In many cases, researchers can only access the data without a solid "ground truth" of their self-reports, interviews, or screening questionnaires. For example, a recent work [9] analyzed profiles of users from public groups about "suicide" and "depression" in a popular Japanese online social network Mixi and compared them with a control group of random users of this network. They used such features as age, sex, number of friends, and local

---

[1] http://www.bcbsm.com/pdf/Depression_CES-D.pdf.

clustering coefficient of their egocentric networks and trained a binary logistic regression model to define key features that distinguish users from suicide-related communities and random users. Their results showed that egocentric networks of users with suicide propensity have smaller local clustering coefficients, which means that their friends mostly do not know each other, and there are too few social groups, or "circles", that the user belongs to. It's noteworthy to say that these results contradict the findings of [5] according to which people who are diagnosed with depression have denser and more clustered ego networks. These difference might be explained by such factors as cultural difference between English-speaking and Japanese users, technological features and different contexts of usage of Twitter's communication networks via replies, retweets and mentions and Mixi "friendship" networks, or by general differences in the audience of these social networks.

Hence we decided to reproduce this line of research on the data from the Russian social network VKontakte, to check if its applicable for conducting such kind of analysis and compare the results with [9]. The rest of the paper is structured the following way. In Sect. 2 we discuss our research design, sampling methods and features, used in the analysis. In Sect. 3 we provide descriptive statistics of the obtained datasets and in Sect. 4 we describe our models and compare their performance. Finally, in the Sect. 5 we summarize the results of our analysis, discuss its limitations and future work.

2   Dataset and Research Design

2.1   User Sampling

We identify depression propensity among social network users based on their demographic features and structural properties of their egocentric networks, introducing several important improvements over the methodology of [9] in the sampling design and methods.

First, we chose people with depression propensity not only by the fact of their membership in the communities related to suicide and depression: obviously, plenty of members of such groups are not really depressed or have already overcome their depression. To gather the data on potentially depressive users, we searched for relevant communities at VK.com. The search request queried for two relevant keywords: "suicide" and "depression"; it returned more than 60 sufficiently large communities with more than 500 users. From this list of results, we excluded blocked or irrelevant ones (with advertisement, inactive etc.). As a result, we came up with a list of 37 communities with more than 110,000 unique members.

In order to select particular users with depression propensity, we searched for discussion threads where users claim to be depressed and "on the edge" of suicide. One such thread, started in January 2013, by January 2015 contained about 30,000 comments from 2550 unique users who shared their personal stories and claims. All those users were selected as our proxy for the target group of people with depression propensity.

Next, we had to select a comparable sample of random users. Note that a random sample out of the entire social network (as was done, e.g., in [9]) will with large probabilities produce many bots, inactive, or otherwise irrelevant accounts. Fortunately, the technical properties of the VKontakte social network itself provide an elegant solution: user ids in VKontakte are generated randomly with respect to chronological order according to which users registered in the network. Hence, we drew a random sample of users of the same size (2550 users) as follows: for each original user id we added random Poisson noise with mean 10, 000. This sampling scheme resulted in a much higher similarity between the demographical distributions of our target and control groups and in a much higher number of active and relevant accounts in the random sample. This brought us to a total of 5100 users for our comparative experiment; there was no intersection between target and control groups.

Due to ethical and privacy reasons [15] we decided not to provide the names of the groups and thread URL in the paper. All the data we gathered were obtained via API and contains only publicly available and officially provided information. All the measures are presented in aggregated form and pose no threat to the privacy of the users, which profiles were analyzed.

2.2 Features

For the 5100 collected users, we extracted both profile data and ego networks, i.e., friendship graphs among friends of a particular user. Following the research design of [9], we used the following variables (that is, their counterparts on VKontakte): age, gender, number of communities, number of friends, local clustering coefficient, and homophily. Age is missing from nearly 60 % of downloaded user profiles, which renders this feature hard to use. Hence we decided to use the median of friends' ages as a proxy for the user's real age, rather than the age itself. Besides that we couldn't find a way to calculate the period since registration, so we didn't use this feature in the analysis.

Due to profile deactivation and banning, not all user profiles were present by the data collection time. We also omitted blank profiles, i.e., users with no friends. This led us to nearly 30 % reduction in sample size, leaving us with 1635 random users and 1909 depression-ideated ones. This class size asymmetry may be caused by the fact that such types of accounts like bots and silent spectators will not write confessions in the depression thread, yet can be revealed via random sampling.

Another problem is that because of privacy reasons, profile information like age and groups is often hidden. Nearly 60 % of our data contains missing values for age variable and independent from them approximately 50 % don't have the number of groups open (doesn't imply 0 groups). Besides, some users set themselves unrealistic age of 90 and even 100 years. Therefore, we dismiss users with ages higher than 70 years due to low reliability. In order to reconstruct previous research results with all the variables, we are left only with 602 users.

In what follows, we refer to the target class of users who left a comment as "depressed", and their opposite as the "random" class. The class distribution

Table 1. Variable descriptive statistics summary.

Variable	"Depressive" group		"Random" group		p-value
	Mean ± sd	Range (min,max)	Mean ± sd	Range (min,max)	
Gender (female)	47.2 %		45 %		
Age	24.8 ± 5.96	(14,51)	28.1 ± 10.9	(14,68)	<0.001
Community number	213 ± 435	(0,3330)	40 ± 109	(0,1450)	<0.001
k_i	185 ± 448	(2,5300)	64 ± 105	(1,1160)	<0.001
C_i	0.066 ± 0.083	(0,0.667)	0.203 ± 0.238	(0,1)	<0.001
Homophily	0.004 ± 0.018	(0,0.222)	0.001 ± 0.014	(0,0.2)	0.122

in our dataset was 32 % depressed users and 68 % random. This asymmetry may be caused by having depressed users being potentially more stealthy or shy and hiding their profile data, whereas their ego network data is open to anyone regardless of privacy policies.

Descriptive statistical summary of the obtained dataset is presented in Table 1. P-values were obtained using the Student's T-test.

3   Exploratory Analysis

3.1   Demographics

We begin with an investigation of the distributions of collected features. First, we study the entire dataset with all considered variables and 602 users.

Figure 1 shows the basic demographic distributions for the collected users; it shows that neither age nor gender distributions have significant deviations in our groups. Both distributions are bimodal and only slightly differ: most depressed users are between 21 and 25 years old while random users are concentrated between 19 and 29 years of age. Note that there are fewer depressive users over 40 years of age compared to the random sample. Both classes contain slightly more males (53 %) than females.Particularly interesting is the fact, that users from "depressive" communities participate in 5 times more groups, and have smaller local clustering coefficient as in previous research.

3.2   Social Features

Next we add features that characterise basic statistics of a user's profile: user friendship degree and the number of communities that user participates in. As both of these features are power-law distributed, we visualize both of them on the logarithmic scale (base 10). The resulting distributions for both classes are shown on Fig. 2. It is important to note that we artificially included zero values as an important part of distribution; to keep the log well-defined we added a small value of 0.1 to the number of groups.

Distributions on Fig. 2 indicate that both of these features show considerable signs of discrimination between classes, unlike user demography. Random users participate in fewer communities and, somewhat surprisingly, have

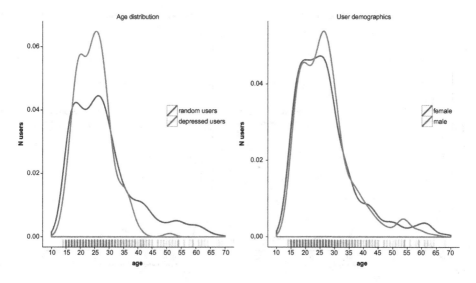

Fig. 1. Demographic distributions: (a) age, (b) gender.

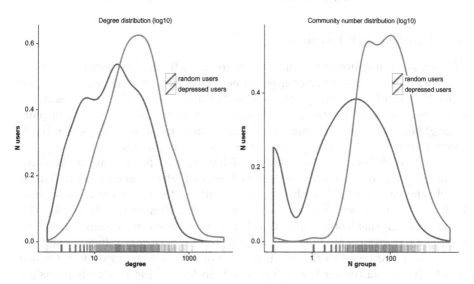

Fig. 2. Distributions of social features: (a) degree, (b) number of communities.

less friends. However, such discrimination can also be caused by other reasons such as the overall user activity and engagement on the social network; after all, our "depressed" sample consists entirely of users who actively cried out for help in a depression community.

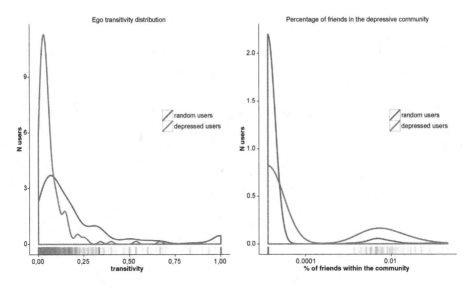

Fig. 3. Distributions of ego network features: (a) transitivity, (b) homophily.

3.3 Ego Network Features

The last set of features are features extracted from friendship connections. First we consider the transitivity of users in ego networks, treating the user as the central vertex of the ego-graph that connects this user's friends. The transitivity, also called the local clustering coefficient of a vertex v is the fraction of pairs of neighbours of v that are connected among all pairs of neighbours of v [7]. It shows how many social groups or "circles" the ego is part of.

Our second feature is the fraction of friends who participate in the same community on depression & suicide, which we will refer to as homophily. This is simply the ratio of friends within the community to the number of friends which indicates the tendency of users from these communities to form friendship ties on vk.com. As this feature involves other features prone to transformation, we use the log transform of homophily as well. To deal with zero values, we shift all values by a bias equal to the minimal value of a feature in our dataset divided by 10 (the base of the log transform). The transformed feature distributions are shown on Fig. 3.

4   Analysis

4.1 Framework

Following the design of [9], we used binary logistic regression to identify the features of profiles that are able to predict a user's depression propensity. We choose the Area Under the Curve (AUC) criterion as our measure of goodness of fit [8]; in particular, this measure is robust to asymmetric class distributions.

In order to compare classification performance of different models, we utilized bootstrap sampling. We consider the classic approach with bootstrap sampling with replacement. At each iteration the model is trained on a bootstrapped data sample comprising of approximately 63.2 % of the original unique values. The model is then validated on the remaining data, called the Out-Of-Bag estimate. The procedure is repeated multiple times (1000 simulations in our case) to evaluate the distribution of the OOB goodness of fit metric (AUC). Finally, the models are compared by their mean performances, with potential inference of the associated variance of models.

4.2   Model Comparison

There were no feature transformations in [9]; in what follows we refer to this model as the "basic" model, meaning that it simply reproduces a linear GLM fit for the original 6 features.

In the previous section, we noted that it is reasonable to keep logarithmic feature transformations in the model design. Hence we will compare the basic model with raw features to a model with our proposed log transforms, in the exact same way as they were outlined earlier in the paper. These transformed features include the number of groups, degree (number of friends), and homophily (percentage of friends in the community). This model will be further referred to as the "transformed" model.

Finally, we compare these two models to the one after the feature selection process. In particular, we perform stepwise feature elimination by Akaike Information Criterion (AIC) to the "transformed" model. We consider the backward stepwise regression, starting with the full model and proceeding backwards by sequentially removing variables that improve the model by being deleted, until no more deletions will improve the model. On our data the result of this procedure is a model with only two remaining variables: the logarithm of the number of groups and transitivity. We will refer to this model as the "selected" one.

With these three models, we performed the bootstrap simulation of their OOB AUC estimates over 1000 trials. The resulting boxplots and ROC curves for these three models are given in Fig. 4. The ROC curves are visualized based on OOB predictions from 100 simulation runs, which leads to higher smoothness of the curves.

Based on these experimental results, we can make three important observations. First of all, all three models achieve a reasonably high AUC, higher than 0.8. This implies that these models consistently capture a reasonable amount of variation in the data and successfully classify our target class. The second observation is that both models with feature transformations yield considerably higher AUC results (approximately 0.84) as compared to the basic model described in the original study [9]. This means that proper model design can potentially improve the performance substantially. Finally, the third point is that a simplified model with only two features achieves the same performance as the complete model; as a result, not only the model becomes simpler and more stable, but also the resulting model becomes easier to tract.

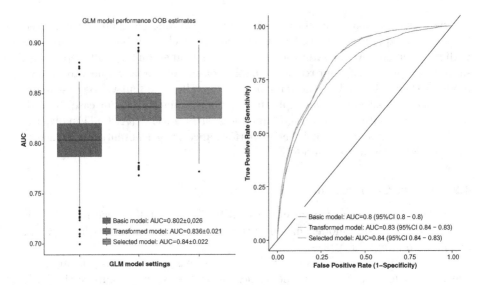

Fig. 4. ROC curves and boxplots for bootstrap simulations.

4.3   Coefficient Analysis

To analyze the resulting coefficients, we trained all models on the entire dataset without bootstrapping. We then compared the three models described above based on their coefficient values, p-values for coefficients, and the variance inflation factor (VIF) that indicates multicollinearity. All of these metrics are listed in Table 2.

Based on Table 2, we distinguish four important points.

First of all, in every model the most significant variables are the transitivity and community count (regardless of transformation). In every case, the community count has a positive coefficient and transitivity has a negative one. This coincides with the previously published results, meaning that transitivity is negatively correlated with the chance of depression propensity, while number of communities are positively correlated with this probability.

The second point is that in the basic model the age variable turns out to be significant with p-value of 0.005, and homophily is likely to be insignificant. On the other hand, in the transformed model logarithms of homophily and friend count turn out to be on the very edge of significance and age is no longer significant for that model. This behaviour can be explained by having the discriminative power of age so little that introduction of proper transformations for homophily and degree dissolves the impact of age in favour of graph features. This also means that the resulting model no longer depends on demography. It is an important result considering that this model also achieves higher AUC values compared to the basic one.

The third interesting phenomenon is the behaviour of the intercept term in these models. Intercept is only significant in the model with feature selection,

Table 2. A comparison of model coefficients.

Variable	Basic model AUC: 0.802 ± 0.026			Transformed model AUC: 0.836 ± 0.021			Selected model AUC: 0.84 ± 0.022		
	Coef. value	P-value	VIF	Coef. value	P-value	VIF	Coef. value	P-value	VIF
Intercept	0.46	0.26	-	−0.18	0.82	-	−2.1	$1.9 \cdot 10^{-11}$	-
Gender (female)	0.16	0.44	1.02	0.17	0.04	1.02	-	-	-
Age	−0.04	0.005	1.03	−0.02	0.23	1.06	-	-	-
Community number	0.005	$1.4 \cdot 10^{-8}$	1.11	1.56	$2.2 \cdot 10^{-14}$	1.52	1.41	$5 \cdot 10^{-18}$	1.01
k_i	0.001	0.25	1.08	−0.46	0.25	1.56	-	-	-
C_i	−7.77	$1.9 \cdot 10^{-10}$	1.1	−6.33	$6.2 \cdot 10^{-6}$	1.06	−6.29	$7.1 \cdot 10^{-6}$	1.01
Homophily	5.40	0.34	1.01	0.23	0.04	1.10	-	-	-

while being most likely insignificant in the other two models. This can imply that all of our four other features collapse into a single bias in the model; this bias plays a significant part but it is independent of the conditioning on the data. The bias value is also reasonable, meaning that *a priori* users are not depressive.

Finally, the Variance Inflation Factor is close to 1 in both basic and selected model, which is good from the model design point of view. For the transformed model, it is around 1.5 for logarithms of degree and group count, suggesting that they share a common source of variation and are connected. Removing redundant variables leads to better model design in the case of the selected model, which also means that both remaining variables represent different sources of variation.

4.4 Results

We investigated the possibility to distinct users with depression propensity in the Russian online social network VKontakte based on demographic and structural features of their profiles. The collected features were sufficient to achieve high AUC result of 0.84, comparable to the results of [9] where the authors achieved AUC of 0.87.

The first result is that in order achieve higher AUC values with GLM one has to consider proper variable transformations. In particular, based on bootstrap simulations, exploiting proper log transforms on variables yields an AUC improvement for GLM from 0.8 to 0.84.

The second result is that the most valuable features that characterize the depression propensity are the (log) number of communities and the user's transitivity within his or her personal ego network, which are sufficient for propensity detection in our setting. In particular, the probability of depression propensity rises with an increase in the communities count and a decrease in transitivity (i.e. local clustering coefficient).

It is worth mentioning that not only age and sex but also the number of friends turned out to be statistically insignificant. According to our simulations, only (log) homophily can be considered significant enough to be added to the model, but its improvement is negligible and thus redundant.

5 Conclusion

We have shown that data from profiles of the VK.com users, who participated in the communities on depression and suicide can be used to distinguish features related to depression propensity via standard GLM models. Moreover, structural features, derived from egocentric networks, such as local clustering coefficient, perform even better than traditional demographic and attributive features from the profiles. It means that the structure of social environment of the users is as significant factor in the problem of depression detection on-line data as it's "offline" [12].

It is likely that introduction of more advanced graph-based measures like ego-network density, dyad and triad census have the potential to improve the model. Another characteristic of users provided by vk.com, called "interests" can also be included to the analysis alongside with their groups via bi- and triclustering, which already showed decent results on this kind of data [6]. However, these and other features derived from the available data have to be thoroughly justified from the psychological perspective.

It is also feasible to fuse text data from the user's wall posts and messages with the profile and graph traits of the user.Sentiment analysis can give quantitative clues about how positive/negative are the current posts and how they change over time. Semantic analysis of topics that the user touches upon, for instance with topic modeling techniques [2,13], is a direct proxy for a rich list of his character traits. In this context, data fusion has the potential to greatly improve the analysis and uncover new insights.

Obviously, even after all this analysis we cannot directly diagnose depression with machine learning techniques: our sample has not been collected via proper clinical screening surveys, the quality of this data is also not perfect and might contain some improper class labels. Moreover, collected user profiles could dramatically change over one or two years, potentially even due to the aftermath of depression. Therefore, further work has to be done with the quality of data; in particular, we would like to obtain ground truth data on the psychological conditions of the users and their informed consent for participation and analysis of their profiles data.

Acknowledgements. The work of Sergey Nikolenko was supported by the Basic Research Program of the National Research University Higher School of Economics, 2015, grant No 78.

References

1. Achrekar, H., Gandhe, A., Lazarus, R., Yu, S.H., Liu, B.: Predicting flu trends using twitter data. In: 2011 IEEE Conference on Computer Communications Workshops (INFOCOM WKSHPS), pp. 702–707. IEEE (2011)
2. Blei, D.M., Ng, A.Y., Jordan, M.I.: Latent dirichlet allocation. J. Mach. Learn. Res. **3**(4–5), 993–1022 (2003)

3. Bollen, J., Mao, H., Zeng, X.J.: Twitter mood predicts the stock market, October 2010. http://arxiv.org/abs/1010.3003
4. Choudhury, M.D., Counts, S., Horvitz, E., Hoff, A.: Characterizing and predicting postpartum depression from shared facebook data. In: Computer Supported Cooperative Work, CSCW 2014, Baltimore, MD, USA, 15–19 February, pp. 626–638 (2014). http://doi.acm.org/10.1145/2531602.2531675
5. Choudhury, M.D., Gamon, M., Counts, S., Horvitz, E.: Predicting depression via social media. In: Proceedings of the Seventh International Conference on Weblogs and Social Media, ICWSM 2013, Cambridge, Massachusetts, USA, 8–11 July 2013. http://www.aaai.org/ocs/index.php/ICWSM/ICWSM13/paper/view/6124
6. Gnatyshak, D., Ignatov, D.I., Semenov, A., Poelmans, J.: Gaining insight in social networks with biclustering and triclustering. In: Aseeva, N., Babkin, E., Kozyrev, O. (eds.) BIR 2012. LNBIP, vol. 128, pp. 162–171. Springer, Heidelberg (2012)
7. Holland, P.W., Leinhardt, S.: Transitivity in structural models of small groups. Comp. Group Stud. **2**, 107–124 (1971)
8. Ling, C.X., Huang, J., Zhang, H.: AUC: a statistically consistent and more discriminating measure than accuracy. In: Proceedings of the International Joint Conference on Artificial Intelligence 2003, pp. 519–526 (2003)
9. Masuda, N., Kurahashi, I., Onari, H.: Suicide ideation of individuals in online social networks. PLoS ONE **8**(4), e62262 (2013). http://dx.doi.org/10.1371%2Fjournal.pone.0062262
10. Porshnev, A., Redkin, I.: Analysis of twitter users' mood for prediction of gold and silver prices in the stock market. In: Ignatov, D.I., Khachay, M.Y., Panchenko, A., Konstantinova, N., Yavorsky, R.E. (eds.) AIST 2014. CCIS, vol. 436, pp. 190–197. Springer, Heidelberg (2014)
11. Romero, D.M., Meeder, B., Kleinberg, J.: Differences in the mechanics of information diffusion across topics. In: Proceedings of the 20th International Conference on World Wide Web - WWW 2011, p. 695. ACM Press, New York (2011). http://portal.acm.org/citation.cfm?doid=1963405.1963503
12. Rosenquist, J.N., Fowler, J.H., Christakis, N.A.: Social network determinants of depression. Mol. Psychiatry **16**(3), 273–281 (2011)
13. Vorontsov, K., Potapenko, A.: Tutorial on probabilistic topic modeling: additive regularization for stochastic matrix factorization. In: Ignatov, D.I., Khachay, M.Y., Panchenko, A., Konstantinova, N., Yavorsky, R.E. (eds.) AIST 2014. CCIS, vol. 436, pp. 29–46. Springer, Heidelberg (2014)
14. Yu, S., Kak, S.: A survey of prediction using social media. arXiv preprint arXiv:1203.1647 (2012)
15. Zimmer, M.: "But the data is already public": on the ethics of research in Facebook. Ethics Inf. Technol. **12**(4), 313–325 (2010). http://www.springerlink.com/index/10.1007/s10676-010-9227-5

Tutorial

Normalization of Non-standard Words with Finite State Transducers for Russian Speech Synthesis

Artem Lukanin[✉]

South Ural State University, Chelyabinsk, Russia
artyom.lukanin@gmail.com

Abstract. This paper describes finite state transducers employed for expansion of numbers, acronyms and graphic abbreviations into full-word numerals and phrases in the task of Russian speech synthesis. The developed finite state transducers cover cardinal and ordinal numbers, convert phone numbers, dates, codes, etc. The developed project is the first Russian open-source normalization system known to the author.

Keywords: Preprocessing · Text-to-speech · Morphology · Numeral · Abbreviation · Acronym

1 Introduction

Text preprocessing for speech synthesis is usually a very complex task. Text normalization is one of the steps in text preprocessing. It usually subsumes sentence segmentation, tokenization, and normalization of non-standard words (NSWs) [1]. Non-standard words such as numbers, abbreviations, and acronyms usually cannot be found in lexicons and their pronunciation cannot be calculated using letter-to-sound rules. They must be expanded into full standard words to be pronounced correctly. The task is even more complex in inflective languages such as Russian where every number, abbreviation, and acronym can be expanded into different word forms depending on the context. For example, ordinal numerals in Russian are similar to adjectives: they have six cases (nominative, genitive, dative, accusative, instrumental and prepositional), two numbers (singular and plural) and three genders (masculine, feminine and neuter). Therefore, every ordinal number can be converted into 36 different word forms. The correct word form can be chosen if the grammatical features of the corresponding nearest nouns, adjectives and prepositions are taken into account. Hence, a basic morphological analysis and a shallow syntactic analysis are required to perform this task.

Most text-to-speech systems do text normalization before applying pronunciation dictionaries and letter-to-sound rules to the input text. There are closed commercial text-to-speech systems and open-source systems such as Festival [2]. Every company builds its own text normalization modules, but the quality of the results is still not perfect. For example, the Russian male voice for Festival provides only digit-by-digit normalization of numbers. If the input texts are normalized beforehand the quality of

© Springer International Publishing Switzerland 2015
M.Y. Khachay et al. (Eds): AIST 2015, CCIS 542, pp. 39–48, 2015.
DOI: 10.1007/978-3-319-26123-2_4

the synthesized speech can be improved. I develop Normatex[1], an open-source Russian text-normalization module, which serves to address this task. This is the first open-source system known to the author of such kind. By providing this resource I invite scientists to collaboratively build a high quality text-normalization system.

To build a normalization grammar a test parallel corpus was developed. It consists of 66 original texts of the official site of South Ural State University[2] and 66 manually preprocessed texts, where all numbers, abbreviations and acronyms were expanded into full words or replaced with pronounceable combination of letters (for example, initials are replaced with the alphabet names of the Russian letters). 66 original texts have 17,180 letter and numeric tokens (38,439 tokens in total), 977 of which are number tokens (2,511 digits). The corpus contains 431 acronyms and initials and 379 graphic abbreviations. A token in Unitex [3] can be: the sentence delimiter {S}; the stop marker {STOP}; a lexical tag, e.g. {ЮУрГУ,.N}; a contiguous sequence of Russian or Latin letters (including Latin letters with diacritics); one (and only one) non-letter character, i.e. all characters not defined in the alphabet file of the current language (if it is a newline, it is replaced by a space) [4]. In this sense a token can be called a *broad segmentation unit* (BSU) [5].

2 Related Work

There are different approaches to automatic text normalization. Today automatic inference from large corpora is applied throughout many areas of natural language processing. There have been several attempts to use it for normalizing texts for the task of text-to-speech synthesis. However, the quality of obtained results is not as perfect as could be [6, 7]. Therefore nowadays it is a norm to build grammars for normalization of non-standard words by hand. For example, the developers of a commercial speech synthesis system VitalVoice apply a partial morphological and syntactic analysis for normalization of NSWs and detection of the correct stress position in Russian words [8]. In [2] the English normalization module employs heuristic disambiguation and expansion rules.

Reichel et al. [1] report 28.9 % word error rate in normalization of English cardinal and ordinal numbers by finite state transducers, spelling unknown abbreviations and pronouncing unknown acronyms as standard words if the latter do not violate phonotactics. Since retrieval of proper names, acronyms and abbreviations is crucial for appropriate sentence segmentation and normalization of non-standard words, they carry out this task prior to text normalization.

3 Finite-State Transducers

I develop context-dependent finite-state transducers (FST) in the graphical interface of Unitex 3.1beta [3] for Russian text normalization. The developed grammars are representations of linguistic phenomena on the basis of recursive transition networks

---

[1] https://github.com/avlukanin/normatex

[2] http://susu.ac.ru

(RTN), a formalism closely related to finite state automata. Numerous studies have shown the adequacy of automata for linguistic problems at all descriptive levels from morphology and syntax to phonetic issues. Grammars created with Unitex carry this approach further by using a formalism even more powerful than automata. These grammars are represented as graphs that the user can easily create and update [4]. The main graph is compiled to an .fst2 file, which contains all the subgraphs that make up a grammar. The grammar is then ready to be used by Unitex programs. The FST2 format conserves the architecture in subgraphs of the grammars, which is what makes them different from strict finite state transducers. The Flatten program allows turning an FST2 grammar into a finite state transducer whenever this is possible, and constructing an approximation if not. This function thus permits to obtain objects that are easier to manipulate and to which all classical algorithms on automata can be applied. [4]. As the main function of graphs in Normatex is transduction, i.e. production of a new output, I will call them finite state transducers in this paper.

The pre-processing is done in several steps. At first, the text is tokenized. Then a morphological dictionary is applied to the tokens to tag each token with all possible grammatical features, producing a tagged corpus. For example, the token *правила* is tagged with five different word forms: править.V+nsv+tr:AeFVi and править.V+intr+nsv:AeFVi (transitive and intransitive variants of "to rule", imperfective active singular feminine, past tense, indicative), as well as правило.N+anim(j)+gen(N):geN:nm:ajm ("a rule", inanimate noun in genitive singular or nominative plural or accusative plural). I use the full version of the Russian computational morphological dictionary, developed at CIS, Munich [9] and I am expanding it with new words from our test corpus.

Finally, an FST is applied to the corpus to produce a preprocessed text, which can be further used by a speech synthesis module.

Unitex allows assigning integer weights to the boxes of a transducer. Thus, when a sequence of tokens is matched by several paths with different outputs (an ambiguous transducer), only a path with the highest weight will produce an output. This helps to assign exceptions a bigger weight. However, this approach is limited to the main graphs, as the weights are ignored in the subgraphs.

Normalization is the process of replacing NSW tokens with standard word tokens. That is why the original text can be normalized in several steps. Unitex has the tool Cassys that provides users with the possibility to create Unitex cascade of transducers, where a set of transducers can be applied to the text sequentially. I have created a Unitex cascade, which applies graphic abbreviation and acronym FSTs after applying the number FST.

3.1 Cardinal Numbers

The cardinal numerals in Russian agree with nouns in case, but the numerals *один* "one" and *два* "two" agree in gender as well ("two" has the same word form in masculine and in neuter). If the numeral is compound, all the constituent words agree with the corresponding noun, for example, *двадцати одного* ("twenty-one" in genitive masculine) and *двадцати одной* ("twenty-one" in genitive feminine), but *двадцать один* ("twenty-one" in nominative masculine). The numeral *одни* ("one" in plural) agrees only with pluralia tantum, e.g. *одни ножницы* "one pair of scissors", *одни брюки* "one pair of pants" [10].

I have developed separate FSTs for conversion of digits into numerals in different cases and for different digit positions. For example, I have a separate FST for numbers 20, 30, …, 90 in the nominative case (see Fig. 1a and 2x-9xncard block in Fig. 2), where "2" is converted into *двадцать* ("twenty"), and a separate FST for numbers 200, 300, …, 900 in the instrumental case, where "2" is converted into *двумястами* ("two hundred"). The same FSTs are used for conversion of the compound numbers 21, 22, …, 29, 31, …, 99 and 101, 102, …, 121, …, 999 by combining them into a separate FST. In Fig. 2 you can see how a subset of these numbers ending in 5–19 is converted into the corresponding numerals.

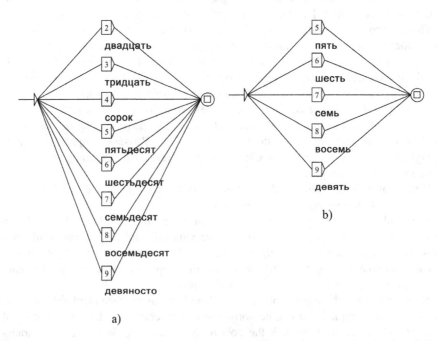

a)

b)

Fig. 1. Simple finite state transducers: (a) 2x-9xncard.grf for conversion the first digit of compound numbers 20, 21, 22, … 99 in the Nominative case; and (b) 5-9ncard.grf for conversion of numbers 5, 6, 7, 8 and 9 in the Nominative case. If the number in the box is matched, then it is substituted by the text under the box

The majority of such combined FSTs have left or right context nodes, which must match to apply the conversion rule. For example, to convert "2" into *две* ("two" in nominative feminine) before the word *ложки* ("spoons", a feminine noun), an FST must have the right context node <N:geF>, which searches for a feminine noun in genitive singular, because the numbers ending with "два/две", "три", "четыре" ("two", "three", "four") require the genitive case in singular [10].

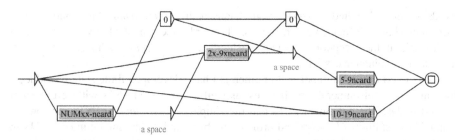

Fig. 2. A combined finite state transducer NUM-5-9-ncard.grf for conversion of compound numbers, ending in "5" to "19", into the corresponding numerals in the Nominative case. The nodes with labels against a dark background denote other FSTs (see corresponding 2x-9xncard and 5-9ncard FSTs in Fig. 1a and b). The nodes labeled "0" remove "0". The nodes with a comment "a space" under them add a space after digit conversion in other FSTs

Cardinal numbers can be used without nouns, e.g. in phone numbers and different codes. In this case the correspondent numeral is inflected in nominative masculine. That is why the combined FST for conversion of numbers into nominative singular does not have context nodes and is used as a fall-back conversion rule for non-matched contexts of other FSTs.

Sometimes a longer context is required to correctly pronounce numbers. For example, recipes have ranges, such as *1-2 столовые ложки растительного масла* "1-2 table spoons of vegetable oil", which must be converted into *одна-две столовые ложки растительного масла* "one-two table spoons of vegetable oil". In this case both numerals of the range agree with the noun *ложки* "spoons" in the feminine gender (the genitive case of nouns is used only with the nominative and accusative cases of numerals corresponding to numbers 2–9, so this rule applies only to the second number of this range). If the noun is in the masculine gender, both numerals in the preceding range must be in the masculine gender as well. For example, *1-2 стакана холодной воды* "1-2 glasses of cold water" must be converted into *один-два стакана холодной воды* "one-two glasses of cold water". To correctly convert both numbers to the corresponding numerals I developed a special FST for ranges before feminine and neuter nouns. The numbers in the ranges before masculine nouns are converted by the default combined FST, which does not require contexts.

3.2   Ordinal Numbers

Simple ordinal numerals in Russian agree with nouns in gender, case and number. In compound ordinal numerals only the last constituent word agrees with the noun [11]. For example, *две тысячи четырнадцатый* ("two thousand fourteenth" in masculine nominative) and *две тысячи четырнадцатом* ("two thousand fourteenth" in masculine prepositional).

Complex ordinal numbers ending in -00, –000, -000000, -000000000 are written without spaces. For example, "153000" is converted into *стопятидесятитрёхтысячный* "one hundred and fifty-three thousandth" in nominative masculine. Only the last constituent words *–сотый* "hundredth", *-тысячный* "thousandth", *-миллионный* "millionth", *-миллиардный* "billionth" agree with the nouns. The words preceding the last word are used in genitive plural (the exceptions are *сто* "one hundred" and *девяносто* "ninety", which are used in the nominative case) [12].

To convert ordinal numbers into numerals I have developed finite state transducers for simple ordinal numerals in all cases, as well as combined FSTs with left or right context nodes. The combined FSTs call the simple ordinal numeral FSTs for conversion of the last digits of compound ordinal numbers, and the cardinal numeral FSTs in nominative masculine for the words preceding the last one. The complex numerals are generated employing cardinal numeral FSTs in genitive masculine (or nominative masculine for words *сто* "one hundred" and *девяносто* "ninety").

These FSTs are used, for example, for converting numbers denoting years: *В 2002, 2007, 2010 годах удостоена Почётных грамот...* "In 2002, 2007, 2010 [she] was awarded Certificates of Merit..." is converted into *В две тысячи втором, две тысячи седьмом, две тысячи десятом годах удостоена Почётных грамот...* "In two thousand [second], two thousand [seventh], two thousand [tenth] [she] was awarded Certificates of Merit..." (literally translated).

3.3   Acronyms

An abbreviation in Russian is a noun consisting of shortened words of a phrase or consisting of shortened components of a compound word. An acronym is an abbreviation formed from the initial letters or sounds in a phrase or a word. Acronyms can be therefore alphabetic acronyms or initialisms, phonetic acronyms and alphabetic-phonetic acronyms [13, 14]. In alphabetic acronyms all the letters are pronounced separately as the letter names in the alphabet, e.g. *РФ* is pronounced as *эр-эф* [er ef]. Phonetic acronyms are pronounced as words, e.g. *ЮУрГУ* [ju:u:rgu:]. In alphabetic-phonetic acronyms some parts of the acronym are pronounced as words and other parts are pronounced letter by letter, e.g. *ЧелГУ* [chel ge u:].

Most acronyms should be converted into full words before speech synthesis, because it is difficult for people to comprehend a letter-by-letter pronunciation in speech and because acronyms are often rare for everybody to know what phrase the acronym corresponds to. For example, in the sentence (1) *ФГБОУ ВПО* should be converted into *Федеральное государственное бюджетное образовательное учреждение высшего профессионального образования* (Higher State State-Financed Educational Institution of Higher Professional Education) and *НИУ* should be converted into *Национальный исследовательский университет* (National Research University).

$$\begin{aligned}&\textit{ФГБОУ ВПО «Южно–Уральский государственный}\\&\textit{университет» } \big(\textit{НИУ}\big),\ \textit{на договорной основе...}\\&\text{HSSFEI HPE "South Ural State University" } \big(\text{NRU}\big)\ \text{on a}\\&\text{contractual basis...}\end{aligned} \qquad (1)$$

Some acronyms are known better than corresponding phrases, that is why they should not be converted into full words. The Russian National Corpus (RNC)[3], for example, has only 43 sentences with the term *дезоксирибонуклеиновая кислота* "deoxyribonucleic acid", 33 of which have the corresponding acronym *ДНК* "DNA". At the same time RNC has 2,313 entries of acronym *ДНК* "DNA". Compare the statistics of the term *Российская Федерация* "Russian Federation" and its acronym *РФ* "RF" in RNC: 19,368 and 18,052 entries correspondingly. Therefore the normalized text should have *Российская Федерация* "Russian Federation", but *ДНК* "DNA".

The main component of an acronym is a noun, that is why there can be 12 possible forms of the converted phrase (six cases and two numbers) in Russian. For example, *ВПО* in (1) is in the genitive singular. There are rules for all six cases in Normatex.

Acronyms can be ambiguous in different corpora. For example, *ВПП* in our corpus should be converted into *Всероссийской политической партии* "of All-Russian political party". Wikipedia lists 2 more interpretations of this acronym: *взлётно-посадочная полоса* "runway" and *Всемирная продовольственная программа* "World Food Programme". For all ambiguous or unknown acronyms I substitute each letter with its alphabet name, e.g. *ЧРО ВПП* in (2) will be replaced with *ЧээРО ВэПэПэ*.

$$\text{Благодарность ЧРО ВПП «Единая Россия».}$$
$$\text{Acknowledgement of ChRD ARPP "United Russia".} \tag{2}$$

3.4 Graphic Abbreviations

Abbreviation as a means of word-formation should not be confused with shortening, which is an elliptic form of speech naturally appearing in some communication situations. A graphic abbreviation is a conditional shortening of a frequently used word or a phrase, which can be found only in written speech and that is deciphered when is read [14].

There are simple graphic abbreviations, which have a single interpretation, for example, *и т.д.* "etc." should be converted into *и так далее* "et cetera", *т.е.* "i.e." should be converted into *то есть* "that is", etc. In other graphic abbreviations the interpretation depends on the context, for instance, *и др.* "et al." can have 3 interpretations: *и другие* "and others" (in nominative plural or accusative plural before inanimate nouns), *и других* "and others" (in genitive plural, accusative plural before animate nouns and in prepositional plural) and *и другим* "and others" (in dative plural). Sometimes, *и др.* "et al." can be converted into *и другое* "and other" (in nominative singular neuter).

When a Graphic Abbreviation Ending in a Period (Full Stop) Is Normalized the Final Period Should be Retained at the End of the Sentence

Graphic abbreviations can be ambiguous. If we take the case of *г.*, it may stand for *год* "year", *город* "city", *грамм* "gram" (every noun can have 12 word forms) or it can just be a letter name in the address (e.g., *Аудитория: 339-г, 339-д.* "Room 339-g,

---

[3] http://ruscorpora.ru

339-d.). To deal with all these cases sufficient left and right contexts should be provided in FSTs as well as FSTs should be applied in a definite order. For example, I normalize dates and measures of weight along with numbers before graphic abbreviations and acronyms, so that graphic abbreviation *г.* is converted into *год* "year" after the numbers denoting years and into *грамм* "gram" after the numbers denoting quantity of grams. In Normatex there are FSTs for normalizing dates in nominative, genitive and prepositional cases at the moment, which cover all the dates in our parallel corpus. Our corpus has 79 matches of token "г", of which there are no interpretations of *грамм* "gram". To test the FSTs of units of measurement an additional set of 269 sentences containing token "г" was taken from the Recipes subcorpus of the Russian National Corpus.

4 Results

Normatex was evaluated on our test parallel corpus. The results are presented in Table 1. 920 of 977 numbers (94.17 %) were normalized correctly. Incorrectly normalized numbers are mostly enumerations, non-standardly written phone numbers (e.g. *(8-35146)-30383*), dates (e.g. *2012.11.07*) and time ranges (*Обед 14.00–15.00*) as well as document numbers (e.g. *Инструкция № 3-1-0 г.*).

The highest precision (98.05 %) as well as the lowest recall (61.21 %) were obtained for conversion of graphic abbreviations. The overall recall is 84.33 % and the overall precision is 93.95 %.

Table 1. Evaluation results

Token type	Tokens	Correct	Errors	Recall	Precision
Numbers	977	920	53	**94.17 %**	94.55 %
Acronyms and initials	431	355	40	82.37 %	89.87 %
Graphic abbreviations	379	232	4	61.21 %	**98.05 %**
Total	1787	1507	97	84.33 %	93.95 %

5 Discussion

The result of the research is Normatex, an open-source normalization system for the Russian language. Normatex consists of 118 finite state transducers for conversion of cardinal and ordinal numbers into the corresponding numerals, which can preprocess different ranges, time, dates, telephone numbers, postal codes, etc. It has additional 33 FSTs for normalization of graphic abbreviations and acronyms to increase the accuracy of the generated text. The work is still in progress. I plan to expand the FSTs to cover different contexts of NSW in different corpora.

I plan to improve FSTs to deal with enumerations, non-standard phone numbers, dates, time ranges and document numbers.

As the reader can see, this preprocessing method depends hugely on the quality and completeness of the morphological dictionary. However, the context nodes of the proposed FSTs can have so called morphological filters, which allow using Perl regular expressions for matching words, which end in some combination of letters. For example, instead of using the lexical mask <N:teF>, which allows searching for feminine nouns in instrumental singular, we can use a combined lexical mask <N: teF>+<!DIC>≪..+ой$≫, which will also find the words, not listed in the dictionary, which end in *ой*. I plan to employ these morphological filters to guess the grammatical features of unknown words.

This developed Russian text normalization system is the first open-source project of such kind known to the author. As a consequence it is very hard to compare Normatex with other normalization systems. The only way to do it is to listen to and compare the synthesized speech outputs of commercial text-to-speech systems, when the original text and the preprocessed text by Normatex are fed into these systems. Random tests on VitalVoice[4] and Google Translate[5] showed improvements, when Normatex is used, but a more systematic comparison is required. By providing this resource I invite scientists to collaboratively build a high quality text-normalization system.

References

1. Reichel, U.D., Pfitzinger, H.R.: Text preprocessing for speech synthesis (2006)
2. The Festival Speech Synthesis System. http://www.cstr.ed.ac.uk/projects/festival/
3. Unitex 3.1beta. http://www-igm.univ-mlv.fr/~unitex/
4. Paumier, S.: Unitex 3.1.beta User Manual. Université Paris-Est Marne-la-Vallée. http://igm.univ-mlv.fr/~unitex/UnitexManual3.1.pdf (2015). Accessed 15 Jan 2015
5. Dutoit, T.: An Introduction to Text-to-Speech Synthesis, vol. 3. Springer Science & Business Media, Berlin (1997)
6. Sproat, R., Black, A., Chen, S., Kumar, S., Ostendorfk, M., Richards, C.: Normalization of non-standard words. Comput. Speech Lang. **15**, 287–333 (2001)
7. Sproat, R.: Lightly supervised learning of text normalization: Russian number names. In: Spoken Language Technology Workshop (SLT), 2010 IEEE, pp. 436–441. IEEE, December 2010
8. Khomitsevich, O.G., Rybin, S.V., Anichkin, I.M.: Linguistic analysis for text normalization and homonymy resolution in a Russian TTS system [Использование лингвистического анализа для нормализации текста и снятия омонимии в системе синтеза русской речи]. Instrument making. Thematic issue "Speech information systems" [Приборостроение. Тематический выпуск «Речевые информационные системы»], vol. 2, pp. 42–46. Izvestija vuzov (2013)
9. Nagel, S.: Formenbildung im Russischen. Formale Beschreibung und Automatisierung für das CISLEX-Wörterbuchsystem (2002)
10. Russian Grammar [Русская грамматика], vol. 1. Nauka, Moscow (1980)
11. Rosental, D.E., Golub, I.B., Telenkova, M.A.: The Modern Russian Language [Современный русский язык]. Airis-Press, Moscow (1997)
12. Rosental, D.E., Djandjakova, E.V., Kabanova, N.P.: Reference Book on Orthography, Pronunciation, Literary Editing [Справочник по правописанию, произношению, литературному редактированию]. CheRo, Moscow (1998)

---

[4] http://cards.voicefabric.ru/

[5] https://translate.google.ru/

13. Linguistics. Big encyclopedic dictionary [Языкознание. Большой энциклопедический словарь]. Big Russian Encyclopedy, Moscow (1998)
14. Akhmanova, O.S.: The Dictionary of Linguistic Terms [Словарь лингвистических терминов]. Editorial URSS, Moscow (2004)

Analysis of Images and Videos

Transform Coding Method for Hyperspectral Data: Influence of Block Characteristics to Compression Quality

Marina Chicheva[1,2(✉)] and Ruslan Yuzkiv[1,2]

[1] Samara State Aerospace University, Samara, Russia
{marina.chicheva, ruslan.yuzkiv}@gmail.com
[2] Image Processing Systems Institute, RAS, Samara, Russia

Abstract. The aim of this paper is to study how block characteristics influence on the compression quality fo hyperspectral data using block transform method. Coordinates in hyperspectral image are not equivalent – two of them are space-based ones and the third coordinate corresponds to a spectral channel. Thus it is necessary to investigate the algorithm implementation with blocks extended along spectral axis. Moreover, it is known that in two-dimensional case, an increasing block size leads to improvement of compression quality. Hence it is useful to investigate the algorithm implementation with cubic blocks of increased size.

Keywords: Hyperspectral data compression · Transform coding method · Discrete cosine transform · Block size

1 Introduction

The aim of the paper is to study the parameters of the earlier proposed method for hyperspectral data compression based on transform coding method [1].

Nowadays there are very urgent problems, which deal with earth remote sensing. So-called hyperspectral data (or images) are especially interesting. They are represented as three-dimensional array consisting of a set of images for the same region obtained in different spectral bands (as illustrated in Fig. 1).

It is obvious that the volume of every hyperspectral image is extremely large. That leads to complications during data storage and transmission. So the community interest in the problem of hyperspectral data compression is high. In recent years many articles were published on this problem, where researchers propose various hyperspectral data compression methods: principal component analysis method [2, 3], difference schemes [4], methods, based on inter-band correlation [5], wavelet methods [6]. There also are some works related to lossless compression [7] and other approaches [8, 9].

It was shown earlier [1] that transform coding method generalized by authors for the case of three-dimensional (hyperspectral) data provides compression quality comparable to state-of-the-art approaches. In this method the fixed block size, that equals to $8 \times 8 \times 8$, was used. However, coordinates in hyperspectral image are not equivalent (two of them are space-based ones and third coordinate corresponds to a spectral channel). So it is necessary to study the algorithm implementation with blocks

© Springer International Publishing Switzerland 2015
M.Y. Khachay et al. (Eds.): AIST 2015, CCIS 542, pp. 51–56, 2015.
DOI: 10.1007/978-3-319-26123-2_5

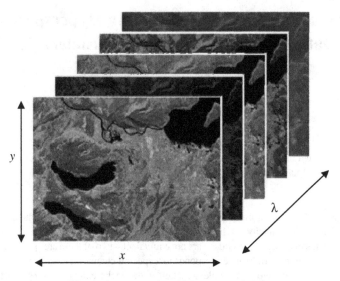

Fig. 1. An example of hyperspectral data

which are extended along spectral axis. Moreover, it is known that in two-dimensional case, an increasing block size leads to improvement of compression quality (see 6.5.2, [10]). So it is useful to study the algorithm implementation with cubic blocks of increased size as well.

Below we describe the considered compression algorithm and show experimental results.

2 Generalized Transform Coding Method

Let us consider the hyperspectral image as three-dimensional array of brightness values:

$$\{f(n_1, n_2, n_3)\}_{n_1, n_2, n_3=0}^{N_1-1, N_2-1, N_3-1},$$

where N_1 and N_2 denote spatial sizes of image (height and width respectively), N_3 denotes the number of spectral channels.

In the beginning, we partition data array into non-overlapping blocks. Then three-dimensional discrete cosine transform (DCT) is performed for each block:

$$\sum_{n_1=0}^{N_1-1} \sum_{n_2=0}^{N_2-1} \sum_{n_3=0}^{N_3-1} f(n_1, n_2, n_3) h_{m_1}(n_1) h_{m_2}(n_2) h_{m_3}(n_3)$$

where

$$h_m(n) = \lambda_m \cos \frac{\pi(2n+1)m}{2N}, \quad \lambda_m = \begin{cases} \sqrt{\frac{1}{N}} \text{ for } m = 0, \\ \sqrt{\frac{2}{N}} \text{ for } m \neq 0. \end{cases}$$

Further, selection and quantization of significant transformants (coefficients) are carried out. Each transformant is quantized and coded by binary word with length $b = \Sigma_m b_m$ where $m = (m_1, m_2, m_3)$ is transformant number in block. Total length of all binary words

$$b = \sum_m b_m$$

is defined by required compression ratio.

Selection of significant transformants can be performed as allocation of total bit number b for each transformant. A suboptimal iterative bit allocation algorithm (see Sect. 6.5.4 [10]), generalized for three-dimensional transformation, is used for bit allocation. Assuming that each transformant $F(m_1, m_2, m_3)$ follows a Gaussian distribution, where (m_1, m_2, m_3) is not equal to $(0, 0, 0)$ we use a Lloyd-Max quantizer [11], which minimizes the mean-squared error. Finally, quantized values for each block are written to archive file with some metadata, such as gauss distribution parameters and binary words lengths.

Note that considered block approach allows to restore not only all data entirely. In addition it allows restoring a certain spatial layer (with fixed band number) as well as certain spectral reflection curves (with fixed spatial coordinates).

3 Experimental Research

The experiments are carried out using public set of hyperspectral images [12], obtained by «Aviris» scanner in 2006. It provides spectral response in the bandwidth from 400 to 2500 nm, with 224 contiguous channels and pixel resolution 680 × 512. Figure 2 shows one image from the set in spectral band No 130.

Fig. 2. «Landscape» image (scene No 18) in spectral band No 130

The major purpose of the current work is to study efficiency of the transform coding method with larger size of cubic and non-cubic blocks. All the results are comparing against the reference algorithm, which uses cubic $8 \times 8 \times 8$ block. Figure 3 shows peak signal-to-noise ratio (PSNR) of the same algorithm with respect to various compression rates for cubic blocks $16 \times 16 \times 16$ and $32 \times 32 \times 32$ Fig. 4 shows dependence of compression quality for the extended blocks. Block size $8 \times 8 \times L$ was considered, where $L = 16, 32, 64, 128, 224$. In this work «noise» means the error, which occurs after compression and decompression procedures.

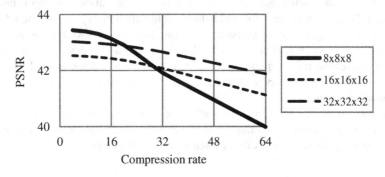

Fig. 3. Influence of cubic block size on compression quality

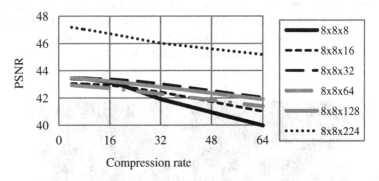

Fig. 4. Influence of blocks size along spectral axe on compression quality

Plots, represented on Fig. 3, show that an increase of cubic block size provides a significant improvement of compression quality only for very large compression rates (more than 32). On low compression rates standard cubic block size $8 \times 8 \times 8$ provides satisfying quality. It can be explained by the fact that we have enough transformants data in archive file for high-quality data reconstruction.

Figure 4 shows, that increasing block size along spectral coordinate can significantly improve the quality of reconstructed hyperspectral data. Increasing L to the number of spectral bands (224) provides the best quality. Disadvantage of such approach is the absence of fast transformations with specified length. However, this problem can be resolved almost for all sizes of hyperspectral images. Some degeneration of compression quality may be found at $L = 64, 128$ in comparison with $L = 32$. It deals with the number of spectral channels in the considered hyperspectral image, which is indivisible to block size. We have to add data and that leads to distortion in statistic.

We report the spatial layers for different block sizes and scenes for visual estimation of recovery quality. Besides, we show the results for another block sizes and the performed comparative analysis with another author's works.

One of the major applied task in hyperspectral imagery is analysis of spectral reflection curves. Figure 5 shows the results of spectral reflection curve recovery for spatial point (280, 340) on scene No 18. It can be noticed, that increasing block size along spectral coordinate provides an improvement of recovery quality. In the comprehensive technical report we will include more detailed information of the obtained results.

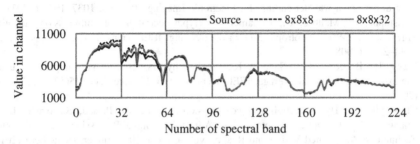

Fig. 5. Spectral reflection curve reconstruction for the point on scene No 18 with compression rate 32

4   Conclusion

This paper shows results of the experimental study of the transform coding method's efficiency for hyperspectral image compression. It has been shown, that simple increase of cubic block size provides a notable improvement of compression quality only for very large compression rates (more than 32). At the same time, using extended blocks along the spectral coordinate can significantly improve recovery quality in case of entirely hyperspectral images and in relation to spectral reflection curves. Transform length along spectral axe, that equals to the number of bands, provides the best quality in PSNR sense.

Acknowledgements. This work was supported by the Russian Scientific Foundation (RSF) grant N14-31-00014 "Establishment of a Laboratory of Advanced Technology for Earth Remote Sensing".

References

1. Chicheva, M.A., Yuzkiv, R.R.: Hyperspectral image compression using transform coding method. Comput. Opt. **38**(4), 798–803 (2014). (In Russian)
2. Smirnov, S.I., Mikhailov, V.V., Ostrikov, V.N.: Application of randomized principal component analysis for compression of hyperspectral data. Sovremennye problem distancionnogo zondirovanija Zemli iz kosmosa – Mod. Probl. Remote Sens. Earth, **11** (**2**), 9–17 (2014). (In Russian)
3. Ramakrishna, B., Wang, J., Chang, C.-I., Plaza, A., Ren, H., Chang, C.-C., Jensen, J.L., Jensen, J.O.: Spectral/spatial hyperspectral image compression in conjunction with virtual dimensionality. In: Proceedings of SPIE, vol. 5806, pp. 772–781 (2005)
4. Sarinova, A.Z., Zamjatin, A.V.: Hyperspectral remote sensing images compression algorithm. In: Sborniktrudov XI Mezhdunarodnoj nauchno-prakticheskoj konferencii studentov, aspirantov I molodyh uchjonyh, Tomsk, 13–16 noyabra (2013). (Proceedings of XI International Scientific-Practical Conference of PhD Students and Young Scientists, Tomsk, pp. 384–386 (2013) (In Russian)
5. Tang, X., Pearlman, W.A., Modestino, J.W.: Hyperspectral image compression using three-dimensional wavelet coding. In: Electronic Imaging 2003, pp. 1037–1047 (2003)
6. Christophe, E., Mailhes, C., Duhamel, P.: Hyperspectral image compression: adapting SPIHT and EZW to anisotropic 3-D wavelet coding. IEEE Trans. Image Process. **17**(12), 2334–2346 (2008)
7. Wang, H., Babacan, S.D., Sayood, K.: Lossless hyperspectral-image compression using context-based conditional average. IEEE Trans. Geosci. Remote Sens. **45**(12), 4187–4193 (2007)
8. Christophe, E.: Hyperspectral data compression tradeoff. In: Prasad, S., Bruce, L.M., Chanussot, J. (eds.) Optical remote sensing. Advances in Signal Processing and Exploitation Techniques. Augmented Vision and Reality, vol. 3, pp. 9–30. Springer, Heidelberg (2011)
9. Nian, Y., He, M., Wan, J.: Low-complexity compression algorithm for hyperspectral images based on distributed source coding. Math. Probl. Eng. **2013**, 7 (2013). Article ID 8256732013
10. Soifer, V.A.: Computer Image Processing. Part II. Methods and Algorithms, p. 568. VDM Verlag Dr. Muller Aktiengesellschaft & Co. KG, Saarbrucken (2010)
11. Pratt, W.K.: Digital Image Processing, p. 750. Wiley, New York (1978)
12. Hyperspectral Image Compression. NASA Jet Propulsion Laboratory. http://compression.jpl.nasa.gov/hyperspectral/

Fréchet Filters for Color and Hyperspectral Images Filtering

Ekaterina Ostheimer[1]([✉]), Valeriy Labunets[2], Denis Komarov[2],
and Tat'yana Fedorova[2]

[1] Capricat LLC, 1340 S., Ocean Blvd., Suite 209, Pompano Beach, FL 33062, USA
katya@capricat.com
[2] Ural Federal University, Pr. Mira, 19, Yekaterinburg 620002, Russian Federation
vlabunets05@yahoo.com

Abstract. Median filtering has been widely used in scalar-valued image processing as an edge preserving operation. The basic idea is that the pixel value is replaced by the median of the pixels contained in a window around it. In this paper, we extend the notion of the Fréchet vector median to the general Fréchet vector median, which minimizes the Fréchet cost function (FCF) in the form of an aggregation function instead of the ordinary sum. Moreover, we propose to use an aggregation distance instead of the classical one. We use the generalized Fréchet median for constructing new nonlinear filters based on an arbitrary pair of aggregation operators that can be changed independently. For each pair of parameters, we get the unique class of new nonlinear filters.

Keywords: Nonlinear filters · Hyperspectral image processing · Generalized aggregation mean

1 Introduction

We develop a conceptual framework and design methodologies for multichannel image median filtering systems with assessment capability. The term multichannel (= multicomponent, multispectral, multicolor, hyperspectral) image is used for an image with more than one component. They are composed of a series of images in different optical bands at wavelengths $\lambda_1, \lambda_2, ..., \lambda_{K-1}$, called the spectral channels: $\mathbf{f}(x, y) = (f_{\lambda_1}, f_{\lambda_2}, ..., f_{\lambda_K})$, where K is the number of different optical channels, *i.e.*, $\mathbf{f}(x, y) : \mathbf{R}^2 \longrightarrow \mathbf{R}^K$, where $\mathbf{R}^K$ is multicolor space.

Let us introduce the observation model and notion used throughout the paper. We consider noise images of the $\mathbf{f}(\mathbf{x}) = \mathbf{s}(\mathbf{x}) + \mathbf{n}(\mathbf{x})$, where $\mathbf{s}(\mathbf{x})$ is the original K-channel image $\mathbf{s}(\mathbf{x}) = (s_1(\mathbf{x}), s_2(\mathbf{x}), ..., s_K(\mathbf{x}))$ and $\mathbf{n}(\mathbf{x})$ denotes the K-channel noise $\eta(\mathbf{x}) = (n_1(\mathbf{x}), n_2(\mathbf{x}), ..., n_K(\mathbf{x}))$ introduced into the image $\mathbf{s}(\mathbf{x})$ to produce the corrupted image $\mathbf{f}(\mathbf{x}) = (f_1(\mathbf{x}), f_2(\mathbf{x}), ..., f_K(\mathbf{x}))$. Here, $\mathbf{x} = (i, j) \in \mathbf{Z}^2$ is a 2D coordinates that belong to the image domain and represent the pixel location. The aim of image enhancement is to reduce the noise as much as possible or to find a method, which, given $\mathbf{s}(\mathbf{x})$, derives an image

© Springer International Publishing Switzerland 2015
M.Y. Khachay et al. (Eds.): AIST 2015, CCIS 542, pp. 57–70, 2015.
DOI: 10.1007/978-3-319-26123-2_6

$\widehat{\mathbf{s}}(\mathbf{x})$ as close as possible to the original $\mathbf{s}(\mathbf{x})$ subjected to a suitable optimality criterion. In a 2D standard linear and median scalar filters with a square N-cellular window $M_{(i,j)}(m,n)$ and located at (i,j), the mean and median replace the central pixel

$$\widehat{s}(i,j) = \underset{(k,l)\in M(i,j)}{\mathbf{Mean}} [f(k,l)], \tag{1}$$

$$\widehat{s}(i,j) = \underset{(k,l)\in M(i,j)}{\mathbf{Med}} [f(k,l)], \tag{2}$$

where $\widehat{s}(i,j)$ is the filtered grey-level image, $\{f(k,l)\}_{(k,l)\in M_{(i,j)}}$ is an image block of the fixed size N extracted from f by moving N-cellular window $M_{(i,j)}$ at the position (i,j), **Mean** and **Med** is the mean (average) and median operators. Median filtering has been widely used in image processing as an edge preserving filter. The basic idea is that the pixel value is replaced by the median of the pixels contained in a window around it. In this work, this idea is extended to vector-valued images, based on the fact that the median is also the value that minimizes the L_1 distance in $\mathbf{R}$ between all the grey-level pixels in the N-cellular window (see Fig. 1).

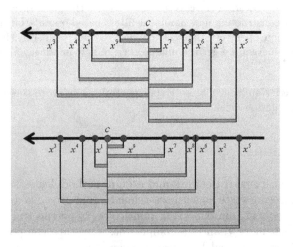

Fig. 1. Distances from an arbitrary point c to each point $x^1, x^2, ..., x^9 \in \mathbf{R}$ from 9-cellular window

In multichannel case, we need to define a distance ρ between pair of objects on the domain $\mathbf{R}^K$. Let $\langle \mathbf{R}^K, \rho \rangle$ be a metric multicolor space, and $\rho(\mathbf{x}, \mathbf{y})$ is a distance function for pair of objects $\mathbf{x}$ and $\mathbf{y}$ in $\mathbf{R}^K$ (that is, $\rho(\mathbf{x}, \mathbf{y}) : \mathbf{R}^K \times \mathbf{R}^K \longrightarrow \mathbf{R}^+$). Let $w^1, w^2, ..., w^N$ be N weights summing to 1 and let $\mathbf{x}^1, \mathbf{x}^2, ..., \mathbf{x}^N \subset \mathbf{R}^K$ be N observations (for example, N pixels in the N-cellular window).

Definition 1. *The optimal Fréchet point associated with the metric $\rho(\mathbf{x}, \mathbf{y})$ is the point $\mathbf{c}_{\mathrm{opt}} \in \mathbf{R}^K$ that minimizes the Fréchet cost function $\sum_{i=1}^{N} w_i \rho(\mathbf{c}, \mathbf{x}^i)$ (the weighted sum distances from an arbitrary point $\mathbf{c}$ to each point $\mathbf{x}^1, \mathbf{x}^2, \ldots, \mathbf{x}^N$) [1–3]. It is formally defined as*

$$\mathbf{c}_{\mathrm{opt}} = \mathbf{FrechPt}\left(\rho | \mathbf{x}^1, \mathbf{x}^2, \ldots, \mathbf{x}^N\right) = \arg\min_{\mathbf{c} \in \mathbf{R}^K} \left(\sum_{i=1}^{N} w_i \rho(\mathbf{c}, \mathbf{x}^i)\right). \quad (3)$$

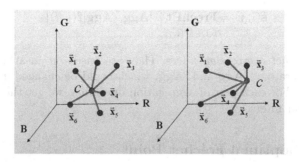

Fig. 2. Distances from an arbitrary point $\mathbf{c}$ to each point $\mathbf{x}^1, \mathbf{x}^2, \ldots, \mathbf{x}^9 \in \mathbf{R}^K$ from 9-cellular window

Note that **arg** min means the argument, for which the sum is minimized. In this case, it is the point $\mathbf{c}_{\mathrm{opt}}$ from $\mathbf{R}^K$, for which the sum of all distances to the $\mathbf{x}^i$s is minimum. So, the optimal Fréchet point of a discrete set of the observations (N pixels) in the metric space $\langle \mathbf{R}^K, \rho \rangle$ is the point minimizing the sum of distances to the N pixels (see Fig. 2). This generalizes the ordinary median, which has the property of minimizing the sum of distances for one-dimensional data. The properties of this point have been extensively studied since the time of Fermat, (this point is often called the *Fréchet point* [1] or *Fermat–Weber point* [4]). In this paper, we extend the notion of the Fréchet point to the generalized Fréchet point which minimizes the aggregation Fréchet cost function (AFCF) in the form of an aggregation function $^{cf}\mathbf{Agg}_{i=1}^{N}\left[w_i \rho(\mathbf{c}, \mathbf{x}^i)\right]$, instead of the sum (3):

$$\mathbf{c}_{\mathrm{opt}} = \mathbf{FrechPt}\left(^{cf}\mathbf{Agg}, \rho | \mathbf{x}^1, \mathbf{x}^2, \ldots, \mathbf{x}^N\right)$$

$$= \arg\min_{\mathbf{c} \in \mathbf{R}^K} \left(^{cf}\mathbf{Agg}_{i=1}^{N}\left[w_i \rho(\mathbf{c}, \mathbf{x}^i)\right]\right). \quad (4)$$

Moreover, we propose use an aggregation distance $^{\rho}\mathbf{Agg}(\mathbf{c}, \mathbf{x})$ instead of the classical distance ρ. It is gives new cost function

$$^{cf}\mathbf{Agg}\left[w_1{}^{\rho}\mathbf{Agg}(\mathbf{c}, \mathbf{x}^1), w_2{}^{\rho}\mathbf{Agg}(\mathbf{c}, \mathbf{x}^2), \ldots, w_N{}^{\rho}\mathbf{Agg}(\mathbf{c}, \mathbf{x}^N)\right]$$

and new Fréchet point associated with the aggregation distance $^\rho\mathbf{Agg}(\mathbf{c}, \mathbf{x})$ and the aggregation Fréchet cost function $^{cf}\mathbf{Agg}$

$$\mathbf{c}_{\text{opt}} = \mathbf{FrechPt}\left(^{cf}\mathbf{Agg}, {}^\rho\mathbf{Agg}|\mathbf{x}^1, \mathbf{x}^2, \ldots, \mathbf{x}^N\right)$$

$$= \arg \min_{\mathbf{c} \in \mathbf{R}^K} \left(^{cf}\mathbf{Agg}_{i=1}^N \left[w_i \cdot {}^\rho\mathbf{Agg}\left(\mathbf{c}, \mathbf{x}^i\right)\right]\right) \tag{5}$$

We use the generalized Fréchet point for constructing new nonlinear filter. When filter (1) is modified as follows:

$$\widehat{\mathbf{s}}(i, j) = \mathbf{FrechPt}_{(k,l) \in M(i,j)}\left[^{cf}\mathbf{Agg}, {}^\rho\mathbf{Agg}|\mathbf{f}(k, l)\right] \tag{6}$$

it becomes *Fréchet aggregation filters*. They are based on an arbitrary pair of aggregation operators $^{cf}\mathbf{Agg}$ and $^\rho\mathbf{Agg}$, which could be changed independently of one another. For each pair of aggregation operators, we get the unique class of new nonlinear filters.

2   The Suboptimal Fréchet Point

In computation point view, it is better to restrict the infinite search domain from $\mathbf{R}^K$ until the finite subset $\mathbf{D} = \left\{\mathbf{x}^1, \mathbf{x}^2, \ldots, \mathbf{x}^N\right\} \subset \mathbf{R}^K$. In this case, we obtain definition of the *suboptimal Fréchet point* or the *optimal Fréchet median*.

Definition 2. *The suboptimal Fréchet point (or optimal Fréchet median) associated with the metric $\rho(\mathbf{x}, \mathbf{y})$ is the point $\widehat{\mathbf{c}} \in \mathbf{D}$ that minimizes the FCF over the restricted search domain $\mathbf{D} \subset \mathbf{R}^K$*

$$\widehat{\mathbf{c}}_{\text{opt}} = \mathbf{FrechMed}\left(\rho|\mathbf{x}^1, \mathbf{x}^2, \ldots, \mathbf{x}^N\right) = \arg \min_{\mathbf{c} \in \mathbf{D}} \left(\sum_{i=1}^N w_i \rho(\mathbf{c}, \mathbf{x}^i)\right). \tag{7}$$

We use the generalized Fréchet point and median for constructing new non-linear filters. When filters (1)–(2) are modified as follows:

$$\widehat{\mathbf{s}}(i, j) = \mathbf{FrechPt}_{(k,l) \in M(i,j)}\left[\rho|\mathbf{f}(k, l)\right], \tag{8}$$

$$\widehat{\mathbf{s}}(i, j) = \mathbf{FrechMed}_{(k,l) \in M(i,j)}\left[\rho|\mathbf{f}(k, l)\right] \tag{9}$$

it becomes *Fréchet mean* and *median filters*, associated with the metric ρ.

Example 1. If observation data are real numbers, i.e., $x^1, x^2, \ldots, x^N \in \mathbf{R}$, and the distance function is the city distance $\rho(x, y) = \rho_1(x, y) = |x - y|$, then the optimal Fréchet point (3) and median (7) for data $x^1, x^2, \ldots, x^N \in \mathbf{R}$ to be the classical *Fréchet point* and *classical median*, respectively. They are associated with the city metric $\rho_1(x, y)$, *i.e.*,

$$\mathbf{c}_{\text{opt}} = \mathbf{FrechPt}\left(\rho_1 | \mathbf{x}^1, \mathbf{x}^2, \ldots, \mathbf{x}^N\right) = \arg \min_{c \in \mathbf{R}} \left(\sum_{i=1}^{N} |c - x^i|\right), \qquad (10)$$

$$\widehat{\mathbf{c}}_{\text{opt}} = \mathbf{FrechMed}\left(\rho_1 | \mathbf{x}^1, \mathbf{x}^2, \ldots, \mathbf{x}^N\right)$$

$$= \arg \min_{c \in \mathbf{D}} \left(\sum_{i=1}^{N} |c - x^i|\right) = \mathbf{Med}\left(x^1, x^2, \ldots, x^N\right). \qquad (11)$$

In this case, filter (8) is *optimal maximum likelihood filter* for Laplace noise, and filter (9) is ordinary *median filter*.

Example 2. If observation data are vectors, *i.e.*, $\mathbf{x}^1, \mathbf{x}^2, \ldots, \mathbf{x}^N \in \mathbf{R}^K$ and the distance function is the city distance $\rho(\mathbf{x}, \mathbf{y}) = \rho_1(\mathbf{x}, \mathbf{y}) = ||\mathbf{x} - \mathbf{y}||_1$, then the Fréchet point (3) and median (7) for vectors $\mathbf{x}^1, \mathbf{x}^2, \ldots, \mathbf{x}^N \in \mathbf{R}^K$ to be the *Fréchet* point (vector) and *vector median*, respectively, associated with the same metric $\rho_1(\mathbf{x}, \mathbf{y})$

$$\mathbf{c}_{\text{opt}} = \mathbf{FrechPt}\left(\rho_1 | \mathbf{x}^1, \mathbf{x}^2, \ldots, \mathbf{x}^N\right) = \arg \min_{c \in \mathbf{R}^K} \left(\sum_{i=1}^{N} ||\mathbf{c} - \mathbf{x}^i||_1\right), \qquad (12)$$

$$\widehat{\mathbf{c}}_{\text{opt}} = \mathbf{FrechMed}\left(\rho_1 | \mathbf{x}^1, \mathbf{x}^2, \ldots, \mathbf{x}^N\right)$$

$$= \arg \min_{c \in \mathbf{D}} \left(\sum_{i=1}^{N} ||\mathbf{c} - \mathbf{x}^i||_1\right) = \mathbf{VecMed}\left(\rho_1 | \mathbf{x}^1, \mathbf{x}^2, \ldots, \mathbf{x}^N\right). \qquad (13)$$

In this case, filter (8) is *optimal maximum likelihood vector filter* for Laplace noise, and filter (9) is *vector median filter* associated with city metric [5,6].

Example 3. If observation data are vectors again: $\mathbf{x}^1, \mathbf{x}^2, \ldots, \mathbf{x}^N \in \mathbf{R}^K$ but the distance function is the Euclidean $\rho(\mathbf{x}, \mathbf{y}) = \rho_2(\mathbf{x}, \mathbf{y}) = ||\mathbf{x} - \mathbf{y}||$, then the Fréchet point (3) and median (7) for vectors $\mathbf{x}^1, \mathbf{x}^2, \ldots, \mathbf{x}^N \in \mathbf{R}^K$ to be the *Fréchet* point (vector) and *vector median*, respectively, associated with the Euclidean metric $\rho_2(\mathbf{x}, \mathbf{y})$

$$\mathbf{c}_{\text{opt}} = \mathbf{FrechPt}\left(\rho_2 | \mathbf{x}^1, \mathbf{x}^2, \ldots, \mathbf{x}^N\right) = \arg \min_{c \in \mathbf{R}^K} \left(\sum_{i=1}^{N} ||\mathbf{c} - \mathbf{x}^i||_2\right), \qquad (14)$$

$$\widehat{\mathbf{c}}_{\text{opt}} = \mathbf{FrechMed}\left(\rho_2 | \mathbf{x}^1, \mathbf{x}^2, \ldots, \mathbf{x}^N\right)$$

$$= \arg \min_{c \in \mathbf{D}} \left(\sum_{i=1}^{N} ||\mathbf{c} - \mathbf{x}^i||_2\right) = \mathbf{VecMed}\left(\rho_2 | \mathbf{x}^1, \mathbf{x}^2, \ldots, \mathbf{x}^N\right). \qquad (15)$$

In this case, filter (8) is *optimal maximum likelihood vector filter* for Gaussian noise, and filter (9) is *vector median filter* associated with Euclidean metric.

3   Generalized Vector Aggregation

In Definitions 1 and 2, the Fréchet point and median are points $\mathbf{c}_{\text{opt}} \in \mathbf{R}^K$, $\widehat{\mathbf{c}}_{\text{opt}} \in \mathbf{D} = \{\mathbf{x}^1, \mathbf{x}^2, \ldots, \mathbf{x}^N\}$ that minimize the Fréchet cost function (FCF) $\sum_{i=1}^{N} w_i \rho(\mathbf{c}, \mathbf{x}^i)$. But this sum up to constant factor is the simplest aggregation function [7–9].

The aggregation problem [7–9] consist in aggregating N-tuples of objects all belonging to a given set $\mathbf{S}$, into a single object of the same set $\mathbf{S}$, *i.e.*, $\mathbf{Agg}$: $\mathbf{S}^N \longrightarrow \mathbf{S}$. In the case of mathematical aggregation operator (AO) the set $\mathbf{S}$, is an interval of the real $\mathbf{S} = [0, 1] \subset \mathbf{R}$, or integer numbers $\mathbf{S} = [0, 255] \subset \mathbf{Z}$. In this setting, an AO is simply a function, which assigns a number y to any N-tuple of numbers $(x_1, x_2, \ldots, x_N)$: $y = \mathbf{Agg}(x_1, x_2, \ldots, x_N)$ that satisfies:

1. $\mathbf{Agg}(x) = x$.
2. $\mathbf{Agg}(a, a, \ldots, a) = a$.
   In particular, $\mathbf{Agg}(0, 0, \ldots, 0) = 0$ and $\mathbf{Agg}(1, 1, \ldots, 1) = 1$ (or $\mathbf{Agg}(255, 255, \ldots, 255) = 255$).
3. $\min(x_1, x_2, \ldots, x_N) \leq \mathbf{Agg}(x_1, x_1, \ldots, x_N)) \leq \max(x_1, x_2, \ldots, x_N$.

Here $\min(x_1, x_2, \ldots, x_N)$ and $\max(x_1, x_2, \ldots, x_N)$ are respectively the *minimum* and the *maximum* values among the elements of $(x_1, x_2, \ldots, x_N)$. All other properties may come in addition to this fundamental group. For example, if for every permutation $\forall \sigma \in S_N$ of $\{1, 2, \ldots, N\}$ the AO satisfies:

$$y = \mathbf{Agg}(x_{\sigma(1)}, x_{\sigma(2)}, \ldots, x_{\sigma(N)}) = \mathbf{Agg}(x_1, x_2, \ldots, x_N),$$

then it is invariant (symmetric) with respect to the permutations of the elements of $(x_1, x_2, \ldots, x_N)$. In other words, as far as means are concerned, the *order* of the elements of $(x_1, x_2, \ldots, x_N)$ is - and must be - completely irrelevant. According to Kolmogorov [9] a sequence of functions $\mathbf{Agg}_N = \mathbf{Agg}(x_1, x_2, \ldots, x_N)$ (for different N) defines a regular type of average if the following conditions are satisfied:

1. $\mathbf{Agg}_N$ is continuous and monotone in each variable.
2. $\mathbf{Agg}_N$ is increasing in each variable.
3. $\mathbf{Agg}_N$ is a symmetric function.
4. The average of identical numbers is equal to their common value:
   $\mathbf{Agg}(x, x, \ldots, x) = x$.
5. A group of values can be replaced by their own average, without changing the overall average: $\mathbf{Agg}_{N+M}(x_1, x_2, \ldots, x_N; y_1, y_2, \ldots, y_M) = \mathbf{Agg}_{N+M}(m, m, \ldots, m; y_1, y_2, \ldots, y_M)$, where $m = \mathbf{Agg}(x_1, x_2, \ldots, x_N)$.

Theorem 1 (Kolmogorov [9]). *If conditions 1–5 are satisfied, the generalized Kolmogorov mean (average) $\mathbf{Agg}(x_1, x_2, \ldots, x_N)$ is of the form:*

$$\mathbf{Agg}(x_1, x_2, \ldots, x_N) = \mathbf{Kol}(K | x_1, x_2, \ldots, x_N) = K^{-1} \left[\frac{1}{N} \sum_{i=1}^{N} K(x_i) \right],$$

where K is a strictly monotone continuous function in the extended real line.

We list below a few particular cases of aggregation means:

1. Arithmetic mean $(K(x) = x)$: $\mathbf{Mean}(x_1, x_2, \ldots, x_N) = \frac{1}{N} \sum\limits_{i=1}^{N} x_i$.

2. Geometric mean $(K(x) = \log(x))$: $\mathbf{Geo}(x_1, x_2, \ldots, x_N) = \sqrt[N]{\left(\prod_{i=1}^{N} x_i \right)}$.

3. Harmonic mean $(K(x) = x^{-1})$: $\mathbf{Harm}(x_1, x_2, \ldots, x_N) = \left(\frac{1}{N} \sum\limits_{i=1}^{N} x_i^{-1} \right)^{-1}$.

4. One-parametric family quasi arithmetic (power or Hólder) means corresponding to the functions $K(x) = x^p$: $\mathbf{Hold}(x_1, x_2, \ldots, x_N) = \sqrt[p]{\left(\frac{1}{N} \sum\limits_{i=1}^{N} x_i^p \right)}$. This family is particularly interesting, because it generalizes a group of common means, only by changing the value of p.

A very notable particular cases correspond to the logic functions $(\min, \max,$ median$)$: $y = \mathbf{Min}(x_1, \ldots, x_N)$, $y = \mathbf{Max}(x_1, \ldots, x_N)$, $y = \mathbf{Med}(x_1, \ldots, x_N)$. When filters (1) and (2) are modified as follows:

$$\widehat{\mathbf{s}}(i,j) = \underset{(k,l) \in M(i,j)}{\mathbf{Agg}} [\mathbf{f}(k,l)], \tag{16}$$

we get the unique class of nonlinear *aggregation filters* proposed in [10–13].

In this work, we are going to use the cost function in the form of an aggregation function

$$^{cf}\mathbf{Agg}_{i=1}^{N} \left[w_i \cdot \rho(\mathbf{c}, \mathbf{x}^i) \right] = {}^{cf}\mathbf{Agg} \left[w_1 \rho(\mathbf{c}, \mathbf{x}^1), w_2 \rho(\mathbf{c}, \mathbf{x}^2), \ldots, w_N \rho(\mathbf{c}, \mathbf{x}^N) \right]$$

instead of $\sum_{i=1}^{N} w_i \rho(\mathbf{c}, \mathbf{x}^i) = \left[w_1 \rho(\mathbf{c}, \mathbf{x}^1) + w_2 \rho(\mathbf{c}, \mathbf{x}^2) + \cdots + w_N \rho(\mathbf{c}, \mathbf{x}^N) \right]$. We obtain the next generalizations of the Fréchet point and median.

Definition 3. *The Fréchet aggregation point and median are the points* $\mathbf{c}_{\mathrm{opt}} \in \mathbf{R}^K$ *and* $\widehat{\mathbf{c}}_{\mathrm{opt}} \in \mathbf{D} = \{ \mathbf{x}^1, \mathbf{x}^2, \ldots, \mathbf{x}^N \}$ *that minimize the aggregation cost function (ACF)* $^{cf}\mathbf{Agg}_{i=1}^{N} \left[w_i \cdot \rho(\mathbf{c}, \mathbf{x}^i) \right]$. *They are formally defined as*

$$\mathbf{c}_{\mathrm{opt}} = \mathbf{FrechPt} \left({}^{cf}\mathbf{Agg}, \rho | \mathbf{x}^1, \mathbf{x}^2, \ldots, \mathbf{x}^N \right)$$

$$= \arg \min_{\mathbf{c} \in \mathbf{R}^K} \left({}^{cf}\mathbf{Agg}_{i-1}^{N} \left[w_i \cdot \rho(\mathbf{c}, \mathbf{x}^i) \right] \right) \tag{17}$$

and

$$\widehat{\mathbf{c}}_{\mathrm{opt}} = \mathbf{FrechMed} \left({}^{cf}\mathbf{Agg}, \rho | \mathbf{x}^1, \mathbf{x}^2, \ldots, \mathbf{x}^N \right)$$

$$= \arg \min_{\mathbf{c} \in \mathbf{D}} \left({}^{cf}\mathbf{Agg}_{i=1}^{N} \left[w_i \cdot \rho(\mathbf{c}, \mathbf{x}^i) \right] \right). \tag{18}$$

Note that **argmin** means the argument, for which the $^{cf}\mathbf{Agg}_{i=1}^{N}\left[w_i\rho(\mathbf{c},\mathbf{x}^i)\right]$ is minimized. In this case, it is the point $\mathbf{c}_{\mathrm{opt}} \in \mathbf{R}^K$ in (17) or point $\widehat{\mathbf{c}}_{\mathrm{opt}} \in \mathbf{D}$ in (18) for which the aggregation of all distances to the $\mathbf{x}^i$s is minimum.

When filters (1) and (2) are modified as follows:

$$\widehat{\mathbf{s}}(i,j) = \mathbf{FrechPt}\left[^{cf}\mathbf{Agg},\rho|\mathbf{f}(k,l)\right], \atop {(k,l)\in M(i,j)} \tag{19}$$

$$\widehat{\mathbf{s}}(i,j) = \mathbf{FrechMed}\left[^{cf}\mathbf{Agg},\rho|\mathbf{f}(k,l)\right] \atop {(k,l)\in M(i,j)} \tag{20}$$

it becomes *Fréchet aggregation mean* and *median filters*. They are based on an aggregation operator $^{cf}\mathbf{Agg}$ and a metric ρ, which could be changed independently of one another. For each pair of aggregation operator and metric, we get the unique class of new nonlinear filters.

Example 4. If observation data are real numbers, *i.e.*, $x^1, x^2, \ldots, x^N \in \mathbf{R}$, the distance function is the city distance $\rho(x,y) = \rho_1(x,y) = |x - y|$ and ACF is quadratic $^{cf}\mathbf{Agg}_{i=1}^{N}\left[w_i \cdot \rho(\mathbf{c},\mathbf{x}^i)\right] = \sum_{i=1}^{N} w_i\rho_1^2(\mathbf{c},\mathbf{x}^i)$, then the optimal Fréchet point (17) and median (18) for grey-level data (numbers) $x^1, x^2, \ldots, x^N \in \mathbf{R}$ to be the ordinary arithmetic mean, and quadratic median, respectively, *i.e.*,

$$\mathbf{c}_{\mathrm{opt}} = \mathbf{FrechPt}\left(^{cf}\mathbf{Agg},\rho_2|x^1,x^2,\ldots,x^N\right)$$

$$= \arg\min_{c\in\mathbf{R}^K}\left(\sum_{i=1}^{N}|c - x^i|^2\right) = \mathbf{Mean}\left(x^1,x^2,\ldots,x^N\right) = \frac{1}{N}\sum_{i=1}^{N}x^i. \tag{21}$$

$$\widehat{\mathbf{c}}_{\mathrm{opt}} = \mathbf{FrechMed}\left(^{cf}\mathbf{Agg},\rho|x^1,x^2,\ldots,x^N\right) = \arg\min_{c\in\mathbf{D}}\left(\sum_{i=1}^{N}|c - x^i|^2\right) \tag{22}$$

In this case, filter (19) is *optimal maximum likelihood vector filter* for Gaussian noise, and filter (20) is *vector median filter* associated with Euclidean metric, because $\rho_2(x,y) = \rho_1^2(x,y) = |x - y|^2$.

Example 5. If observation data are vectors, *i.e.*, $\mathbf{x}^1, \mathbf{x}^2, \ldots, \mathbf{x}^N \in \mathbf{R}^K$, the distance function is the city distance $\rho_1(x,y) = ||x - y||_1$, and ACF is the Kolmogorov mean $^{cf}\mathbf{Agg}_{i=1}^{N}\left[w_i\rho_1\left(\mathbf{c},\mathbf{x}^i\right)\right] = K^{-1}\left[\sum_{i=1}^{N}w_iK\left(||\mathbf{c} - \mathbf{x}||_1\right)\right]$ then Fréchet aggregation point and median for vectors $\mathbf{x}^1, \mathbf{x}^2, \ldots, \mathbf{x}^N \in \mathbf{R}^K$ to be the following Kolmogorov–Fréchet aggregations

$$\mathbf{c}_{\mathrm{opt}} = \mathbf{FrechPt}\left(K,\rho_1|\mathbf{x}^1,\mathbf{x}^2,\ldots,\mathbf{x}^N\right)$$

$$= \arg\min_{\mathbf{c}\in\mathbf{R}^K}\left(K^{-1}\left[\sum_{i=1}^{N}w_iK\left(||\mathbf{c} - \mathbf{x}||_1\right)\right]\right), \tag{23}$$

$$\widehat{c}_{opt} = \textbf{FrechMed}\left(K, \rho_1 | \mathbf{x}^1, \mathbf{x}^2, \ldots, \mathbf{x}^N\right)$$

$$= \arg \min_{\mathbf{c} \in D}\left(K^{-1}\left[\sum_{i=1}^{N} w_i K\left(\|\mathbf{c} - \mathbf{x}\|_1\right)\right]\right), \tag{24}$$

respectively.

In this case, filter (19) is the *Kolmogorov–Fréchet vector mean filter*, and filter (20) is the *Kolmogorov–Fréchet vector median filter* associated with sity metric.

In Definitions 1 and 2 we used a distance function $\rho(\mathbf{x}, \mathbf{y})$. But all known metrics have the aggregation form. By this reason, we can use an aggregation function $^{\rho}\textbf{Agg}\left(\left|c_1 - x_1^i\right|, \ldots, \left|c_K - x_K^i\right|\right)$ instead of $\rho\left(\left|c_1 - x_1^i\right|, \ldots, \left|c_K - x_K^i\right|\right)$.

Definition 4. *The Fréchet aggregation point and median are the points* $\mathbf{c}_{opt} \in \mathbf{R}^K$ *and* $\widehat{\mathbf{c}}_{opt} \in \mathbf{D} = \left\{\mathbf{x}^1, \mathbf{x}^2, \ldots, \mathbf{x}^N\right\}$ *that minimize the aggregation cost function* (ACF) $^{cf}\textbf{Agg}\left(w_1{}^{\rho}\textbf{Agg}\left(\mathbf{c}, \mathbf{x}^1\right), \ldots, w_N{}^{\rho}\textbf{Agg}^{\rho}\left(\mathbf{c}, \mathbf{x}^N\right)\right)$ *(=the weighted aggregation mean of all aggregation distances* $w_1{}^{\rho}\textbf{Agg}\left(\mathbf{c}, \mathbf{x}^1\right), \ldots, w_N{}^{\rho}\textbf{Agg}\left(\mathbf{c}, \mathbf{x}^N\right)$ *from an arbitrary point* $\mathbf{c} \subset \mathbf{R}^K$ *to each point* $\mathbf{x}^1, \mathbf{x}^2, \ldots, \mathbf{x}^N \in \mathbf{R^K}$*). They are formally defined as*

$$\mathbf{c}_{opt} = \textbf{FrechPt}\left(^{cf}\textbf{Agg}, {}^{\rho}\textbf{Agg} | \mathbf{x}^1, \mathbf{x}^2, \ldots, \mathbf{x}^N\right)$$

$$= \arg \min_{\mathbf{c} \in \mathbf{R}^K}\left(^{cf}\textbf{Agg}_{i=1}^{N}\left[w_i \cdot {}^{\rho}\textbf{Agg}\left(\mathbf{c}, \mathbf{x}^i\right)\right]\right) \tag{25}$$

and

$$\widehat{\mathbf{c}}_{opt} = \textbf{FrechMed}\left(^{cf}\textbf{Agg}, {}^{\rho}\textbf{Agg} | \mathbf{x}^1, \mathbf{x}^2, \ldots, \mathbf{x}^N\right)$$

$$= \arg \min_{\mathbf{c} \in \mathbf{D}}\left(^{cf}\textbf{Agg}_{i=1}^{N}\left[w_i \cdot {}^{\rho}\textbf{Agg}\left(\mathbf{c}, \mathbf{x}^i\right)\right]\right). \tag{26}$$

When filters (1) and (2) are modified as follows:

$$\widehat{s}(i,j) = \textbf{FrechPt}_{(m,n) \in M(i,j)}\left[^{cf}\textbf{Agg}, {}^{\rho}\textbf{Agg} | \mathbf{f}(m,n)\right], \tag{27}$$

$$\widehat{s}(i,j) = \textbf{FrechMed}_{(m,n) \in M(i,j)}\left[^{cf}\textbf{Agg}, {}^{\rho}\textbf{Agg} | \mathbf{f}(m,n)\right] \tag{28}$$

it becomes Fréchet aggregation mean and median filters. They are based on an arbitrary pair of aggregation operators $^{cf}\textbf{Agg}$ and $^{\rho}\textbf{Agg}$, which could be changed independently of one another. For each pair of aggregation operators, we get the unique class of new nonlinear filters.

Example 6. For vector observation data $\mathbf{x}^1, \mathbf{x}^2, \ldots, \mathbf{x}^N \in \mathbf{R}^K$, for L–Kolmogorov aggregation distance function $L^{-1}\left(\sum_{k=1}^{K} L\left(|c_k - x_k^i|\right)\right)$, and for ACF in the form of the K–Kolmogorov mean $K^{-1}\left(\sum_{i=1}^{N} w_i K\left[{}^{\rho}\mathbf{Agg}\left(\mathbf{c}, \mathbf{x}^i\right)\right]\right)$, we have

$$\mathbf{c}_{\mathrm{opt}} = \mathbf{FrechPt}\left(K, L | \mathbf{x}^1, \mathbf{x}^2, \ldots, \mathbf{x}^N\right)$$

$$= \arg\min_{\mathbf{c} \in \mathbf{R}^K} \left({}^{cf}\mathbf{Agg}_{i=1}^{N}\left[w_i {}^{\rho}\mathbf{Agg}\left(\mathbf{c}, \mathbf{x}^i\right)\right]\right)$$

$$= \arg\min_{\mathbf{c} \in \mathbf{R}^K} \left\{ K^{-1}\left(\sum_{i=1}^{N} w_i K\left[L^{-1}\left(\sum_{k=1}^{K} L(|c_k - x_k^i|)\right)\right]\right)\right\}, \quad (29)$$

$$\widehat{\mathbf{c}}_{\mathrm{opt}} = \mathbf{FrechMed}\left(K, L | \mathbf{x}^1, \mathbf{x}^2, \ldots, \mathbf{x}^N\right)$$

$$= \arg\min_{\mathbf{c} \in \mathbf{D}} \left({}^{cf}\mathbf{Agg}_{i=1}^{N}\left[w_i {}^{\rho}\mathbf{Agg}\left(\mathbf{c}, \mathbf{x}^i\right)\right]\right)$$

$$= \arg\min_{\mathbf{c} \in \mathbf{D}} \left\{ K^{-1}\left(\sum_{i=1}^{N} w_i K\left[L^{-1}\left(\sum_{k=1}^{K} L(|c_k - x_k^i|)\right)\right]\right)\right\}, \quad (30)$$

where L and K are two Kolmogorov functions.

In this case, filter (27) is the *Kolmogorov–Fréchet vector mean filter*, and filter (28) is the *Kolmogorov–Fréchet vector median filter* associated with sity metric and with a pair of Kolmogorov K–, L–functions.

4   Experiments

Generalized vector aggregation filtering with ${}^{cf}\mathbf{Agg} = \mathbf{Mean}, \mathbf{Med}, \mathbf{Geo}$ and Euclidean metric ${}^{\rho}\mathbf{Agg} = \rho_2$ has been applied to noised 256×256 image "Dog": Figs. 3(b), 4(b), 5(b). We use window with size 3×3. The denoised images are shown in Figs. 3, 4 and 5. All filters have very good denoised properties.

5   Conclusions

A new class of nonlinear generalized MIMO-filters (vector median filters or Frèchet filters) for multichannel image processing is introduced in this paper. These filters are based on an arbitrary pair of aggregation operators, which could be changed independently of one another. For each pair of parameters, we get the unique class of new nonlinear filters. The main goal of the work is to show that generalized Frèchet aggregation means can be used to solve problems of image filtering in a natural and effective manner.

Acknowledgements. This work was supported by grants the RFBR Nos.13-07-12168 and 13-07-00785.

Appendix. Figures

a) Original image

b) Noised image, $PSNR = 21.83$

c) $^{cf}\mathbf{Agg} = \mathbf{Mean}$, $^{\rho}\mathbf{Agg} = \rho_2$,
$PSNR = 32.52$

d) $^{cf}\mathbf{Agg} = \mathbf{Med}$, $^{\rho}\mathbf{Agg} = \rho_2$,
$PSNR = 31.79$

e) $^{cf}\mathbf{Agg} = \mathbf{Min}$, $^{\rho}\mathbf{Agg} = \rho_2$,
$PSNR = 28.29$

f) $^{cf}\mathbf{Agg} = \mathbf{Geo}$, $^{\rho}\mathbf{Agg} = \rho_2$,
$PSNR = 30.52$

Fig. 3. Original (a) and noised (b) images; noise: Salt-Pepper; denoised images (c)–(f)

a) Original image

b) Noised image, $PSNR = 17.19$

c) $^{cf}\mathbf{Agg} = \mathbf{Mean}$, $^{\rho}\mathbf{Agg} = \rho_2$,
$PSNR = 21.83$

d) $^{cf}\mathbf{Agg} = \mathbf{Med}$, $^{\rho}\mathbf{Agg} = \rho_2$,
$PSNR = 20.84$

e) $^{cf}\mathbf{Agg} = \mathbf{Min}$, $^{\rho}\mathbf{Agg} = \rho_2$,
$PSNR = 19.05$

f) $^{cf}\mathbf{Agg} = \mathbf{Geo}$, $^{\rho}\mathbf{Agg} = \rho_2$,
$PSNR = 20.51$

Fig. 4. Original (a) and noised (b) images; noise: Gaussian PDF; denoised images (c)–(f)

a) Original image

b) Noised image, $PSNR = 28.24$

c) $^{cf}\mathbf{Agg} = \mathbf{Mean}$, $^{\rho}\mathbf{Agg} = \rho_2$,
$PSNR = 30.68$

d) $^{cf}\mathbf{Agg} = \mathbf{Med}$, $^{\rho}\mathbf{Agg} = \rho_2$,
$PSNR = 29.61$

e) $^{cf}\mathbf{Agg} = \mathbf{Min}$, $^{\rho}\mathbf{Agg} = \rho_2$,
$PSNR = 27.77$

f) $^{cf}\mathbf{Agg} = \mathbf{Geo}$, $^{\rho}\mathbf{Agg} = \rho_2$,
$PSNR = 30.14$

Fig. 5. Original (a) and noised (b) images; noise: Laplasian PDF; denoised images (c)–(f)

References

1. Frèchet, M.: Les elements aleatoires de nature quelconque dans un espace distancie. Ann. Inst. Henri Poincare **10**(3), 215–310 (1948)
2. Bajaj, C.: Proving geometric algorithms nonsolvability: an application of factoring polynomials. J. Symbolic Comput. **2**, 99–102 (1986)
3. Bajaj, C.: The algebraic degree of geometric optimization problems. Discrete Comput. Geom. **3**, 177–191 (1988)
4. Chandrasekaran, R., Tamir, F.: Algebraic optimization: the fermat-weber problem. Math. Program. **46**, 219–224 (1990)
5. Astola, J., Haavisto, P., Neuvo, Y.: Vector median filters. Proc. IEEE **78**, 678–689 (1990)
6. Tang, K., Astola, J., Neuvo, Y.: Nonlinear multivariate image filtering techniques. IEEE Trans. Image Process. **4**, 788–798 (1996)
7. Mayor, G., Trillas, E.: On the representation of some aggregation functions. Proc. ISMVL **20**, 111–114 (1986)
8. Ovchinnikov, S.: On robust aggregation procedures. In: Bouchon-Meunier, B. (ed.) Aggregation Operators for Fusion under Fuzziness, pp. 3–10. Physica, Heidelberg (1998)
9. Kolmogorov, A.: Sur la notion de la moyenne. Atti Accad. Naz. Lincei **12**, 388–391 (1930)
10. Labunets V. G.: Filters based on aggregation operators Part 1. Aggregation operators. In: 24th International Crimean Conference Microwave and Telecommunication Technology (CriMiCo2014), Sevastopol, Crimea, Russia, vol. 24, pp. 1239–1240 (2014)
11. Labunets, V.G., Gainanov, D.N., Ostheimer, E.: Filters based on aggregation operators. Part 2. The kolmogorov filters. In: 24th International Crimean Conference Microwave and Telecommunication Technology (CriMiCo2014), Sevastopol, Crimea, Russia, vol. 24, pp. 1241–1242 (2014)
12. Labunets, V.G., Gainanov, D.N., Tarasov A.D., Ostheimer E.: Filters based on aggregation operators. Part 3. The heron filters. In: 24th International Crimean Conference "Microwave and Telecommunication Technology" (CriMiCo2014), Sevastopol, Crimea, Russia, vol. 24, pp. 1243–1244 (2014)
13. Labunets, V.G., Gainanov, D.N., Arslanova, R.A., Ostheimer, E.: Filters based on aggregation operators. Part 4. Generalized vector median filters. In: 24th International Crimean Conference Microwave and Telecommunication Technology (CriMiCo2014), Sevastopol, Crimea, Russia, vol. 24, pp. 1245–1246 (2014)

Fast Global Image Denoising Algorithm on the Basis of Nonstationary Gamma-Normal Statistical Model

Inessa Gracheva$^{(\boxtimes)}$, Andrey Kopylov, and Olga Krasotkina

Tula State University, Tula, Russian Federation
gia1509@mail.ru, And.Kopylov@gmail.com, ko180177@yandex.ru

Abstract. We consider here a Bayesian framework and the respective global algorithm for adaptive image denoising which preserves essential local peculiarities in basically smooth changing of intensity of reconstructed image. The algorithm is based on the special nonstationary gamma-normal statistical model and can handle both Gaussian noise, which is an ubiquitous model in the context of statistical image restoration, and Poissonian noise, which is the most common model for low-intensity imaging used in biomedical imaging. The algorithm being proposed is simple in tuning and has linear computation complexity with respect to the number of image elements so as to be able to process large data sets in a minimal time.

Keywords: Image denoising · Bayesian framework · Nonstationary gamma-normal statistical model

1 Introduction

Image analysis and information extraction are often used in many imaging systems. General degradation of image quality during acquisition and transmission process in many cases is described by an additive Gaussian noise model. At the same time, variations in low-intensity levels in quantum noise caused by fluctuations in either the number of detected photons or the inherent limitation of the discrete nature for photon detection can be described by a Poissonian noise model, which is the most popular in biomicroscopy. The primary aim of the image denoising technique is to remove the noisy observations while reconstructing a satisfying estimation of the original image, thereby enhancing the performance of these applications in imaging systems.

The most popular methods are good in case when we have only one type of noise. For example, for denoising of Gaussian used the Steins unbiased risk estimate from the concept of linear expansion of thresholds (SURE-LET) [1], where directly the denoising process is parametrized as a sum of elementary nonlinear processes with unknown weights; block-matching 3-D (BM3D) algorithm [2] based on enhanced sparse representation in transform domain; fast bilateral filter (FBF) [3] was first termed by Tomasi and Manduchi [4] based on

© Springer International Publishing Switzerland 2015
M.Y. Khachay et al. (Eds.): AIST 2015, CCIS 542, pp. 71–82, 2015.
DOI: 10.1007/978-3-319-26123-2_7

the work [5,6], and later modified and improved in [7]. A widespread alternative to the direct handling of Poisson statistics is to apply variance-stabilizing transforms (VSTs) with the underlying idea of exploiting the broad class of denoising methods which are based on a Gaussian noise model [8], such as the HaarFisz transform [9]. Platelet approach [10], which stands among the state-of-the-art algorithms for Poisson intensity estimation. Poisson unbiased risk estimate from the concept of linear expansion of thresholds (PURE-LET) [11] is based on the minimization of an unbiased estimate of the MSE for Poisson noise, a linear parametrization of the denoising process and the preservation of Poisson statistics across scales within the Haar VST.

The purpose of this article is to present an algorithm that would satisfy compromise between restoration quality and computational complexity. Firstly, we want a method which is able to effectively remove Gaussian noise as well as Poissonian noise. In the second place, the method should satisfy strict constraints in terms of computational cost and memory requirements, so as to be able to process large data sets. To achieve these purpose a nonstationary gamma-normal noise model has been proposed in the framework of Bayesian approach to the problem of image processing. This model allows us to develop a fast global algorithm on the basis of Gauss-Seudel procedure and Kalman filter-interpolator.

2   Gamma-Normal Model of Hidden Random Field

The task of image reconstruction within the Bayesian approach can be expressed as the problem of estimating a hidden Markov component $X = (x_t, t = 1, ..., N)$ $T = \{t = (t_1, t_2) : t_1 = 1, ..., N_1, t_2 = 1, ..., N_2\}$ of a two-component random field, where observed component $Y = (y_t, t \in T)$ is the analyzed image.

Probabilistic properties of a two-component random field (X, Y) are completely determined by the joint conditional probability density $\Phi(Y|X, \delta)$ of original functions $Y = (y_t, t \in T)$ with respect to the secondary data $X = (x_t, t \in T)$, and the a prior joint distribution $\Psi(X|\Lambda, \delta)$ of hidden component $X = (x_t, t \in T)$.

Let the joint conditional probability density $\Phi(Y|X, \delta)$ be in the form of Guassian distribution:

$$\Phi(Y|X, \delta) = \frac{1}{\delta^{N/2}(2\pi)^{N/2}} \exp(-\frac{1}{2\delta} \sum_{t \in T} (y_t - x_t)^2), \tag{1}$$

where $E(e_t^2) = \delta$ is the variance of the observation noise, which to be unknown.

Just as in [12,13], we will express our prior knowledge about sought for estimates of a hidden component in the form of Markov random field. Let the conditional probability densities of hidden variables with respect to their neighbors be also Gaussian with some variance $E(\xi_t^2)$ and the conditional mathematical expectation equal to the value of the adjacent variable.

In the case of Poissonian noise the variance of corresponding random variables is determined by the intensity of photons emission and takes different values in different parts of an image. We do not use here VST to reforms the data so that the noise approximately becomes Gaussian with a constant variance. On the contrary, we do not assume the common variance of hidden components to remain the same in an image plane, and suppose $E(\xi_t^2) = r_t$. The unknown variances $(r_t, t \in T)$ are considered as proportional to the variance of the observation noise $r_t = \lambda_t \delta$ with the proportionality coefficients λ_t acting as factors of the unknown instantaneous volatility of hidden field $X = (x_t, t \in T)$, unknown as well.

Under this assumption, the a priori joint distribution of $X = (x_t, t \in T)$ is conditionally normal with respect to the field of the factors $\Lambda = (\lambda_t, t \in T)$. So, we come to the improper a priori density:

$$\Psi(X|\Lambda,\delta) \propto \frac{1}{\left(\prod_{t \in T} \delta\lambda_t\right)^{1/2} (2\pi)^{n(N-1)/2}}$$

$$\times \exp\left(-\frac{1}{2} \sum_{t',t'' \in V} \frac{1}{\delta\lambda_t}(x_{t'} - x_{t''})^2\right), \qquad (2)$$

where V is the neighborhood graphs of image elements having the form of a lattice.

Finally, we assume the inverse factors $1/\lambda_t$ to be a priori independent and identically gamma-distributed $\gamma(1/\lambda_t|\alpha,\vartheta) \propto (1/\lambda_t)^{\alpha-1} \exp(-\vartheta(1/\lambda_t))$ on the positive half-axis $\lambda_t \geq 0$. The mathematical expectation and variance of gamma-distribution are ratios α/ϑ, and α/ϑ^2. The a priori distribution density of the entire field of the factors:

$$G(\Lambda|\alpha,\vartheta) = \prod_{t \in T} \gamma(\lambda_t|\alpha,\vartheta) \propto \left(\prod_{t \in T} \frac{1}{\lambda_t}\right)^{\alpha-1} \exp\left(-\vartheta \sum_{t \in T} \frac{1}{\lambda_t}\right)$$

$$= \exp\left[-(\alpha - 1)\sum_{t \in T} \ln\lambda_t - \vartheta\sum_{t \in T} \frac{1}{\lambda_t}\right]. \qquad (3)$$

We redefine the parameters α and ϑ through new parameters μ and λ

$$\alpha = \frac{1}{2}\left[\frac{1}{\delta}\left(1 + \frac{1}{\mu}\right) + 1\right], \vartheta = \frac{\lambda}{2\delta\mu}, \qquad (4)$$

assuming thereby the parametric family of gamma distributions of inverse factors $1/\lambda_t$

$$\gamma(1/\lambda_t|\delta,\lambda,\mu) \propto (1/\lambda_t)^{\frac{2\mu+1}{2\delta\mu}} \exp\left(-\frac{\lambda}{2\delta\mu}(1/\lambda_t)\right), \qquad (5)$$

with mathematical expectations and variances

$$E(1/\lambda_t) = \frac{(1+\delta)\mu + 1}{\lambda}, Var(1/\lambda_t) = 2\delta\mu\frac{(1+\delta)\mu + 1}{\lambda^2}.$$

In terms of this parameterization, the independent prior distribution of each instantaneous inverse factors $1/\lambda_t$ is almost completely concentrated around the mathematical expectation $1/\lambda$ if $\mu \to 0$. On the contrary, with $\mu \to \infty$ coefficient $1/\lambda$ have tends to the almost uniform distribution.

In this way, we come to the prior density:

$$G(\Lambda|\delta,\lambda,\mu) = \exp\left[-\frac{1}{2\delta\mu}\sum_{t\in T}\left(\lambda\frac{1}{\lambda_t} + \frac{1}{\lambda}\ln\lambda_t\right)\right], \qquad (6)$$

and, so, have completely defined the joint prior normal gamma-distribution of both hidden fields $X = (x_t, t \in T)$ and $\Lambda = (\lambda_t, t \in T)$:

$$H(X,\Lambda|\delta,\lambda,\mu) = \Psi(X|\Lambda,\delta)G(\Lambda|\delta,\lambda,\mu).$$

Coupled with the conditional density of the observable field (2), it makes basis for Bayesian estimation of the field $X = (x_t, t \in T)$.

3   The Bayesian Estimate of the Hidden Random Field

Bayesian reasoning makes it possible to reduce a wide class of image analysis problems to the problem of maximum a posteriori (MAP) estimation.

The joint a posteriori distribution of hidden elements, namely, those of field $X = (x_t, t \in T)$ and its instantaneous factors $\Lambda = (\lambda_t, t \in T)$, is completely defined by (1) and (2) and, in terms of the original parameters (α, ϑ), by (3):

$$P(X,\Lambda|Y,\delta,\alpha,\vartheta) = \frac{\Psi(X|\Lambda,\delta)G(\Lambda|\alpha,\vartheta)\Phi(Y|X,\delta)}{\int\int\Psi(X'|\Lambda',\delta)G(\Lambda'|\alpha,\vartheta)\Phi(Y|X',\delta)dX'd\Lambda'}.$$

The Bayesian estimate of (X, Λ) is the maximum point of the numerator

$$\begin{cases} (\hat{X},\hat{\Lambda}|\delta,\alpha,\vartheta) = \arg\max_{X,\Lambda}[\ln\Phi(Y|X,\delta) + \ln\Psi(X|\Lambda,\delta) + \ln G(\Lambda|\alpha,\vartheta)] \\ = \arg\max_{X,\Lambda}\left(-\frac{1}{2\delta}\sum_{t\in T}(y_t - x_t)^2 - \frac{1}{2}\sum_{t\in T}\ln\delta\lambda_t - \frac{1}{2\delta}\sum_{t',t''\in V}\frac{1}{\lambda_t}(x_{t'} - x_{t''})^2 \\ - (\alpha - 1)\sum_{t\in T}\ln\lambda_t - \vartheta\sum_{t\in T}\frac{1}{\lambda_t}\right), \end{cases}$$

or, what is equivalent,

$$\begin{cases} (\hat{X},\hat{\Lambda}|\delta,\alpha,\vartheta) = \arg\min_{X,\Lambda}J(X,\Lambda|Y,\delta,\alpha,\vartheta), \\ J(X,\Lambda|Y,\delta,\alpha,\vartheta) = \sum_{t\in T}(y_t - x_t)^2 \\ + \sum_{t',t''\in V}\left\{\frac{1}{\lambda_t}\left[(x_{t'} - x_{t''})^2 + 2\delta\vartheta\right] + \delta(2\alpha - 1)\ln\lambda_t\right\}. \end{cases}$$

The substitution of the new parameters (4) makes the Bayesian estimate independent of the observation noise variance δ:

$$\begin{cases} (\hat{X}, \hat{\Lambda} | \lambda, \mu) = \arg\min_{X,\Lambda} J(X, \Lambda | Y, \lambda, \mu), \\ J(X, \Lambda | Y, \lambda, \mu) = \sum_{t \in T} (y_t - x_t)^2 \\ + \sum_{t',t'' \in V} \left\{ \frac{1}{\lambda_t} \left[(x_{t'} - x_{t''})^2 + \lambda/\mu \right] + (1 + 1/\mu) \ln \lambda_t \right\}. \end{cases} \quad (7)$$

As it will be shown, the growing value of parameter μ endows this criterion with a pronounced tendency to keep the majority of estimated volatility factors $\hat{\lambda}_t$ close to the basic low value λ and to allow single large outliers, revealing thereby hidden events in the primarily smooth original field.

4 The Gauss-Seudel Procedure of Edge-Preserving Image Denoising

The conditionally optimal factors $\hat{\Lambda}(X, \lambda, \mu) = [\hat{\lambda}_t(X, \lambda, \mu), t \in T]$ are defined independently of each other:

$$\hat{\Lambda}(X, \lambda, \mu) = \arg\min_{\Lambda} J(\Lambda | X, \lambda, \mu) :$$

$$\frac{\partial}{\partial \lambda_t} \left\{ \frac{1}{\lambda_t} [(x_{t'} - x_{t''})^2 + \lambda/\mu] + (1 + 1/\mu) \ln \lambda_t \right\} = 0. \quad (8)$$

The zero conditions for the derivatives, excluding the trivial solutions $\lambda_t \to \infty$, lead to the equalities

$$\frac{1}{\lambda_t} \left[(x_{t'} - x_{t''})^2 + \lambda/\mu \right] = (1 + 1/\mu),$$

and, hence,

$$\hat{\lambda}_t(X, \lambda, \mu) = \lambda \frac{(1/\lambda)(x_{t'} - x_{t''})^2 + 1/\mu}{1 + 1/\mu}, \quad (9)$$

Substitution of (9) into (7) gives the equivalent form, which avoids immediate finding the factors themselves:

$$\begin{cases} (\hat{X}, \hat{\Lambda} | \lambda, \mu) = \arg\min_{X,\Lambda} J(X, \Lambda | Y, \lambda, \mu), \\ J(X, \Lambda | Y, \lambda, \mu) = \sum_{t \in T} (y_t - x_t)^2 \\ + \sum_{t',t'' \in V} \left\{ (1 + 1/\mu) \ln \frac{(1/\lambda)(x_{t'} - x_{t''})^2 + 1/\mu}{1 + 1/\mu} \right\}. \end{cases} \quad (10)$$

It is almost quadratic function in a vicinity of the zero point $(x_{t'} - x_{t''})^2 = 0$ and remains being so practically over the entire number axis if μ is small (Fig. 1).

But as μ grows, the originally quadratic penalty undergoes more and more marked effect of saturation at some distance from zero. This means that the

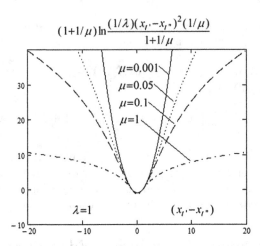

Fig. 1. The "saturation effect" of the unsmoothness penalty for sufficiently large values of the parameter μ and fixed value $\lambda=1$.

criterion strongly penalizes estimates of original function but becomes more and more indulgent to sharp discontinuities.

For finding the minimum point of the objective function with fixed structural parameters μ and λ, we apply the Gauss-Seidel iteration to both groups of variables $X = (x_t, t \in T)$ and $\Lambda = (\lambda_t, t \in T)$ starting with the initial values $\hat{\Lambda}^0 = (\hat{\lambda}_t^0 = \lambda, t \in T)$. At each iteration, current values of variables $\hat{\Lambda}^k = (\hat{\lambda}_t^k = 1, t \in T)$ according to (7) give calculate of the field $X = (x_t, t \in T)$, whose minimum point of the objective function gives the new approximation to the estimate of the mutually agreed field $\hat{X}^k = (\hat{x}_t^k, t \in T)$:

$$\hat{X}^k = (\hat{x}^k, t \in T) = \arg\min_X J(X, \Lambda^k | Y, \lambda, \mu)$$

$$= \arg\min_X \left\{ \sum_{t \in T} (y_t - x_t)^2 + \sum_{t', t'' \in V} \frac{1}{\lambda_t} (x_{t'} - x_{t''})^2 \right\}. \tag{11}$$

It is clear that there is no way to replace the lattice-like neighborhood graph (Fig. 2a) by a tree-like one without loss of the crucial property to ensure smoothness of the hidden secondary data field in all directions from each point of the image plane.

To avoid this obstacle, for finding the values of the hidden field at each vertical row of the picture, we use a separate pixel neighborhood tree which is defined, nevertheless, on the whole pixel grid and has the same horizontal branches as the others (Fig. 2b). The resulting image processing procedure is aimed at finding optimal values only for the hidden variables at the stem nodes in each tree [14]. For this combination of partial pixel neighborhood trees, the algorithm of finding the optimal values of the stem node variables boils down into a combination of two usual Kalman filtration-interpolation procedures, each dealing with single, respectively, horizontal and vertical image rows considered as signals on the one-dimensional argument axis.

First, such a one-dimensional procedure is applied to the horizontal rows $t_1 = 1, ..., N_1$ independently for each $t_2 = 1, ..., N_2$. The resulting marginal node functions $\hat{J}_{t_1,t_2}(x)$ should be stored in the memory. Then, the procedure is applied to the vertical rows $t_2 = 1, ..., N_2$ independently for each $t_1 = 1, ..., N_1$ with the only alteration: the respective marginal node functions $\hat{J}_t(x)$, obtained at the first step, are taken instead of the image-dependent node functions. In the case of real-valued variables x_t and quadratic pair-wise separable objective function, these elementary procedures applied to single horizontal and vertical rows are nothing else than Kalman filters-interpolators of special kind.

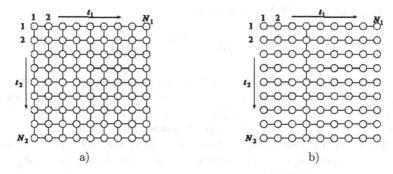

Fig. 2. Neighborhood graphs of image elements: (a) rectangular lattice; (b) simplest tree.

Each iteration of this algorithm has linear computational complexity with respect to size of the original image. Once the estimates $\hat{X}^k = (\hat{X}_t^k = \hat{x}_t^k, t \in T)$ are found, the next approximation to the estimates of factors $\hat{\Lambda}^k = (\hat{\lambda}_t^k = \lambda, t \in T)$ is defined by the rule (9):

$$\hat{\lambda}_t^{k+1} = \frac{(\hat{x}_{t'}^k - \hat{x}_{t''}^k)^2 + \lambda/\mu}{1 + 1/\mu}, t', t'' \in T, \tag{12}$$

which, in accordance with (8), gives the solution of the conditional optimization problem

$$\hat{\Lambda}^{k+1} = \arg\min_\Lambda J(\hat{X}^k, \Lambda | Y, \lambda, \mu)$$

$$= \arg\min_\Lambda \left\{ \sum_{t \in T} (y_t - \hat{x}_t)^2 + \sum_{t',t'' \in V} \frac{1}{\lambda_t} (\hat{x}_{t'}^k - \hat{x}_{t''}^k)^2 \right\}.$$

The very structure of the iterative procedure (11) and (12) provides objective function satisfying the inequality

$$J(\hat{X}^{k+1}, \hat{\Lambda}^{k+1} | Y, \lambda, \mu) < J(\hat{X}^k, \hat{\Lambda}^k | Y, \lambda, \mu),$$

which, as it is shown previously, the equality holds only at the stationary point.

It remains only to specify the way of choosing the values of the structural parameters μ and λ, which control, respectively, the basic average factors and the ability of instantaneous volatility factors to change along the image plane.

5   Experimental Results

We use here one of the common measures of image distortion, namely peak signal-to-noise ratio (PSNR). PSNR is an engineering term for the ratio between the maximum possible power of a signal and the power of corrupting noise that affects the fidelity of its representation. PSNR is most easily defined via the mean squared error (MSE). Given a noise-free mn monochrome image x and its noisy approximation $\hat{x}$, MSE is defined as:

$$PSNR = 10\log_{10}\left(\frac{\max(x^2)}{\langle|\hat{x}-x|^2\rangle}\right)$$

We have tested the various denoising methods for a representative set of standard 8-bit grayscale images such as Barbara, Moon (size 512×512) and Peppers, Cameraman (size 256×256), corrupted by simulated additive Gaussian white noise and Poissonian noise at six different power levels, which corresponds to PSNR decibel values. The parameters of each method have been set according to the values given by their respective authors in the corresponding referred papers. Variations in output PSNRs are, thus, only due to the denoising techniques themselves.

Table 1 summarizes the PSNRs obtained by the various algorithms for denoising of Poisson. We can see that the PURE-based approach clearly outperforms

Table 1. Comparison of Poissonian noise removal algorithms

Images	Cameraman 256 × 256						Moon 512 × 512					
Peak intensity	120	60	30	20	10	5	120	60	30	20	10	5
Input PSNR	24.08	21.07	18.05	16.29	13.28	10.27	26.27	23.25	20.23	18.48	15.47	12.46
Haar-Fisz	28.49	26.40	24.36	23.69	22.39	20.93	29.03	26.89	25.01	24.33	23.55	22.78
Platelet	28.29	26.79	25.44	24.60	23.24	21.49	29.16	26.01	25.05	24.60	23.96	23.63
PURE-LET	30.07	28.28	26.54	25.55	23.94	22.42	29.62	27.97	26.56	25.87	24.92	24.23
Our algorithm	29.03	26.95	25.67	24.70	23.54	21.86	29.56	27.64	25.98	25.05	24.39	23.79

Table 2. Comparison of Gaussian noise removal algorithms

Images	Papers 256 × 256						Barbara 512 × 512					
Peak intensity	120	60	30	20	10	5	120	60	30	20	10	5
Input PSNR	8.13	14.15	18.59	22.11	28.13	34.15	8.34	14.35	19.23	22.48	28.47	34.46
FBF	21.39	24.39	27.57	29.60	32.62	35.72	22.66	24.21	26.43	28.32	32.92	36.51
SURE-LET	21.32	24.43	27.13	29.33	33.18	37.17	21.76	23.70	25.83	27.98	32.18	36.71
BM3D	21.49	24.40	27.70	30.16	34.68	37.93	23.03	24.59	26.99	28.72	33.98	37.78
Our algorithm	21.42	24.40	27.47	29.50	33.54	37.76	22.56	23.34	25.98	28.05	33.39	36.79

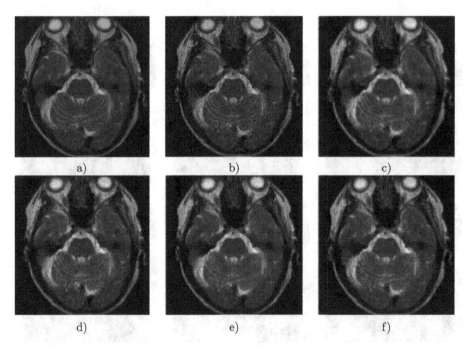

Fig. 3. (a) Part of the original MRI slice. (b) Noisy version with simulated Poissonian noise: PSNR = 22.31 dB. (c) Denoised with our algorithm ($\lambda = 10$, $\mu = 0.5$): PSNR = 29.03 dB. (d) Denoised with our algorithm ($\lambda = 0.1$, $\mu = 10$): PSNR = 29.54 dB. (e) Denoised with our algorithm ($\lambda = 0.0001$, $\mu = 0.1$): PSNR = 29.32 dB. (f) Denoised with our algorithm ($\lambda = 0.1$, $\mu = 0.5$): PSNR = 29.89 dB.

the standard VST-based wavelet denoisier applied in an orthonormal wavelet basis. Our solution also gives significantly better PSNRs than the non-redundant version of the Platelet approach and Haar-Fisz algorithm.

Table 2 summarizes the results obtained for denoising of Gaussian. Our results are already competitive with the best techniques available such as SURE-LET, FBF and BM3D.

Figures 3 and 4 present the results of our algorithm for different images. The values of the parameters (the basic average factors) and (the ability of instantaneous volatility factors to change in time) are chosen different that show their influence on the PSNRs value.

It is also interesting to evaluate the various denoising methods from a practical point of view: the computation time. Indeed, the results achieved by algorithms used for comparison in this paper are superiorly than many other algorithms, but their weakness is the time they require (on a Core i3 workstation with 1.6 GHz for 256 × 256 and 512 × 512 images to obtain the redundant results reported in Table 3. With our method, the whole denoising process lasts approximately 0.16 s for 256 × 256 images (0.53 s for 512 × 512 images), using a similar workstation. To compare with, FBF lasts approximately 1.8 s for

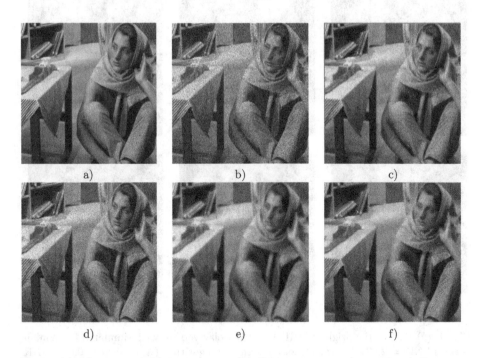

Fig. 4. (a) The original Barbara image. (b) Noisy version with simulated Gaussian noise: PSNR = 19.323 dB. (c) Denoised with our algorithm ($\lambda = 0.1$, $\mu = 1$): PSNR = 21.34 dB. (d) Denoised with our algorithm ($\lambda = 0.1$, $\mu = 0.5$): PSNR = 23.65 dB. (e) Denoised with our algorithm ($\lambda = 0.0001$, $\mu = 0.05$): PSNR = 22.59 dB. (f) Denoised with our algorithm ($\lambda = 0.01$, $\mu = 0.8$): PSNR = 25.98 dB.

Table 3. Relative computation time of various denoising techniques (seconds)

Methods	Image size	
	256×256	512×512
Platelet	856	1112
PURE-LET	4.6	10.2
BM3D	4.1	9.2
Haar-Fisz	1.3	3.2
FBF	0.5	1.8
SURE-LET	0.4	1.6
Our algorithm	0.16	0.53

512×512 images, PURE-LET lasts approximately $10.2\,\mathrm{s}$ for 512×512 images. Besides giving competitive results, our method is also much faster.

6   Conclusion

The gamma-normal model of the image and the expected result of processing, proposed in this paper in a combination with computationally effective Kalman filtration-interpolation procedure, allows us to develop fast global image denoising algorithm which is able to preserve substantial local image features, in particular edges of objects.

The comparison of the denoising results, obtained with our algorithm with respect to the other methods, demonstrates the efficiency of our approach for most of the images. The visual quality of our denoised images is moreover characterized by fewer artifacts than the other methods.

However, the most important advantage of the algorithm is low computation time in comparison with other algorithms. With this algorithm you can handle large images in a relatively short time.

Acknowledgements. This research is funded by RFBR, grant #13-07-00529.

References

1. Luisier, F., Blu, T., Unser, M.: A new SURE approach to image denoising: interscale orthonormal wavelet thresholding. IEEE Trans. Image Process. **16**(3), 593–606 (2007)
2. Dabov, K., Foi, A., Katkovnik, V., Egiazarian, K.: Image denoising by sparse 3-D transform-domain collaborative ltering. IEEE Trans. Image Process. **16**(8), 2080–2095 (2007)
3. Yang Q., Tan K.H., Ahuja N.: Real-time O(1) bilateral filtering. In: IEEE Conference on Computer Vision and Pattern Recognition, pp. 557–564, Miami (2009)
4. Tomasi C., Manduchi R.: Bilateral filtering for gray and color images. In: 6th International Conference on Computer Vision, pp. 839–846, Bombay (1998)
5. Aurich V., Weule J.: Non-linear gaussian filters performing edge preserving diffusion. In: DAGM Symposium, pp. 538–545, Bielefeld (1995)
6. Smith, S.M., Brady, J.M.: SUSANA new approach to low level image processing. Int. J. Comput. Vis. **23**(1), 45–78 (1997)
7. Elad, M.: On the origin of the bilateral filter and ways to improve it. IEEE Trans. Image Process. **11**(10), 1141–1151 (2002)
8. Donoho D.L.: Nonlinear wavelet methods for recovery of signals densities and spectra from indirect and noisy data. In: Daubechies, I. (ed.) Different Perspectives on Wavelets, Proceedings of Symposia in Applied Mathematics, vol. 47, pp. 173–205. American Mathematical Society, Providence (1993)
9. Fryzlewicz, P., Nason, G.P.: A HaarFisz algorithm for poisson intensity estimation. J. Computat. Graph. Stat. **13**(3), 621–638 (2004)
10. Willett, R.M., Nowak, R.D.: Multiscale poisson intensity and density estimation. IEEE Trans. Inf. Theory **53**(9), 3171–3187 (2007)

11. Luisier, F., Vonesch, C., Blu, T., Unser, M.: Fast interscale wavelet denoising of poisson-corrupted images. J. Sig. Process. **90**, 415–427 (2010)
12. Gracheva I., Kopylov A.: Adaptivnyj parametricheskij algoritm sglazhivanija izobrazhenij. Izvestija TulGU, ser. "Tehnicheskie nauki", Tula: Izd-vo TulGU **9**(2), 61–67 (2013) (in Russian)
13. Markov, M., Mottl, V., Muchnik, I.: Principles of nonstationary regression estimation: a new approach to dynamic multi-factor models in finance. DIMACS Technical report, Rutgers University, USA (2004)
14. Mottl, V., Blinov, A., Kopylov, A., Kostin, A.: Optimization techniques on pixel neighborhood graphs for image processing. In: Jolion, J.-M., Kropatsch, W.G. (eds.) Graph-Based Representations in Pattern Recognition. Computing Supplement, vol. 12, pp. 135–145. Springer, Wien (1998)

Theoretical Approach to Developing Efficient Algorithms of Fingerprint Enhancement

Mikhail Yu. Khachay[1,2](✉) and Maxim Pasynkov[1]

[1] Krasovsky Institute of Mathematics and Mechanics, 16 S. Kovalevskoy St., Ekaterinburg, Russia
pmk0690@gmail.com
[2] Ural Federal University, 19 Mira St., Ekaterinburg, Russia
mkhachay@imm.uran.ru

Abstract. A new theoretical approach to construction of efficient algorithms for fingerprint image enhancement is proposed. The approach comprises novel modifications of advanced orientation field estimation techniques such as the method of fingerprint core extraction based on Poincaré indexes and model-based smoothing for the gradient-based approximation of an orientation field by Legendre polynomials, and new adaptive Gabor filtering technique based on holomorphic transformations of coordinates.

Keywords: Fingerprint image enhancement · Orientation field · Poincaré index · Model-based approximation · Gabor filter · Holomorphic function · Conformal map

1 Introduction

For the last decades, the world has seen an increasing interest to biometrics-equipped authentication systems in both security and commerce applications. Among other biometric sources, fingerprint images have the most valuable place due to their individuality, which is commonly regarded by criminologists, forensic experts, anthropologists and ordinary people. The related problems of developing the reliable and efficient algorithms for fingerprint verification and identification became a great challenge for many specialists in computer science [3,5,13–17].

Since 2000, the international Fingerprint Verification Competition (FVC) was established as a challenging benchmark for the best fingerprint analysis algorithms [1]. These competitions provide the solid experimental proof for the optimality of minutiae-based fingerprint recognition algorithms w.r.t. their performance and efficiency. The general scheme of any such an algorithm consists of the following stages: preprocessing, minutiae extraction, secondary features construction (deep learning), and matching. Unfortunately, the conventional minutiae-extraction algorithms are poorly reliable to possible defects of the enrolled fingerprints, which decreases the overall performance of a recognition system. Therefore, the value of the preliminary enhancement procedures for improving the fingerprint images to be analyzed can hardly be overestimated.

© Springer International Publishing Switzerland 2015
M.Y. Khachay et al. (Eds.): AIST 2015, CCIS 542, pp. 83–95, 2015.
DOI: 10.1007/978-3-319-26123-2_8

Fingerprint enhancement is a young actively developing topic in image processing, which attracts many researchers and numerous publications. A variety of practical results are obtained in this field, some of them seem to be very promising. On the other hand, acquaintance with these results suggests that different applied problems can often be solved by near identical mathematical methods. However, researchers do not always use the existing or even classical mathematical formalisms, preferring to develop new and new heuristics. In this paper, we try to make such a gap between theory and practice a slightly narrower.

2   Proposed Approach

The proposed scheme of fingerprint enhancement extends the general framework introduced in [9]. The scheme consists of several stages, each of them (maybe except the last one) was considered separately in cited works, but, to the best of our knowledge, there are no papers presenting them together as entire approach. According to the proposed scheme, the enhancement procedure of a fingerprint image can be considered as a sequence of the following steps.

The initial step of this procedure deals with 'a coarse' estimation of the orientation field (OF) using Sobel approximation for gradients of the gray-scale image intensity function. This popular algorithm (a brief overview is provided in Sect. 3) was proposed for the first time in [12] and was adopted by many authors (see, e.g. [6,8,27]). The gradient-based algorithm has two main advantages: it can be implemented efficiently and performs well in several typical cases. Unfortunately, the resulting OF suffers from interference of scars, dirt, moisture or dryness of finger and other fingerprint defects produced at the enrollment phase (Fig. 1(a)). Therefore, this field is called *coarse* should be significantly improved, and can be used only as a starting point for the subsequent processing.

As we mentioned above, there is a variety of methods for such an improvement, among them blurring of the obtained OF, line and Bézier sensors [10,18], multi-scale analysis [15,19], etc. These approaches can perform well for many images, but have predominantly heuristic nature. Following to [2,22], we propose the improved model-based OF approximation method leveraging the previously obtained information on *singular points* (*cores* and *deltas*) of the fingerprint under consideration. Thus, in Sect. 4, we provide an explanation of the core detection technique based on Poincaré indexes of closed curves estimations over vector fields. Further, in Sect. 5, we use this additional information at the stage of model selection for the subsequent global approximation of the OF to be estimated. After that, the processed orientation field is fitted by two-variate Legendre polynomials. The resulting approximated OF (Fig. 1(b)) seems to be rather smoother and easily explainable than its initial gradient-based estimate and can be utilized in the following adaptive Gabor filtering stage.

The conventional way (see, e.g. [3,11,24,26]) to fingerprint image enhancement using regular Gabor filter (to distinguish it from the proposed below modification, we call the former *standard* and the latter *curved*), for a given

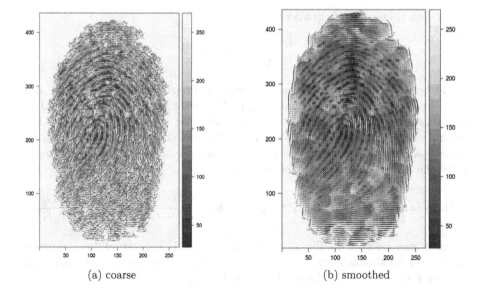

(a) coarse   (b) smoothed

Fig. 1. Orientation field estimates

rectangular image region, assumes rotating of this rectangle around its center up to the angle of average local orientation and subsequent tuning other filter parameters, frequency and sigmas. Such an approach performs well during the enhancement of regions with near flat papillary ridges. Unfortunately, in the most interesting parts of fingerprint images neighboring to their cores and deltas, the ridges curvature is high, which leads to significant distortions during such a filtering.

Developing the idea introduced in [7], we propose (in Sect. 6) the novel kernel-based approach to Gabor filtering. According to this approach, any region of interest (ROI) of the analyzed fingerprint image is previously mapped onto some rectangular window by means of the appropriate holomorphic complex-valued function. After such a transformation all papillary lines become near horizontal and can be processed successfully by regular Gabor filter. Then, using the corresponding inverse mapping, the filtered data is transferred to their initial location in the analyzed image.

3   Coarse Field Estimation

Hereinafter, we use the following notation: $I : D \to Y$ denotes an intensity function of the image in question (or just *an image*), where D is a (rectangular) pixel-domain and Y is a segment of gray values ($[0, 255]$ in our study). In theoretical reasoning, it is convenient to treat I as a smooth real-valued function (of two real or one complex argument). Certainly, for real fingerprints, this function has integer values and its gradient

$$\begin{bmatrix} G_x \\ G_y \end{bmatrix} = \nabla I(x, y)$$

should be approximated by some discrete counter part (Sobel operator in our case). In ideal case, for any point (x, y), the vector $\nabla I(x, y)$ is orthogonal to the tangent vector of a papillary ridge (valley) and can be used to reveal the orientation field at this point. To overcome the equivalence orientation issue for angles θ and $\theta + \pi$ we estimate (see, e.g. [6]) not the angle θ itself but its image under the following analytic function $U : D \to \mathbb{C}$:

$$U(x, y) = \cos 2\theta(x, y) + i \cdot \sin 2\theta(x, y), \quad \cos 2\theta = \frac{G_y^2 - G_x^2}{G_x^2 + G_y^2}, \quad \sin 2\theta = \frac{2G_x G_y}{G_x^2 + G_y^2}.$$

As it can be seen in Fig. 1(a), the obtained OF has rather low accuracy and should be significantly refined. To assess its quality, we use the *coherence* function $Coh : D \to [0, 1]$ defined by the following equation:

$$Coh(x, y) = \frac{|\sum_{(\xi, \eta) \in W(x, y)} U(\xi, \eta)|}{\sum_{(\xi, \eta) \in W(x, y)} |U(\xi, \eta)|},$$

where function U is averaged over some nonuniform window $W(x, y)$ centered at the point $(x, y) \in D$. Further, we use $Coh(x, y)$ in Sect. 5 as a weight function at the stage of model-based OF approximation. The result of such an approximation is presented in Fig. 1(b).

4 Extraction of Cores and Deltas

Cores and deltas (see, e.g. Fig. 2) are generally recognized as main features of a fingerprint image. Since their locations and types define the global structure of a fingerprint, knowledge of these parameters affects fundamentally the overall accuracy of any fingerprint analysis technique.

Mathematically, cores and deltas are singular points (SPs) of an orientation field, i.e. the points where the field is discontinuous. Indeed, OF can be treated as a phase portrait of an appropriate dynamic system. Moreover, it is useful to consider the problem of fingerprint analysis as a special case of the identification problem for a certain system of nonlinear ordinary differential equations (see, e.g. [25])

$$\frac{dx}{dt} = P(x, y), \quad \frac{dy}{dt} = Q(x, y)$$

for which papillary ridges play a role of uncertain measured phase trajectories (and the following approximate equalities $P(x, y) \approx G_y(x, y)$ and $Q(x, y) \approx G_x(x, y)$ are valid). Therefore, to find cores and deltas on a given fingerprint, one can use (see, e.g. [2,22]) a variety of singular point locating techniques developed

Fig. 2. Core and delta

for such dynamic systems, among them Poincaré index technique [20] appears to be the most famous.

Basically, for any simply connected domain B with a border ∂B, to answer the question on existence of a singular point of the given type (at this domain) it is sufficient to calculate the number J of turns made by the tangent vector (around its origin) while moving counterclockwise along the closed contour ∂B. This number is called *Poincaré index* and is defined [20] by the following equation

$$J = \frac{1}{2\pi} \oint_{\partial B} d\left(\arctan \frac{Q(x,y)}{P(x,y)}\right) = \frac{1}{2\pi} \oint_{\partial B} \frac{PdQ - QdP}{P^2 + Q^2}.$$

It is known, that, for any domain B containing no singular points, $J = 0$. Since J depends exclusively on number of singular points contained in B and is independent on the form of ∂B, for all known types of singular points[1], the corresponding values of J can be easily calculated [13] as well. In Fig. 3 we present such values for circular domains containing a single core (a), delta (b), and no singular points (c).

To estimate locations of singular points for a fingerprint under consideration efficiently, we follow the approach proposed in [2].

Indeed, representing J in the form

$$J = \frac{1}{2\pi} \oint_{\partial B} L\,dx + M\,dy,$$

where

$$L = \frac{P\frac{\partial Q}{\partial x} - Q\frac{\partial P}{\partial x}}{P^2 + Q^2}, \quad M = \frac{P\frac{\partial Q}{\partial y} - Q\frac{\partial P}{\partial y}}{P^2 + Q^2},$$

---

[1] Which can be found in fingerprint images.

by Green's Theorem, we have

$$J = \frac{1}{2\pi} \iint_B \left(\frac{\partial M}{\partial x} - \frac{\partial L}{\partial y} \right) dx dy.$$

Therefore, to find locations and types of all singular points it is sufficient to compute J for a sliding small square window B (we use the window of size 3×3).

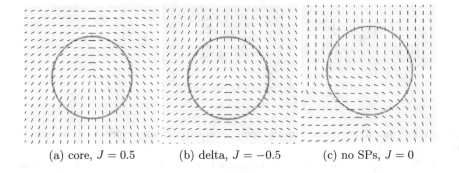

(a) core, $J = 0.5$ (b) delta, $J = -0.5$ (c) no SPs, $J = 0$

Fig. 3. Poincaré indexes

5   Model-Based of Approximation

To the starting point of this stage we have the complex-valued function U and real-valued function Coh defining a coarse orientation field and its coherence at any pixel (x, y), respectively, which are augmented by a list of locations and types of cores and deltas extracted from the fingerprint in question. Given by this input information, we construct the following model-based approximation of the orientation field. Indeed, it is known (see, e.g. [27]) that, in a small neighborhood of a core or a delta located at the point (x_0, y_0) and rotated clockwise to the angle φ, the induced orientation field is perfectly approximable by the following functions U_{core} and U_{delta} respectively

$$U_{core}(x', y') = \frac{y' - y_0}{r} - i\frac{x' - x_0}{r}, \quad U_{delta}(x', y') = -\frac{y' - y_0}{r} - i\frac{x' - x_0}{r},$$

where $r = \sqrt{(x' - x_0)^2 + (y' - y_0)^2}$ and

$$\begin{bmatrix} x' - x_0 \\ y' - y_0 \end{bmatrix} = \begin{pmatrix} \cos\varphi & \sin\varphi \\ -\sin\varphi & \cos\varphi \end{pmatrix} \cdot \begin{bmatrix} x - x_0 \\ y - y_0 \end{bmatrix}.$$

On the other hand, away from singularities, an orientation field is rather smooth and can be approximated by an analytic function $U_0(x, y)$, whose real and imagine parts are polynomials up to the given order [8,27]. Taking into account this argument, we approximate not the initial coarse field U but the biased field

$$\tilde{U} = \frac{1}{\Gamma - \sum_{i=1}^{k_c} \gamma_i' - \sum_{j=1}^{k_d} \gamma_j''} \left(U - \sum_{i=1}^{k_c} \gamma_i' U_{core,i} - \sum_{j=1}^{k_d} \gamma_j'' U_{delta,j} \right),$$

where k_c and k_d are numbers of cores and deltas retrieved from the analyzed fingerprint at the previous stage, $U_{core,i}$ and $U_{delta,j}$ are fields induced by the i-th core and j-th delta respectively, γ_i' and γ_j'' are the appropriate Gaussian kernels, and Γ is a tunable outer parameter.

To increase the overall efficiency of the polynomial fitting procedure, we use (as in [21]) an approximation approach based on two-variate Legendre polynomials [23]. This polynomial family can be easily constructed by means of the appropriate univariate Legendre polynomials defined on the segment $[-1,1]$, $P_n(x)$, $n = 0, 1, \ldots$ having the *orthogonality* property

$$\int_{-1}^{1} P_n(x) P_m(x) dx = \frac{2}{2n+1} \delta_{n,m},$$

where $\delta_{n,m}$ is the Kronecker symbol, which is equal to 1 if $m = n$ and to 0 otherwise. There are many ways to construct the Legendre polynomials, e.g. Gram-Schmidt orthogonalization process applied to monomials $1, x, x^2, \ldots$ Among them the well-known *Bonnet's recursion formula*

$$P_0(x) = 1, \ P_1(x) = x, \ (n+1)P_{n+1}(x) = (2n+1)xP_n(x) - nP_{n-1}(x)$$

is generally regarded to be the most efficient. Further, for a given k, to compute the set of basis functions for the k-th order two-variate Legendre polynomial expansion, the following simple equation

$$P_{n,m}(x, y) = P_{n-m}(x) P_m(y), \quad (n = 0, \ldots, k, \ m = 0, \ldots, n)$$

can be used. It is useful to establish a linear order on the set of polynomials $P_{n,m}$, e.g. the polynomials can be ordered as follows

$$P_{0,0}, P_{1,0}, P_{1,1}, P_{2,0}, P_{2,1}, P_{2,2}, \ldots$$

Denote ordered in such a way polynomials as $\psi_1(x, y), \ldots, \psi_K(x, y)$ for $K = (k+2)(k+1)/2$.

To apply this approximation technique to our problem, we rescale coordinates to transform the initial domain D to $[-1, 1]^2$. Let all pixels of the fingerprint in question be mapped to points $z_1 = (\xi_1, \eta_1), \ldots, z_l = (\xi_l, \eta_l) \in [-1, 1]^2$. We find an approximation for real and imagine parts of the function $\tilde{U}$ separately using the well-known least squares fitting method. To proceed, we compute the $l \times K$

matrix

$$\Psi = \begin{pmatrix} \Psi(z_1) \\ \vdots \\ \Psi(z_l) \end{pmatrix} = \begin{pmatrix} \psi_1(\xi_1,\eta_1) \ \dots \ \psi_K(\xi_1,\eta_1) \\ \vdots \\ \psi_1(\xi_l,\eta_l) \ \dots \ \psi_K(\xi_l,\eta_l) \end{pmatrix}.$$

Let $F = [f(z_1),\dots,f(z_l)]^T$ be the vector of observed values (of $Re(\tilde{U})$ or $Im(\tilde{U})$), $C = diag(Coh(z_1),\dots,Coh(z_l))$ be the diagonal $l \times l$ matrix with coarse OF coherence values estimated at Sect. 3, and $a = [a_1,\dots,a_K]^T$ be the vector of polynomial coefficients to be fitted by minimizing the weighted sum

$$\sum_{i=1}^{l} Coh(z_i)(\Psi(z_i)a - f(z_i))^2 \to \min_{a}. \tag{1}$$

It is known that the optimal solution of (1) is defined by the equation

$$\hat{a} = (\Psi^T C \Psi)^+ \Psi^T C F,$$

where, as usual, A^+ denotes the pseudoinverse of a matrix A. Finally, for the estimated coefficient vectors $\hat{a}_{Re}$ and $\hat{a}_{Im}$ the value of the approximated field U_{appr} at any point $z = (\xi,\eta)$ has the following value

$$U_{appr}(z) = \left(\Gamma - \sum_{j=1}^{k_c} \gamma_j' - \sum_{j=1}^{k_d} \gamma_j'' \right) \Psi(z)\hat{a}_{Re}$$

$$+ \sum_{j=1}^{k_c} \gamma_j' Re(U_{core,j}(z)) + \sum_{j=1}^{k_d} \gamma_j'' Re(U_{delta,j}(z)) + i \cdot \left[\left(\Gamma - \sum_{j=1}^{k_c} \gamma_j' - \sum_{j=1}^{k_d} \gamma_j'' \right) \Psi(z)\hat{a}_{Im} \right.$$

$$\left. + \sum_{j=1}^{k_c} \gamma_j' Im(U_{core,j}(z)) + \sum_{j=1}^{k_d} \gamma_j'' Im(U_{delta,j}(z)) \right]$$

and the orientation angle value θ_{appr} can be calculated by the formula

$$\theta_{appr} = \frac{\pi}{2} - \frac{1}{2} \text{atan2}(Im(U_{appr}), Re(U_{appr}))$$

An example of the final OF estimation result is presented in Fig. 1(b).

6   Curved Gabor Filtering

Gabor filters are generally believed to be the most useful tool in fingerprint enhancement (see, e.g. [3,11,24,26]). The impulse response function of the filter has the following form

$$g(x,y;\theta,f,\sigma_x,\sigma_y) = \exp\left(-\frac{x_\theta^2}{2\sigma_x^2}\right) \exp\left(-\frac{y_\theta^2}{2\sigma_y^2}\right) \exp(i \cdot 2\pi f x_\theta), \tag{2}$$

where $x_\theta = x\cos\theta + y\sin\theta$, $y_\theta = -x\sin\theta + y\cos\theta$, θ is an orientation angle and f is a frequency. Conventional approach to fingerprint enhancement by means of adaptive Gabor filtering is based on local features of the image (an average of the orientation field, an estimate of frequency of papillary ridges, etc.). According to this approach, the fingerprint image under consideration is partitioned into small rectangular regions; then, for each of these regions, local values of averaged orientation $\bar{\theta}$ of papillary ridges and their frequency $\bar{f}$ are estimated, after that the appropriate Gabor filter is constructed (or taken from a bank of precomputed filters) and used for the subsequent enhancing of this region. It is known that this method performs well in case when papillary ridges behave like near parallel strait lines. But, in real fingerprint images, ridges can be very curvy, especially in the vicinity of singular points (cores and deltas), which are of the most interest for the subsequent analysis. Unfortunately, in these cases, the conventional Gabor filtering method produces numerous artifacts, which lead to spurious minutiae at the minutiae detection phase and increase misclassification rate at the phase of fingerprint matching.

Therefore, in [7] the new approach to Gabor filtering is introduced. The author proposes to curve the Gabor filter according to the curvature of papillary lines (Fig. 4). This method seems to be very fruitful, but the writing style of the paper [7] appears to be lax.

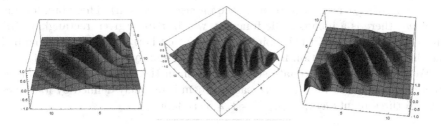

Fig. 4. Examples of a curved Gabor filter

To make the story more rigorous, we use the dual approach. Instead of curving the graph of Gabor filter's impulse response function, we propose to warp coordinate system of the initial image using the appropriate non-singular transformation. Classic complex analysis provides the useful technique for construction of such a transformation on the basis of *holomorphic* functions $w : \mathbb{C} \to \mathbb{C}$ or *conformal maps* (see, e.g. [4]). It is known that a differentiable function $w(x, y) = u(x, y) + i \cdot v(x, y)$ is holomorphic on a given simple connected domain $D \subset \mathbb{C}$, if w satisfies the Cauchy-Riemann conditions

$$\frac{\partial u}{\partial x} = \frac{\partial v}{\partial y}, \quad \frac{\partial u}{\partial y} = -\frac{\partial v}{\partial x}.$$

Any conformal map preserves several geometric features of the mapped domain: border, angles between curves, etc. In particular, at any point $z = x_0 + i \cdot y_0 \in D$, the level contours $u(x, y) = u(x_0, y_0)$ and $v(x, y) = v(x_0, y_0)$ are orthogonal. As it follows from the famous Riemann Mapping Theorem, any simple connected domain (with piecewise-smooth border) can be mapped by some holomorphic function to the given rectangle such a way that lines of the Cartesian grid on the rectangle are mapped by the inverse function to a family of mutually orthogonal curves (Fig. 5).

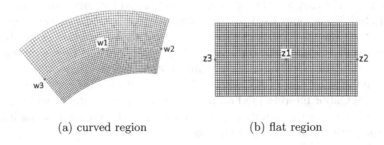

(a) curved region   (b) flat region

Fig. 5. Example of holomorphic transformation

We use this fact in the following way. For any fingerprint, at any point, papillary lines are near parallel to the direction defined by the retrieved orientation field (and orthogonal to a gradient of the intensity function). Therefore, for any point w_1, there is a holomorphic function, which maps a given rectangle to the curved region centered at this point such a way that horizontal lines are mapped onto lines of the orientation field and vertical lines onto lines of gradients.

We propose to approximate such a function in the class of *fractional-linear* functions $w(z) = (az+b)/(cz+d)$, each of them is uniquely defined by its values in given three points $w(z_i) = w_i$ $(i = 1, 2, 3)$ as follows

$$\frac{w - w_1}{w - w_3} : \frac{w_2 - w_1}{w_2 - w_3} = \frac{z - z_1}{z - z_3} : \frac{z_2 - z_1}{z_2 - z_3}. \tag{3}$$

In our study, we locate these points as shown in Fig. 5. W.l.o.g. we assume that $z_1 = w_1 = 0$, $z_2 = -z_3 = A$ for some parameter $A > 0$. In this case, Eq. (3) reduces to

$$w(z) = w_3 \left(1 + \frac{(z + A)(w_2 - w_3)}{z(w_2 + w_3) - A(w_2 - w_3)} \right). \tag{4}$$

Locating a curved ROI in a fingerprint (Fig. 6(a)), we use the function $w(z)$ defined by formula (4) to fill the rectangular window (Fig. 6(c)). Then, using the standard Gabor filter (2) for $\theta = \pi/2$, an estimated (on this window) value of frequency f, and the values $\sigma_x = 3.4/(9f)$, $\sigma_y = 4/3\sigma_x$ (as proposed in [7]), we produce the improved window (Fig. 6(d)), which is mapped into the initial place by the inverse function w^{-1} and simple approximation trick in the rectangular window (Fig. 6(b)).

Indeed, suppose, we need to determine the intensity value at some point w_0 located in the selected ROI[2]. First, using the inverse function w^{-1} we find its

---

[2] Of the fingerprint to be analyzed.

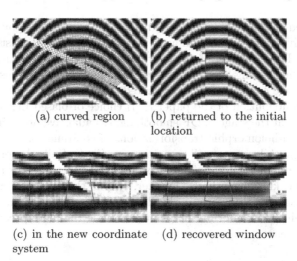

(a) curved region   (b) returned to the initial
                               location

(c) in the new coordinate (d) recovered window
system

Fig. 6. Filtering process

preimage $z_0 = w^{-1}(w_0)$. To this moment, we know intensity values (improved by Gabor filter) in any integer point (pixel) of our rectangular window. To interpolate the required value $I(z_0)$ we use the standard bilinear interpolation algorithm. And, voila! The required intensity value $I(w_0)$ is found, since $I(w_0) = I(z_0)$.

(a) initial image   (b) processed image

Fig. 7. Result of the proposed approach

7 Conclusion

In the paper the novel theoretical approach to fingerprint enhancement, comprising both the well-known approaches (gradient-based OF estimation, core extraction based on Poincaré indexes, and Gabor filtering) and significantly modified or partially brand-new techniques, such as model-based least-squares OF smoothing and local warping of the coordinates (by the appropriate conformal maps) during the filtering. An example of application of the proposed approach to fingerpint image enhancement is presented at Fig. 7.

The main contribution of the paper is (i) choice of biased Legendre polynomials as a model at the stage of OF smoothing, (ii) adaptive procedure of Gabor filtering based on holomorphic transformations of coordinates.

Certainly, multiple issues still remain open. First of all, the proposed approach should be numerically evaluated over some public fingerprint dataset in comparison with known methods. Further, to put the explanation simple, we construct holomorphic transform in the class of fractional-linear functions. Using the more complicated conformal maps may provide the more precise results at the filtering stage.

Acknowledgements. This research was supported by Russian Science Foundation, grant no. 14-11-00109.

References

1. FVC ongoing. https://biolab.csr.unibo.it/fvcongoing/UI/Form/Home.aspx. 20 March 2015
2. Bazen, A., Gerez, S.: Systematic methods for the computation of the directional fields and singular points of fingerprints. IEEE Trans. Pattern Anal. Mach. Intell. **24**(7), 905–919 (2002)
3. Bhanu, B., Tan, X.: Computational Algorithms for Fingerprint Recognition (Kluwer International Series on Biometrics, 1). Kluwer Academic Publishers, Norwell (2003)
4. Cohn, H.: Conformal Mapping on Riemann Surfaces. Dover London, New York (1967)
5. Dremin, A., Khachay, M.Y., Leshko, A.: Fingerprint identification algorithm based on delaunay triangulation and cylinder codes. In: AIST 2014, pp. 128–139 (2014)
6. Feng, J., Zhou, J., Jain, A.: Orientation field estimation for latent fingerprint enhancement. IEEE Trans. Pattern Anal. Mach. Intell. **35**(4), 925–940 (2013)
7. Gottschlich, C.: Curved-region-based ridge frequency estimation and curved gabor filters for fingerprint image enhancement. IEEE Trans. Image Process. **21**(4), 2220–2227 (2012)
8. Gu, J., Zhou, J., Zhang, D.: A combination model for orientation field of fingerprints. Pattern Recogn. **37**(3), 543–553 (2004)
9. Hong, L., Wan, Y., Jain, A.: Fingerprint image enhancement: algorithm and performance evaluation. IEEE Trans. Pattern Anal. Mach. Intell. **20**(8), 777–789 (1998)
10. Jiang, X., Yau, W.Y., Wee, S.: Detecting the fingerprint minutiae by adaptive tracing the gray-level ridge. Pattern Recogn. **34**(5), 999–1013 (2001)
11. Karimimehr, N., Shirazi, A., Keshavars Bahaghighat, M.: Fingerprint image enhancement using gabor wavelet transform. In: 2010 18th Iranian Conference on Electrical Engineering (ICEE), pp. 316–320, May 2010
12. Kass, M., Witkin, A.: Analyzing oriented patterns. Comput. Vis. Graph. Image Process. **37**(3), 362–385 (1987)

13. Kawagoe, M., Tojo, A.: Fingerprint pattern classification. Pattern Recogn. **17**(3), 295–303 (1984)
14. Khachai, M.Y., Leshko, A.S., Dremin, A.V.: The problem of fingerprint identification: a reference database indexing method based on delaunay triangulation. Pattern Recogn. Image Anal. **24**(2), 297–303 (2014)
15. Liu, M., Chen, X., Wang, X.: Latent fingerprint enhancement via multi-scale patch based sparse representation. IEEE Trans. Inf. Forensics Secur. **10**(1), 6–15 (2015)
16. Maltoni, D., Cappelli, R.: Advances in fingerprint modeling. Image Vis. Comput. **27**(3), 258–268 (2009)
17. Maltoni, D., Maio, D., Jain, A.K., Prabhakar, S.: Handbook of Fingerprint Recognition, 2nd edn. Springer Publishing Company, Incorporated, London (2009)
18. Mihăilescu, P., Mieloch, K., Munk, A.: Fingerprint classification using entropy sensitive tracing. In: Bonilla, L.L., Moscoso, M., Platero, G., Vega, J.M. (eds.) Progress in Industrial Mathematics at ECMI 2006, pp. 928–932. Springer, Heidelberg (2008)
19. Oliveira, M., Leite, N.: A multiscale directional operator and morphological tools for reconnecting broken ridges in fingerprint images. Pattern Recogn. **41**(1), 367–377 (2008)
20. Poincaré, H.: Mémoire sur les courbes définies par une équation différentielle. Journal de mathématiques pures et appliquées **7**, 375–422 (1881)
21. Ram, S., Bischof, H., Birchbauer, J.: Modelling fingerprint ridge orientation using Legendre polynomials. Pattern Recogn. **43**(1), 342–357 (2010)
22. Sherlock, B., Monro, D.: A model for interpreting fingerprint topology. Pattern Recogn. **26**(7), 1047–1055 (1993)
23. Suetin, P.: Orthogonal Polynomials in Two Variables. Gordon and Breach, Amsterdam (1999)
24. Turroni, F., Cappelli, R., Maltoni, D.: Fingerprint enhancement using contextual iterative filtering. In: 2012 5th IAPR International Conference on Biometrics (ICB), pp. 152–157 (2012)
25. Voss, H.U., Timmer, J., Kurths, J.: Nonlinear dynamical system identification from uncertain and indirect measurements. Int. J. Bifurcat. Chaos **14**, 1905–1933 (2004)
26. Wang, W., Li, J., Huang, F., Feng, H.: Design and implementation of log-gabor filter in fingerprint image enhancement. Pattern Recogn. Lett. **29**(3), 301–308 (2008)
27. Zhou, J., Gu, J.: Modeling orientation fields of fingerprints with rational complex functions. Pattern Recogn. **37**(2), 389–391 (2004)

Remote Sensing Data Verification Using Model-Oriented Descriptors

Andrey Kuznetsov[1,2(✉)] and Vladislav Myasnikov[1,2]

[1] Samara State Aerospace University (SSAU), Samara, Russia
kuznetsoff.andrey@gmail.com, vmyas@rambler.ru
[2] Image Processing Systems Institute of the Russian Academy of Sciences
(IPSI RAS), Samara, Russia

Abstract. This paper presents a solution of remote sensing data verification problem. Remote sensing data includes digital image data and metadata, which contains parameters of satellite image shooting process (Sun and satellite azimuth and elevation angles, shooting time, etc.). The solution is based on the analysis of special numerical characteristics, which directly depend on the shooting parameters: sun position, satellite position and orientation. We propose two fully automatic algorithms for remote sensing data analysis and decision-making based on data compatibility: the first one uses vector data of the shooting territory, the second doesn't.

Keywords: Satellite image · Vector map · Buffer zone · Model-oriented descriptor · Canny edge detector · Border tracing

1 Introduction

Widely used in the modern world remote sensing data (RSD) consist of two main components: a digital image and its metadata, which describe the process and the shooting conditions of an image. During RSD transmission from source to destination, these data can be distorted accidentally (due to errors) and intentionally (by hackers). When this happens, the satellite image itself or/and its metadata can be changed. The problem of forgery detection in digital images is being solved in [1–5] when shooting parameters and image metadata are not used or unknown.

Nowadays, there are papers devoted to the analysis of light parameters inconsistency for local parts of a single object in digital images [6]. These algorithms use only image data during analysis since additional information about shooting conditions is absent (the research is carried out for digital images obtained by ordinary cameras that do not store shooting information). Due to the lack of this data, there is nothing to compare with angles and lengths of shadows in the analyzed image. Metadata of satellite images and vector maps of images territory allow to analyze the consistency of objects and their shadows. During literature analysis there were not found any papers aimed at the detection of inconsistency in shadows and objects in satellite images.

In this paper we propose a new solution for detection of inconsistency of digital satellite image and its metadata shooting parameters using model-oriented descriptors, proposed by V.V. Myasnikov in papers [7, 8].

© Springer International Publishing Switzerland 2015
M.Y. Khachay et al. (Eds.): AIST 2015, CCIS 542, pp. 96–101, 2015.
DOI: 10.1007/978-3-319-26123-2_9

2 Problem Definition

To identify irrelevance between an image and its metadata, we will analyze the shadows of high objects on the image. There will be used high buildings with height of at least 12 m (for example, houses with 5 floors and more), which have a simple rectangular form on a satellite image received by nadir shooting. The length of the analyzed shadows of such building is 10–15 m – in the length exceeds this value the shadows may be imposed on neighboring buildings (in dense urban areas), which may impair analysis quality. It is better to identify objects and their shadows with such linear characteristics on high-resolution images (0.5–1 m). This is why we will use Geoeye-1 satellite images (spatial resolution – 0.5 m). The above parameters characterize the restrictions wherein the performed algorithms will work correctly.

Image metadata contains the following parameters of the satellite image, which are used in the proposed algorithms:

1. shooting coordinates $\mathbf{s} = (s_1, s_2, \ldots, s_k)^T$, where $s_i = (x_i, y_i)$ is a reference point of a satellite image, k is a number of reference points;
2. satellite position coordinates $\mathbf{p} = (\varphi_{az}, \varphi_{el}, \varphi_{alt})$, where φ_{az} is the azimuth incidence angle of a satellite sensor, φ_{el} is the elevation incidence angle of a satellite angle, h_{alt} corresponds to the altitude value of a satellite;
3. Sun position coordinates $\boldsymbol{\alpha} = (\alpha_{az}, \alpha_{el})^T$, where α_{az}, α_{el} are azimuth and elevation incidence Sun angles respectively.

Figure 1 shows the relative position of azimuth and zenith angles of Sun and spacecraft.

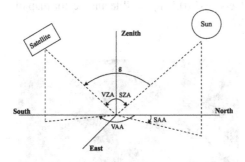

Fig. 1. Arrangements of the angles for Sun and spacecraft (VAA – azimuth shooting angle, VZA – zenith shooting angle, SAA – azimuth Sun angle, SZA – zenith Sun angle, g – phase angle)

3 Proposed Solution

The base of a model-oriented descriptor constructing is the use of probability distribution of the gradient field, which characterizes the model of the analyzed image fragment. Values of descriptor components for a particular image fragment are calculated as the values of probability density of the argument in the form of a specific gradient field or some of its components.

For a formal definition of this descriptor, we introduce some notation. Let D be an analyzed image area (area of some real object shadow), for which the function $\varphi(t_1, t_2)$ is defined. The values of this function define orientation (angle) of a brightness difference line (along shadow boundaries) in the corresponding position (t_1, t_2).

Amplitude-phase mismatch (APM) ζ for an image area D is defined as follows:

$$\zeta = \frac{SGD}{SGM}, \ \zeta \in [0, 1], \tag{1}$$

where SGD and SGM are represented in the form:

$$SGM = \sum_{(t_1, t_2) \in D} |g(t_1, t_2)|, \ SGD$$
$$= \sum_{(t_1, t_2) \in D} |g(t_1, t_2)| \left(\frac{\cos(\varphi(t_1, t_2) - \arg(g(t_1, t_2))) + 1}{2} \right).$$

At this point $g(t_1, t_2)$ is a concrete implementation of the gradient field for the given image fragment, $|g(t_1, t_2)|$ and $\arg(g(t_1, t_2))$ are its modulus and direction (phase) respectively. It is obvious that the closer APM value ζ to 1, the more image area D matches a template, represented by $\varphi(t_1, t_2)$ function.

Let us consider a situation when there is a priori information about the shooting territory – a vector map of this area. By carrying out a geometric calibration of a snapshot and putting it on a vector map of the shooting area, it is possible to determine positions of physical objects in the space image. The example of combining space image received from Geoeye-1 (0.5 m) satellite and vector map of the shooting territory is presented in Fig. 2.

Fig. 2. Combination of satellite image and vector map, $\varphi_{el} = 35$

Using semantic data of the buildings vector layer we select only those buildings, which height is more than 10 m. The first step of the proposed algorithm of space image verification consists of building shadow buffer zone detection. Taking into account the metadata of the analyzed satellite image, we calculate this buffer zone for all the buildings, which meet the requirements (Fig. 3).

Fig. 3. Shadow buffer zone detection for a building

For each of the buffer zones we then calculate APM value (1), which characterizes correspondence of the real object shadow in the satellite image (according to the orientation of the buffer zone) to the value, calculated using metadata parameters.

There is another situation when we have no vector map of the shooting territory. If it is so, we need to detect the correspondence of shadows parameters and shooting parameters using only image analysis. We will use high-resolution snapshots for image analysis as in the previous algorithm. There is proposed another algorithm that identify the corresponding angles of the buildings and their shadows using Canny detector [9, 10]. The advantages of this edge detection method include better detection in noisy images (due to the use of a threshold value) and width of the detected edges (it does not exceed one pixel) – this allows to use edge tracing further [11]. The algorithm of edge detection result analysis consists of 2 steps: detection of corresponding right angles and detection of shadow borders, codirectional with shadow incidence angle, calculated using metadata parameters.

As a result of the algorithm the list of corresponding right angles and the list of shadows borders, which are the closest to these points, there is formed a geometric model of the object shadow. Then the value of APM is calculated for this model. The speed of proposed algorithms enables their use in real-time analysis of satellite images.

4   Experiments

During the algorithm research we define APM threshold values, which will be used for a decision making of satellite image correspondence to shooting conditions. To conduct an experiment we take Geoeye-1 satellite images (0.5 m) and a set of 26 vector objects, randomly selected among the objects belonging to the snapshot territory. Random selection leads to the fact, that some shadow buffer zones boundaries may appear in the shadow region of other vector objects, or may be blocked by other objects in the image.

We will calculate APM values for any of the analyzed satellite image channels in two ways:

(1) for each side of the shadow buffer zone of the object – 104 objects in the train sample (4 values for each vector object);
(2) for the whole shadow buffer zone of the object – 26 objects in the train sample.

Shadow buffer zone boundaries are calculated for a given correct shadow inclination angle $\alpha_s = 75°$, defined in satellite metadata.

The APM threshold value for i-th train sample method is defined as follows:

$$t_i = \min_{k \in L}\{\zeta_k\} \cdot 0.9, \quad i \in \{1,2\},$$

where ζ_k is the APM value for k-th object of a train sample L, 0.9 is a constant defined experimentally.

The distribution of APM values for both techniques of creating a train sample lead to the following threshold values $t_1 = 0.34$ and $t_2 = 0.6$.

Decision making of satellite image to shooting conditions correspondence is performed as follows. There is selected a test sample of 20 buildings (vector objects) for the analyzed satellite image. On the first stage APM values are calculated for each element of the shadow buffer zone and the object doesn't pass a test if $\exists\, j \in \overline{0,3}$, $\zeta_j < t_1$. On the second stage APM values are calculated for the entire shadow buffer zone and the decision is made in a similar way: $\zeta < t_2$. Satellite image will not pass the validation test, if at least one test sample object does not pass this two-stage test procedure.

In order to confirm the APM threshold values t_1, t_2 selection we take a satellite image and a test sample of 20 vector objects, which belong to the territory of the snapshot. We then construct a relationship between the values of shadow inclination angles and the number of objects that have not passed the two-stage procedure of satellite image validation (see Fig. 4).

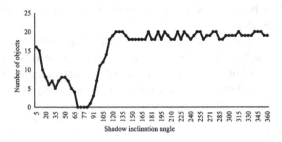

Fig. 4. Dependency of test sample objects number that failed validation test from shadow inclination angle

According to the results of conducted experiments it can be concluded that developed algorithms detect inconsistency of a satellite image and its shooting conditions when the deviation of shadow inclination angle from its correct value is more than 5° for t_1, t_2. This is acceptable for the analysis of satellite images.

5   Conclusion

We have presented two new algorithms for detection of inconsistencies of a satellite image and its shooting conditions: when we have and don't have vector information about the shooting area. The proposed solution allows to detect inconsistencies of shadows and shooting conditions when deviation of shadow incidence angle from correct value is higher than 5.

Acknowledgements. This work was supported by the Russian Science Foundation grant №14-31-00014 «Establishment of a Laboratory of Advanced Technology for Earth Remote Sensing».

References

1. Glumov, N., Kuznetsov, A.: Detection of local artificial changes in images. Optoelectron. Instrum. Data Process. **47**(3), 4–12 (2011)
2. Glumov, N., Kuznetsov, A.: Image copy-move detection. Comput. Opt. **35**(4), 508–512 (2011)
3. Glumov, N., Kuznetsov, A., Myasnikov, V.: The algorithm for copy-move detection on digital images. Comput. Opt. **37**(3), 360–367 (2013)
4. Kuznetsov, A., Myasnikov, V.: Efficient linear local features based copy-move detection algorithm. Comput. Opt. **37**(4), 489–495 (2013)
5. Vladimirovich, K.A., Valerievich, M.V.: A fast plain copy-move detection algorithm based on structural pattern and 2D Rabin-Karp rolling hash. In: Campilho, A., Kamel, M. (eds.) ICIAR 2014, Part I. LNCS, vol. 8814, pp. 461–468. Springer, Heidelberg (2014)
6. Farid, H.: Image forgery detection. IEEE Sig. Process. Mag. **26**, 16–25 (2009)
7. Myasnikov, V.: Method for detection of vehicles in digital aerial and space remote sensed images. Comput. Opt. **36**(3), 429–438 (2012)
8. Myasnikov, V.: Model-based gradient field descriptor as a convenient tool for image recognition and analysis. Comput. Opt. **36**(4), 596–604 (2012)
9. Gashnikov, M., Glumov, N., Ilyasova, N., Myasnikov, V., Popov, S., Sergeyev, V., et al.: In: Soifer, V.A. (ed.) Computer Image Processing, Part II: Methods and Algorithms. VDM Verlag, Saarbrücken (2009)
10. Canny, J.: A computational approach to edge detection. Pattern Anal. Mach. Intell. IEEE Trans. PAMI. **8**(6), 679–698 (1986)
11. Ren, M., Yang, J., Sun, H.: Tracing boundary contours in a binary image. Image Vis. Comput. **20**(2), 125–131 (2002)

New Bi-, Tri-, and Fourlateral Filters for Color and Hyperspectral Images Filtering

Ekaterina Ostheimer[1], Valeriy Labunets[2], Andrey Kurganski[2]([✉]),
Denis Komarov[2], and Ivan Artemov[2]

[1] Capricat LLC, 1340 South Ocean Boulevard Suite 209,
Pompano Beach, FL 33062, USA
katya@capricat.com
[2] Ural Federal University, pr. Mira, 19, Yekaterinburg 620002, Russia
vlabunets05@yahoo.com, k-and92@mail.ru

Abstract. In the paper, we investigate effectiveness of modified bilateral and new tri-, and fourlateral denoising filters for grey, color, and hyperspectral image procession. Conventional bilateral filter performs merely weighted averaging of the local neighborhood pixels. The weight includes two components: spatial and radiometric ones. The first component measures the geometric distances between the center pixel and local neighborhood ones. The second component measures the radiometric distance between the values of the center pixel and local neighborhood ones. Noise affects all pixels even onto the centre one used as a reference for the tonal filtering. Thus, the noise affecting the centre pixel has a disproportionate effect onto the result. This suggests the first modification: the center pixel is replaced by the weighted average (with some estimate of the true value) of the neighborhood pixels contained in a window around it. The second modification uses the matrix-valued weights. They include four components: spatial, radiometric, inter-channel weights, and radiometric inter-channel ones. The fourth weight measures the radiometric distance (for grey-level images) between the inter-channel values of the center scalar-valued channel pixel and local neighborhood channel ones.

Keywords: Bilateral filters · Image processing · Hyperspectral images

1 Introduction

We develop a conceptual framework and design methodologies for multi-channel image bilateral aggregation filtering systems with assessment capability. The term multichannel image is used for an image with more than one component. They are composed of a series of images in different optical bands at wavelengths $\lambda_1, \lambda_2, ..., \lambda_K$, called the spectral channels

$$\mathbf{f}(x, y) = (f_{\lambda_1}(x, y), f_{\lambda_2}(x, y), ..., f_{\lambda_K}(x, y))$$

where K is the number of different optical channels, *i.e.*, $\mathbf{f} : \mathbf{R}^2 \to \mathbf{R}^K$, where $\mathbf{R}^K$ is multicolor space. The bold font for $\mathbf{f}$ emphasizes the fact that images may

© Springer International Publishing Switzerland 2015
M.Y. Khachay et al. (Eds.): AIST 2015, CCIS 542, pp. 102–113, 2015.
DOI: 10.1007/978-3-319-26123-2_10

be multichannel. Each pixel in $\mathbf{f}(x, y)$, therefore, represents the spectrum at the wavelengths λ_1, λ_2, ..., λ_K of the observed scene at point $\mathbf{x} = (x, y)$.

Let us introduce the observation model and notion used throughout the paper. We consider noised image in the form $\mathbf{f}(\mathbf{x}) = \mathbf{s}(\mathbf{x}) + \mathbf{n}(\mathbf{x})$, where $\mathbf{s}(\mathbf{x})$ is the original grey-level image and $\mathbf{n}(\mathbf{x})$ denotes the noise introduced into $\mathbf{s}(\mathbf{x})$ to produce the corrupted image $\mathbf{f}(\mathbf{x})$, and $\mathbf{x} = (i, j) \in \mathbf{Z}^2$ is a 2D coordinates that represent the pixel location. The aim of image enhancement is to reduce the noise as much as possible or to find a method, which, given $\mathbf{s}(\mathbf{x})$, derives an image $\widehat{\mathbf{s}}(\mathbf{x})$ as close as possible to the original $\mathbf{s}(\mathbf{x})$ subjected to a suitable optimality criterion.

The standard bilateral filter (BF) [2–12] with a square N-cellular window $M(\mathbf{x})$ is located at $\mathbf{x}$, the weighted average of pixels in the moving window replaces the central pixel

$$\widehat{\mathbf{s}}(\mathbf{x}) = \underset{\mathbf{p} \in M(\mathbf{x})}{\mathbf{BilMean}} \left[w(\mathbf{x}, \mathbf{p}) \cdot \mathbf{f}(\mathbf{p}) \right] = \frac{1}{k(\mathbf{x})} \sum_{\mathbf{p} \in M(\mathbf{x})} w(\mathbf{x}, \mathbf{p}) \mathbf{f}(\mathbf{p}), \tag{1}$$

where $\widehat{\mathbf{s}}(\mathbf{x})$ is the filtered image and $k(\mathbf{x})$ is the normalization factor

$$k(\mathbf{x}) = \sum_{\mathbf{p} \in M(\mathbf{x})} w(\mathbf{x}, \mathbf{p}). \tag{2}$$

Equation (1) is simply a normalized weighted average of a neighbourhood of the N-cellular window $M(\mathbf{x})$ (*i.e.*, the mask around pixel $\mathbf{x}$, consisting of N pixels).

The scalar-valued weights $w(\mathbf{x}, \mathbf{p})$ are computed based on the content of the neighbourhood. For pixels $\{\mathbf{f}(\mathbf{p})\}_{\mathbf{p} \in M(\mathbf{x})}$ around the centroid $\mathbf{f}(\mathbf{x})$, the weights $\{w(\mathbf{x}, \mathbf{p})\}_{\mathbf{p} \in M(\mathbf{x})}$ are computed by multiplying the following two factors:

$$w(\mathbf{x}, \mathbf{p}) = w_{Sp}(\mathbf{p}) \cdot w_{Rn}(\mathbf{x}, \mathbf{p}) = w_{Sp}(\|\mathbf{p}\|) \cdot w_{Rn}(\|\mathbf{f}(\mathbf{x}) - \mathbf{f}(\mathbf{p})\|_2).$$

The weight includes two ingredients: *spatial* $w_{Sp}(\|\mathbf{p}\|)$ and *radiometric* weights $w_{Rn}(\mathbf{x}, \mathbf{p}) = w_{Rn}(\|\mathbf{f}(\mathbf{x}) - \mathbf{f}(\mathbf{p})\|_2)$. The first weight measures the geometric distance $\|\mathbf{p}\|$ between the center pixel $\mathbf{f}(\mathbf{x})$ and the pixel $\mathbf{f}(\mathbf{p})$ (note, the centroid $\mathbf{x}$ has the position $\mathbf{0} \in M(\mathbf{x})$ inside of the mask $M(\mathbf{x})$). Here, the Euclidean metric $\|\mathbf{p}\| = \|\mathbf{p}\|_2$ is applied. This way, close-by pixels influence the final result more than distant ones. The second weight measures the radiometric distance between the values of the center pixel $\mathbf{f}(\mathbf{x})$ and all N pixels $\mathbf{f}(\mathbf{p})$, $\mathbf{p} \in M(\mathbf{x})$ and again, the Euclidean metric $\|\mathbf{f}(\mathbf{x}) - \mathbf{f}(\mathbf{p})\|_2$ is chosen, too. Therefore, pixels with close-by values tend to influence the final result more than those having distant value.

This paper considers two natural extensions to the bilateral filter. Firstly, instead of the center pixel $\mathbf{f}(\mathbf{x})$ in $w_{Rn}(\|\mathbf{f}(\mathbf{x}) - \mathbf{f}(\mathbf{p})\|)$, we use the *Fréchet median* $\bar{\mathbf{f}}(\mathbf{x})$ for calculating of weighs $w_{Rn}(\mathbf{x}, \mathbf{p}) = w_{Rn}(\|\bar{\mathbf{f}}(\mathbf{x}) - \mathbf{f}(\mathbf{p})\|_2)$. Secondly, instead of the scale-valued weighs, we use a matrix-valued ones

$$\widehat{\mathbf{s}}(\mathbf{x}) = \underset{\mathbf{p} \in M(\mathbf{x})}{\mathbf{BilMean}} \left[\mathbf{W}(\mathbf{x}, \mathbf{p}) \cdot \mathbf{f}(\mathbf{p}) \right], \tag{3}$$

where $\mathbf{W}(\mathbf{x}, \mathbf{p})$ are the matrix-valued weighs.

2   The First Modification of Bilateral Filter

In this modification we use the *Fréchet median* $\bar{\mathbf{f}}(\mathbf{x})$ for calculating of weighs $w_{Rn}(\mathbf{x}, \mathbf{p})$ instead of the center pixel $\mathbf{f}(\mathbf{x})$ in $w_{Rn}(\|\mathbf{f}(\mathbf{x}) - \mathbf{f}(\mathbf{p})\|)$. Let $\langle \mathbf{R}^K, \rho \rangle$ be a metric space, where ρ is a distance function. Let $w_1, w_2, \ldots, w_N$ be N weights summing to 1 and let $\{\mathbf{f}^1, \mathbf{f}^2, \ldots, \mathbf{f}^N\} = \mathbf{D} \subset \mathbf{R}^K$ be N pixels in the N-cellular window $\mathrm{M}(\mathbf{x})$.

Definition 1. *The optimal Fréchet point associated with the metric ρ, is the point, $\bar{\mathbf{f}}_{\mathrm{opt}} \in \mathbf{D}$, that minimizes the Fréchet cost function $\sum_{i=1}^{N} w_i \rho(\bar{\mathbf{f}}, \mathbf{f}^i)$ (the weighted sum distances from an arbitrary point $\bar{\mathbf{f}}$ to each point $\mathbf{f}^1, \mathbf{f}^2, \ldots, \mathbf{f}^N \in \mathbf{R}^K$). It is formally defined as [1]:*

$$\bar{\mathbf{f}}_{\mathrm{opt}} = \mathbf{FrechPt}\left(\rho|\mathbf{f}^1, \mathbf{f}^2, \ldots, \mathbf{f}^N\right) = \arg \min_{\mathbf{f} \in \mathbf{R}^K} \sum_{i=1}^{N} w_i \rho(\mathbf{f}, \mathbf{f}^i). \tag{4}$$

Note that **argmin** means the argument, for which the sum is minimized. So, the vector-valued median of a discrete set of sample points in a Euclidean space $\mathbf{R}^K$ is the point $\bar{\mathbf{f}}$ minimizing the sum of distances to the pixels $\mathbf{f}^1, \mathbf{f}^2, \ldots, \mathbf{f}^N$. This generalizes the ordinary median, which has the property of minimizing the sum of distances for one-dimensional data, and provides a central tendency higher dimensions.

In computation point of view, it is better to restrict the search domain from $\mathbf{R}^K$ until the finite set $\mathbf{D} = \{\mathbf{f}^1, \mathbf{f}^2, \ldots, \mathbf{f}^N\} \subset \mathbf{R}^K$. In this case, we obtain definition of the *suboptimal Fréchet point* or the *optimal vector Fréchet median*.

Definition 2. *The suboptimal weighted Fréchet point or optimal vector Fréchet median associated with the metric $\rho(\mathbf{x}, \mathbf{y})$ is the point $\bar{\mathbf{f}} \in \{\mathbf{f}^1, \mathbf{f}^2, \ldots, \mathbf{f}^N\}$ that minimizes the FCF over the restrict search domain $\mathbf{D} \subset \mathbf{R}^K$*

$$\bar{\mathbf{f}}_{\mathrm{opt}} = \mathbf{FrechMed}\left(\rho|\mathbf{f}^1, \mathbf{f}^2, \ldots, \mathbf{f}^N\right) = \arg \min_{\mathbf{f} \in \mathbf{D}} \sum_{i=1}^{N} w_i \rho(\mathbf{f}, \mathbf{f}^i). \tag{5}$$

Example 1. If observation data are real numbers, *i.e.*, $f^1, f^2, \ldots, f^N \in \mathbf{R}$ and the distance function is the city distance $\rho(f, g) = \rho_1(f, g) = |f - g|$, then the Fréchet point and median (4), (5) for pixels $f^1, f^2, \ldots, f^N \in \mathbf{R}$ associated with the metric $\rho_1(f, g)$ to be *classical the Fréchet point* and *median*, respectively, *i.e.*,

$$\bar{\mathbf{f}}_{\mathrm{opt}} = \mathbf{FrechPt}\left(\rho_1|\mathbf{f}^1, \mathbf{f}^2, \ldots, \mathbf{f}^N\right) = \arg \min_{\mathbf{f} \in \mathbf{R}} \left(\sum_{i=1}^{N} |f - x^i|\right), \tag{6}$$

$$\bar{\mathbf{f}}_{\mathrm{opt}} = \mathbf{FrechMed}\left(\rho_1|\mathbf{f}^1, \mathbf{f}^2, \ldots, \mathbf{f}^N\right)$$

$$= \arg \min_{\mathbf{f} \in \mathbf{D}} \left(\sum_{i=1}^{N} |f - f^i|\right) = \mathbf{Med}\left(f^1, f^2, \ldots, f^N\right). \tag{7}$$

Example 2. If observation data are vectors, *i.e.*, $\mathbf{f}^1, \mathbf{f}^2, \ldots, \mathbf{f}^N \in \mathbf{R}^K$, and distance function is the city distance $\rho(\mathbf{f}, \mathbf{g}) = \rho_1(\mathbf{f}, \mathbf{g})$, then the Fréchet point (4) and median (5) for vectors $\mathbf{f}^1, \mathbf{f}^2, \ldots, \mathbf{f}^{N^2} \in \mathbf{R}^K$ to be the *Fréchet point* and the *Fréchet vector median*, respectively associated with the same metric $\rho_1(\mathbf{x}, \mathbf{y})$

$$\bar{\mathbf{f}}_{\text{opt}} = \mathbf{FrechPt}\left(\rho_1|\mathbf{f}^1, \mathbf{f}^2, \ldots, \mathbf{f}^N\right) = \arg\min_{\mathbf{f}\in\mathbf{R}^K}\left(\sum_{i=1}^N \|\mathbf{f} - \mathbf{f}^i\|_1\right), \quad (8)$$

$$\bar{\mathbf{f}}_{\text{opt}} = \mathbf{FrechMed}\left(\rho_1|\mathbf{f}^1, \mathbf{f}^2, \ldots, \mathbf{f}^N\right)$$

$$= \arg\min_{\mathbf{f}\in D}\left(\sum_{i=1}^N \|\mathbf{f} - \mathbf{f}^i\|_1\right) = \mathbf{VecMed}\left(\rho_1|\mathbf{f}^1, \mathbf{f}^2, \ldots, \mathbf{f}^N\right). \quad (9)$$

Now we use Fréchet median $\bar{\mathbf{f}}_{\text{opt}}$ for calculating radiometric weights $w_{Rn}(\mathbf{x}, \mathbf{p}) = w_{Rn}(\|\bar{\mathbf{f}}_{\text{opt}}(\mathbf{x}) - \mathbf{f}(\mathbf{p})\|_2)$. The modified bilateral filter (MBF) is given as

$$\hat{s}(\mathbf{x}) = \mathbf{BilMean}_{\mathbf{p}\in M(\mathbf{x})}\left[w(\mathbf{x}, \mathbf{p}) \cdot \mathbf{f}(\mathbf{p})\right]$$

$$= \frac{1}{k(\mathbf{x})}\sum_{\mathbf{p}\in M(\mathbf{x})} w_{Sp}(\|\mathbf{p}\|) \cdot w_{Rn}(\|\bar{\mathbf{f}}_{\text{opt}}(\mathbf{x}) - \mathbf{f}(\mathbf{p})\|_2) \cdot \mathbf{f}(\mathbf{p}), \quad (10)$$

where $\hat{s}(\mathbf{x})$ is the filtered image.

3   Vector Fourlateral (4-Factor) Filters

In the case of the color, multispectral, and hyperspectral images, processed data are vector-valued $\mathbf{f}(\mathbf{x}) : \mathbf{R}^2 \to \mathbf{R}^K$

$$\mathbf{f}(\mathbf{x}) = \left(f_1(\mathbf{x}), f_2(\mathbf{x}), \ldots, f_K(\mathbf{x})\right) = [f_c(\mathbf{x})]_{c=1}^K.$$

By this reason, we must use matrix-valued weights $\left\{\mathbf{W}(\mathbf{x}, \mathbf{p})\right\}_{\mathbf{p}\in M(\mathbf{x})}$, where $\mathbf{W}(\mathbf{x}, \mathbf{p})$ is a $(K \times K)$-matrix, and K is the number of different optical channels in $\mathbf{f}(\mathbf{x}) : \mathbf{R}^2 \to \mathbf{R}^K$. The *4-factor MIMO-filter* (V4FF) suggests a weighted average of pixels in the given image $\mathbf{f}(\mathbf{p})$

$$\hat{s}(\mathbf{x}) = \mathbf{VecFourLatMean}_{\mathbf{p}\in M(\mathbf{x})}\left[\overline{\mathbf{W}}(\mathbf{x}, \mathbf{p}) \cdot \mathbf{f}(\mathbf{p})\right]$$

$$= \frac{1}{\mathbf{diag}\{k_1(\mathbf{x}), k_2(\mathbf{x}), \ldots, k_K(\mathbf{x})\}}\sum_{\mathbf{p}\in M(\mathbf{x})} \mathbf{W}(\mathbf{x}, \mathbf{p})\mathbf{f}(\mathbf{p}), \quad (11)$$

or in component-wise form

$$\hat{s}_a(\mathbf{x}) = \mathbf{VecFourLatMean}_{\mathbf{p}\in M(\mathbf{x})}\left[\overline{\mathbf{W}}(\mathbf{x}, \mathbf{p}) \cdot \mathbf{f}(\mathbf{p})\right]$$

$$= \sum_{\mathbf{p}\in M(\mathbf{x})}\sum_{b=1}^K \frac{w^{ab}(\mathbf{x}, \mathbf{p})}{k_a(\mathbf{x})} f_b(\mathbf{p}) = \sum_{\mathbf{p}\in M(\mathbf{x})}\sum_{b=1}^K \overline{w}^{ab}(\mathbf{x}, \mathbf{p}) f_b(\mathbf{p}), \quad (12)$$

where $\hat{\mathbf{s}}(\mathbf{x})$ is the filtered hyperspectral image, $\hat{s}_a(\mathbf{x})$ is its ath channel, $\bar{w}^{ab} = w^{ab}/k_a$, $\bar{\mathbf{W}} = \mathbf{diag}\left\{k_1^{-1}, k_2^{-1}, \ldots, k_K^{-1}\right\} \cdot \bar{\mathbf{W}}$, $k_a(\mathbf{x})$ is the normalization factor in the ath channel

$$k_a(\mathbf{x}) = \sum_{\mathbf{p}\in M(\mathbf{x})} \sum_{b=1}^{K} w^{ab}(\mathbf{x},\mathbf{p}) \tag{13}$$

and $\mathbf{diag}\{k_1(\mathbf{x}), k_2(\mathbf{x}), \ldots, k_K(\mathbf{x})\}$ is a diagonal matrix with channel normalization factors. Note, that

$$\frac{1}{\mathbf{diag}\{k_1(\mathbf{x}), k_2(\mathbf{x}), \ldots, k_K(\mathbf{x})\}}\mathbf{W}(\mathbf{x},\mathbf{p})\mathbf{f}(\mathbf{p}) = \overline{\mathbf{W}}(\mathbf{x},\mathbf{p})\mathbf{f}(\mathbf{p})$$

$$= \begin{bmatrix} \overline{w}^{11}(\mathbf{x},\mathbf{p}) & \overline{w}^{12}(\mathbf{x},\mathbf{p}) & \cdots & \overline{w}^{1K}(\mathbf{x},\mathbf{p}) \\ \overline{w}^{21}(\mathbf{x},\mathbf{p}) & \overline{w}^{22}(\mathbf{x},\mathbf{p}) & \cdots & \overline{w}^{2K}(\mathbf{x},\mathbf{p}) \\ \vdots & \vdots & \vdots & \vdots \\ \overline{w}^{K1}(\mathbf{x},\mathbf{p}) & \overline{w}^{K2}(\mathbf{x},\mathbf{p}) & \cdots & \overline{w}^{KK}(\mathbf{x},\mathbf{p}) \end{bmatrix} \begin{bmatrix} f_1(\mathbf{p}) \\ f_2(\mathbf{p}) \\ \vdots \\ f_K(\mathbf{p}) \end{bmatrix}.$$

The matrix–valued weights $\overline{\mathbf{W}}(\mathbf{x},\mathbf{p})$ are computed based on the content of the neighbourhood. For pixels $\mathbf{f}(\mathbf{p}), \mathbf{p} \in M(\mathbf{x})$ around the Fréchet centroid $\bar{\mathbf{f}}_{\mathrm{opt}}(\mathbf{x})$, the weights $\bar{w}(\mathbf{x},\mathbf{p}), \mathbf{p} \in M(\mathbf{x})$ are computed by multiplying the following four factors:

$$\overline{w}^{cd}(\mathbf{x},\mathbf{p})$$
$$= \overline{w}_{Sp}(||\mathbf{p}||) \cdot \overline{w}_{Ch}(|c-d|) \cdot \overline{w}_{Rn}(||\bar{\mathbf{f}}_{\mathrm{opt}}(\mathbf{x}) - \mathbf{f}(\mathbf{p})||_2) \cdot \overline{w}_{Rn}(|\bar{f}_{c,\mathrm{opt}}(\mathbf{x}) - f_d(\mathbf{p})|).$$

The weight includes four factors: spatial $\overline{w}_{Sp}(||\mathbf{p}||)$, inter-channels $\overline{w}_{Ch}(|c-d|)$, global radiometric $\overline{w}_{Rn}(||\bar{\mathbf{f}}_{\mathrm{opt}}(\mathbf{x}) - \mathbf{f}(\mathbf{p})||_2)$, and radiometric inter–channels weights $\overline{w}_{Rn}(|\bar{f}_{c,\mathrm{opt}}(\mathbf{x}) - f_d(\mathbf{p})|)$. The first factor $\overline{w}_{Sp}(||\mathbf{p}||)$ measures the geometric distance between the center pixel $f(\mathbf{x})$ and the neighbourhood pixel $\mathbf{f}(\mathbf{p}), \mathbf{p} \in M(\mathbf{x})$. The second factor $\overline{w}_{Ch}(|c-d|)$ measures the spectral (inter-channel) distance. The third factor $\overline{w}_{Rn}(||\bar{\mathbf{f}}_{\mathrm{opt}}(\mathbf{x}) - \mathbf{f}(\mathbf{p})||_2)$ measures the global radiometric distance between the values of the Fréchet centroid $\bar{\mathbf{f}}_{\mathrm{opt}}(\mathbf{x})$ and the pixel $\mathbf{f}(\mathbf{p})$. The fourth factor $\overline{w}_{Rn}(|\bar{f}_{c,\mathrm{opt}}(\mathbf{x}) - f_d(\mathbf{p})|)$ measures the radiometric distance between the values of the center sample $\bar{f}_{c,\mathrm{opt}}(\mathbf{x})$ of the c–channel and the pixel $f_d(\mathbf{p})$ of the d–channel. All weights $\overline{w}_{Rn}^{cd}(\mathbf{x},\mathbf{p}) = \overline{w}_{Rn}(|\bar{f}_{c,\mathrm{opt}}(\mathbf{x}) - f_d(\mathbf{p})|)$ form N radiometric inter-channel $K \times K$-matrices

$$\left\{\overline{\mathbf{W}}_{Rn}(\mathbf{x},\mathbf{p})\right\}_{\mathbf{p}\in M(\mathbf{x})} = \left\{\left[\overline{w}_{Rn}^{cd}(\mathbf{x},\mathbf{p})\right]_{c,d=1}^{K}\right\}_{\mathbf{p}\in M(\mathbf{x})}$$

$$= \left\{\left[\overline{w}_{Rn}(|\bar{f}_{c,\mathrm{opt}}(\mathbf{x}) - f_d(\mathbf{p})|)\right]_{c,d=1}^{K}\right\}_{\mathbf{p}\in M(\mathbf{x})}$$

if N-cellular window $M(\mathbf{x})$ is used. Obviously, mean filter (11) is fourlateral (4-factor). If three ingredients are used, for example,

$$\overline{w}^{cd}(\mathbf{x},\mathbf{p}) = \overline{w}_{Sp}(||\mathbf{p}||) \cdot \overline{w}_{Rn}(||\bar{\mathbf{f}}_{\mathrm{opt}}(\mathbf{x}) - \mathbf{f}(\mathbf{p})||_2) \cdot \overline{w}_{Rn}(|\bar{f}_{c,\mathrm{opt}}(\mathbf{x}) - f_d(\mathbf{p})|),$$

a) Original image

b) Noised image PSNR = 25.36

c) Noised image PSNR = 18.34

d) Noised image PSNR = 15.41

e) Noised image PSNR = 12.64

f) Noised image PSNR = 9.22

Fig. 1. The results for "Salt-Pepper" noise and bilateral filters with Laplacian weights for $\alpha = 0.035$

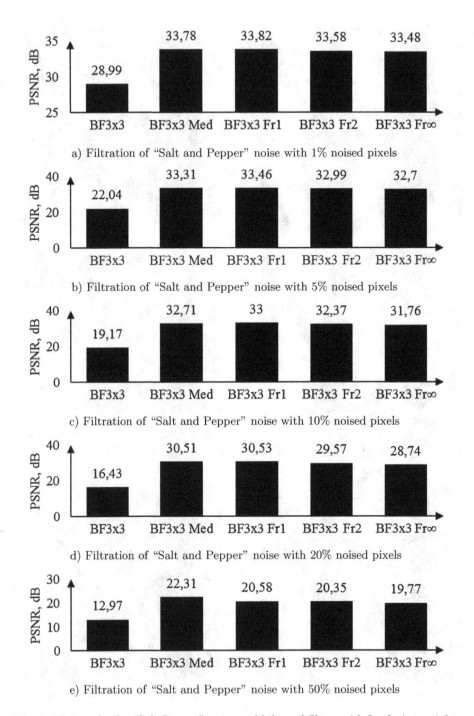

a) Filtration of "Salt and Pepper" noise with 1% noised pixels

b) Filtration of "Salt and Pepper" noise with 5% noised pixels

c) Filtration of "Salt and Pepper" noise with 10% noised pixels

d) Filtration of "Salt and Pepper" noise with 20% noised pixels

e) Filtration of "Salt and Pepper" noise with 50% noised pixels

Fig. 2. The results for "Salt-Pepper" noise and bilateral filters with Laplacian weights for $\alpha = 0.07$

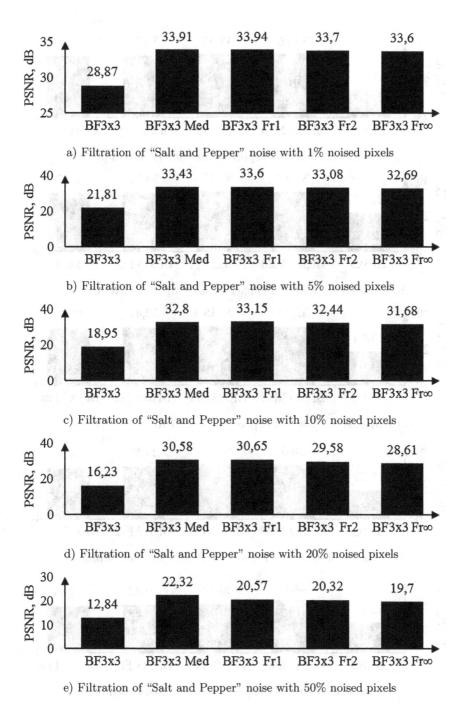

Fig. 3. The results for "Salt-Pepper" noise and bilateral filters with Laplacian weights for $\alpha = 0.1$

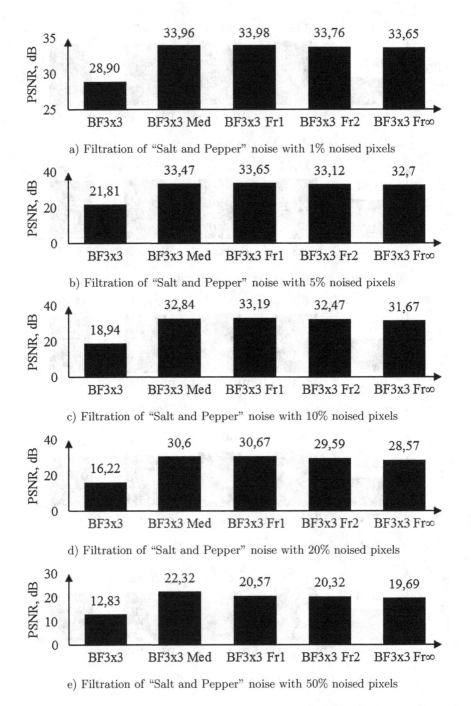

a) Filtration of "Salt and Pepper" noise with 1% noised pixels

b) Filtration of "Salt and Pepper" noise with 5% noised pixels

c) Filtration of "Salt and Pepper" noise with 10% noised pixels

d) Filtration of "Salt and Pepper" noise with 20% noised pixels

e) Filtration of "Salt and Pepper" noise with 50% noised pixels

Fig. 4. Original and noised images; noise: "Salt-Pepper"

or

$$\overline{w}^{cd}(\mathbf{x}, \mathbf{p}) = \overline{w}_{Sp}(||\mathbf{p}||) \cdot \overline{w}_{Ch}(|c - d|) \cdot \overline{w}_{Rn}(|\overline{f}_{c,\text{opt}}(\mathbf{x}) - f_d(\mathbf{p})|),$$

then we obtain threelateral (3-factor) filters.

4   Simulation Experiments

Some variants of the proposed filters are tested. They are compared on real image "LENA". Noise is added (see Fig. 4) with different the Peak Signal to Noise Ratios (PSNRs). The noised images has 1 % noised pixels (PSNR = 25.36 dB), 5 % noised pixels (PSNR = 18.34 dB), 10 % noised pixels (PSNR = 15.41 dB), 20 % noised pixels (PSNR = 12.64 dB), 50 % noised pixels (PSNR = 9.22 dB).

Figures 1, 2 and 3 summarize the results for the "Salt and Pepper" noise and bilateral filters with Laplacian weights $w(f, g) = \exp(-\alpha|f - g|)$ for different α: $\alpha = 0.035$ (Fig. 1), $\alpha = 0.07$ (Fig. 2), $\alpha = 0.1$ (Fig. 3). Figures 1, 2 and 3 show the results obtained by the following bilateral filters with $\boxplus_3$-mask

- the classical bilateral filter (1) (**BF3x3**),
- modified bilateral filter (10) (**BF3x3Med**), where $\overline{f}(\mathbf{x})$ are calculated as classical median in each channel,
- modified bilateral filter (10) (**BF3x3Fr1**), where $\overline{f}(\mathbf{x})$ are calculated as Fréchet median with distance $\rho(\mathbf{f}, \mathbf{g}) = \rho_1(\mathbf{f}, \mathbf{g})$,
- modified bilateral filter (10) (**BF3x3Fr2**), where $\overline{f}(\mathbf{x})$ are calculated as Fréchet median with distance $\rho(\mathbf{f}, \mathbf{g}) = \rho_2(\mathbf{f}, \mathbf{g})$,
- modified bilateral filter (10) (**BF3x3Fr∞**), where $\overline{f}(\mathbf{x})$ are calculated as Fréchet median with distance $\rho(\mathbf{f}, \mathbf{g}) = \rho_\infty(\mathbf{f}, \mathbf{g})$.

It is easy to see that results for all modified bilateral filters are better, compared to the classical bilateral filter **BF3x3**.

5   Conclusion and Future Work

A new class of nonlinear generalized vector-valued bi-, three-, and fourlateral filters for multichannel image processing is introduced in this paper. Weights in fourlateral filters include three components: spatial, radiometric, interchannel, and interchannel radiometric weights. The fourth weight measures the radiometric distance (for grey-level images) between the interchannel values of the center scalar-valued channel pixel and local neighborhood channel pixels. Here, the 1D Euclidean metric is used, too. We are going to use generalized average (aggregation) [16, 17] in (11) instead of ordinary mean.

The aggregation problem [16, 17] consist in aggregating N-tuples of objects all belonging to a given set $\mathbf{S}$, into a single object of the same set $\mathbf{S}$, $i.e.$, $\mathbf{Agg}$: $\mathbf{S}^N \longrightarrow \mathbf{S}$. In the case of mathematical aggregation operator (AO) the set $\mathbf{S}$, is an interval of the real $\mathbf{S} = [0, 1] \subset \mathbf{R}$, or integer numbers $\mathbf{S} = [0, 255] \subset \mathbf{Z}$. In this setting, an AO is simply a function, which assigns a number y to any N-tuple of numbers $(x_1, x_2 \dots, x_N)$: $y = \mathbf{Agg}(x_1, x_2, \dots, x_N)$ that satisfies:

1. $\mathbf{Agg}(x) = x$.
2. $\mathbf{Agg}(a, a, \ldots, a) = a$.

   In particular, $\mathbf{Agg}(0, 0, \ldots, 0) = 0$ and $\mathbf{Agg}(1, 1, \ldots, 1) = 1$
   (or $\mathbf{Agg}(255, 255, \ldots, 255) = 255$).
3. $\min(x_1, x_2, \ldots, x_N) \leq \mathbf{Agg}(x_1, x_1, \ldots, x_N)) \leq \max(x_1, x_2, \ldots, x_N)$.

Here $\min(x_1, x_2, \ldots, x_N)$ and $\max(x_1, x_2, \ldots, x_N)$ are respectively the *minimum* and the *maximum* values among the elements of $(x_1, x_2, \ldots, x_N)$. All other properties may come in addition to this fundamental group. For example, if for every permutation $\forall \sigma \in S_N$ of $\{1, 2, \ldots, N\}$ the AO satisfies:

$$y = \mathbf{Agg}(x_{\sigma(1)}, x_{\sigma(2)}, \ldots, x_{\sigma(N)}) = \mathbf{Agg}(x_1, x_2, \ldots, x_N),$$

then it is invariant (symmetric) with respect to the permutations of the elements of $(x_1, x_2, \ldots, x_N)$. In other words, as far as means are concerned, the *order* of the elements of $(x_1, x_2, \ldots, x_N)$ is - and must be - completely irrelevant.

We list below a few particular cases of aggregation means:

1. Arithmetic mean ($K(x) = x$): $\mathbf{Mean}(x_1, x_2, \ldots, x_N) = \frac{1}{N} \sum_{i=1}^{N} x_i$.

2. Geometric mean ($K(x) = \log(x)$): $\mathbf{Geo}(x_1, x_2, \ldots, x_N) = \sqrt[N]{\left(\prod_{i=1}^{N} x_i \right)}$.

3. Harmonic mean ($K(x) = x^{-1}$): $\mathbf{Harm}(x_1, x_2, \ldots, x_N) = \left(\frac{1}{N} \sum_{i=1}^{N} x_i^{-1} \right)^{-1}$.

4. One-parametric family quasi arithmetic (power or Hólder) means corresponding to the functions $K(x) = x^p$: $\mathbf{Hold}(x_1, x_2, \ldots, x_N) = \sqrt[p]{\left(\frac{1}{N} \sum_{i=1}^{N} x_i^p \right)}$. This family is particularly interesting, because it generalizes a group of common means, only by changing the value of p.

A very notable particular cases correspond to the logic functions ($\min, \max$, median): $y = \mathbf{Min}(x_1, \ldots, x_N)$, $y = \mathbf{Max}(x_1, \ldots, x_N)$, $y = \mathbf{Med}(x_1, \ldots, x_N)$.

When filter (11) is modified as follows:

$$\hat{\mathbf{s}}(\mathbf{x}) = \underset{\mathbf{p} \in M(\mathbf{x})}{\mathbf{VecFourLatAgg}} \left[\overline{\mathbf{W}}(\mathbf{x}, \mathbf{p}) \cdot \mathbf{f}(\mathbf{p}) \right] \tag{14}$$

or in component-wise form

$$\hat{s}_a(\mathbf{x}) = \underset{\mathbf{p} \in M(\mathbf{x})}{\mathbf{VecFourLatAgg}} \left[\overline{\mathbf{W}}(\mathbf{x}, \mathbf{p}) \cdot \mathbf{f}(\mathbf{p}) \right]$$

$$= \underset{\mathbf{p} \in M(\mathbf{x})}{\mathbf{Agg}} \left\{ \mathbf{Agg}_{b=1}^{K} \left[\overline{w}^{a1}(\mathbf{x}, \mathbf{p}) f_1(\mathbf{p}), \overline{w}^{a2}(\mathbf{x}, \mathbf{p}) f_2(\mathbf{p}), \ldots, \overline{w}^{aK}(\mathbf{x}, \mathbf{p}) f_K(\mathbf{p}) \right] \right\}$$

$$\tag{15}$$

we get the unique class of nonlinear *4–factor aggregation MIMO-filters* that we are going to research in future works.

Acknowledgment. This work was supported by the Russian Foundation for Basic Research (grants Nos. 13-07-12168, 13-07-00785).

References

1. Ostheimer, E., Labunets, V., Komarov, D., Fedorova, T.: Fréchet filters for color and hyperspectral images filtering. In: Khachay, M.Y., Konstantinova, N., Panchenko, A., Ignatov, D.I., Labunets, V.G. (eds.) AIST 2015. CCIS, vol. 542, pp. 57–70. Springer, Heidelberg (2015)
2. Tomasi, C., Manduchi, R.: Bilateral filtering for gray and color images. In: Proceedings of the 6th International Conference on Computer Vision, New Delhi, India, pp. 839–846 (1998)
3. Astola, J., Haavisto, P., Neuvo, Y.: Vector median filters. Proc. IEEE **78**, 678–689 (1990)
4. Tang, K., Astola, J., Neuvo, Y.: Nonlinear multivariate image filtering techniques. IEEE Trans. Image Process. **4**, 788–798 (1996)
5. Barash, D.: Bilateral filtering and anisotropic diffusion: towards a unified viewpoint. In: Kerckhove, M. (ed.) Scale-Space 2001. LNCS, vol. 2106, pp. 273–280. Springer, Heidelberg (2001)
6. Durand, F., Dorsey, J.: Fast bilateral filtering for the display of high-dynamic-range images. In: Proceedings of ACM SIGGRAPH, pp. 257–266 (2002)
7. Durand, F., Dorsey, J.: Fast bilateral filtering for the display of high dynamic range images. In: Proceedings of SIGGRAPH, 844–847 (2002)
8. Elad, M.: Analysis of the bilateral filter. In: The 36th Asilomar on Signals, Systems and Computers, Pacific Grove, CA (2002)
9. Elad, M.: On the origin of the bilateral filter and ways to improve it. IEEE Trans. Image Process. **11**(10), 1141–1151 (2002)
10. Fleishman, S., Drori, I., Cohen, D.: Bilateral mesh filtering. In: Proceedings of ACM SIGGRAPH, San Diego, TX, pp. 950–953 (2003)
11. Barash, D.: A fundamental relationship between bilateral filtering, adaptive smoothing and the non-linear diffusion equation. PAMI **24**(6), 844–847 (2002)
12. Labunets, V.G.: Filters based on aggregation operators. Part 1. Aggregation Operators. In: 24th International Crimean Conference on Microwave & Telecommunication Technology (CriMiCo2014), 7–13 September, Sevastopol, Crimea, Russia, vol. 24, pp. 1239–1240 (2014)
13. Labunets, V.G., Gainanov, D.N., Ostheimer, E.: Filters based on aggregation operators. Part 2. The Kolmogorov filters. In: 24th International Crimean Conference on Microwave & Telecommunication Technology (CriMiCo2014), 7–13 September, Sevastopol, Crimea, Russia, vol. 24, pp. 1241–1242 (2014)
14. Labunets, V.G., Gainanov, D.N., Tarasov, A.D., Ostheimer, E.: Filters based on aggregation operators. Part 3. The Heron filters. In: 24th International Crimean Conference on Microwave & Telecommunication Technology (CriMiCo2014), 7–13 September, Sevastopol, Crimea, Russia, vol. 24, pp. 1243–1244 (2014)
15. Labunets, V.G., Gainanov, D.N., Arslanova, R.A., Ostheimer E.: Filters based on aggregation operators. Part 4. Generalized vector median filters. In: 24th International Crimean Conference on Microwave & Telecommunication Technology (CriMiCo2014), 7–13 September, Sevastopol, Crimea, Russia, vol. 24, pp. 1245–1246 (2014)
16. Mayor, G., Trillas, E.: On the representation of some aggregation functions. Proc. ISMVL **20**, 111–114 (1986)
17. Ovchinnikov, S.: On robust aggregation procedures. In: Bouchon-Meunier, B. (ed.) Aggregation Operators for Fusion under Fuzziness, vol. 12, pp. 3–10. Physica, Heidelberg (1998)

Frequency Analysis of Gradient Descent Method and Accuracy of Iterative Image Restoration

Artyom Makovetskii[1], Alexander Vokhmintsev[1(✉)], Vitaly Kober[1,2], and Vladislav Kuznetsov[3]

[1] Research Laboratory, Chelyabinsk State University, Chelyabinsk, Russia
artemmac@mail.ru, vav@csu.ru, vkober@cicese.mx
[2] Department of Computer Science, CICESE, Ensenada, B.C., Mexico
[3] Institute of Informatics Problems of RAS, Moscow, Russia
k.v.net@rambler.ru

Abstract. For images with sharp changes of intensity, the appropriate regularization is based on variational functionals. In order to minimize such a functional, the gradient descent approach can be used. In this paper, we analyze the performance of the gradient descent method in the frequency domain and show that the method converges to the sum of the original undistorted function and the kernel function of a linear distortion operator.

Keywords: Image restoration · Denoising · Deblurring

1 Introduction

In many applications observed images are often degraded owing to atmospheric turbulence, relative motion between a scene and a camera, nonuniform illumination, wrong focus, etc. Many different restoration techniques (linear, nonlinear, iterative, noniterative, deterministic, stochastic, etc.) optimized with respect to various criteria have been introduced [1–7, 12]. The amount of a priori information about degradation, i.e., the size or shape of blurs, and the noise level, determines mathematically the ill-posed problem. The blind and nonblind deconvolutions have been extensively studied and many techniques have been proposed for their solution [2–4]. They usually involve some regularization which assures various statistical properties of the image or constrains the estimated image and restoration filter according to some assumptions. This regularization is required to guarantee a unique solution and stability against noise and some model discrepancies. One of the most popular fundamental techniques is a linear minimum mean square error method. It finds the linear estimate of the ideal image for which the mean square error between the estimate and the ideal image is minimum. The linear operator acting on the observed image to determine the estimate is obtained on the basis of a priori second order statistical information about the image and noise processes. For images with sharp changes of intensity, the appropriate regularization is based on variational functionals. Minimization of the variational

M.Y. Khachay et al. (Eds.): AIST 2015, CCIS 542, pp. 114–122, 2015.
DOI: 10.1007/978-3-319-26123-2_11

functionals preserves edges and fine details in the image and it was applied to blind image restoration [6, 7].

For the numerical solution of the functionals minimization problems various versions of the gradient descent method are widely used. A natural question arises about the asymptotic behavior of these methods [8–10]. In particular, the asymptotic behavior was studied for the s-step method of steepest descent in a Hilbert space. It is proved that the iterative process converges to a plane. This plane depends on the operator which generates the considered functional [8]. In this paper we consider the asymptotic behavior of the gradient descent method for the special kind functionals in a Hilbert space. The influence of the linear variation to the restoration results is also discussed.

In image restoration it is often necessary to solve the following inverse problem [11–13]:

$$A\tilde{u} = u_0, \tag{1}$$

where A is a known linear operator, u_0 is an observed image, and $\tilde{u}$ and $u_0 \in U(\Omega)$. Here $U(\Omega)$ is a Banach space on the bounded set Ω. The function $\tilde{u}$ is called the exact solution of the problem given in Eq. (1). Usually an inverse operator A^{-1} does not exist and the observed image u_0 can be additionally distorted by additive noise n,

$$v_0 = u_0 + n \tag{2}$$

such that Eq. (1) takes the following form:

$$Au = v_0. \tag{3}$$

A common way to solve the problems in Eqs. (1) and (3) is to use the variational functional,

$$J(u) = H(Au) + \lambda R(u), \tag{4}$$

where $H(Au)$ and $R(u)$ are fidelity and regularization terms, respectively, $\lambda > 0$ is a parameter.

The fidelity term often takes the following form:

$$H(Au) = \| Au - u_0 \|_{L_2}^2. \tag{5}$$

We consider the following variational problem to solve Eq. (3):

$$u = \arg\min \| Au - u_0 \|_{L_2}^2 + \lambda R(u). \tag{6}$$

One of the most widely used regularization term is the total variation [14–17],

$$R(u) = \int |\nabla u|, \ u \in BV(\Omega) \tag{7}$$

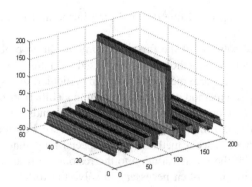

Fig. 1. Two-dimensional discrete function w_k^1.

Note that the variational problem in Eq. (6) without the regularization term corresponds to the problem in Eq. (1),

$$u = \arg\min \| Au - u_0 \|_{L_2}^2. \tag{8}$$

Next, we analyze the gradient descent method [18] in the frequency domain. We show that the method converges to the sum of an original undistorted function and the kernel of a linear distortion operator. So the restoration accuracy of linearly degraded image depends on the kernel function. The accuracy of restoration can be improved by consequent smoothing using the total variation method. However, in this work we use metrical and topological characteristics to improve the restoration quality. The topological characteristics of a function of two variables are linear variations [19]. The proposed topological method modifies the restoration result obtained with the gradient decent method in order to get a given number of linear variations at the output.

2   Example of the Distorting Linear Operator

Let $v \in L_2(\Omega)$ and $u_0 = Av$ be original and degraded functions, respectively. Here the operator A is a centered uniform blurring,

$$Av(t) = \frac{1}{2\Delta} \int_{t-\Delta}^{t+\Delta} v(\tau)d\tau, \tag{9}$$

where t is an arbitrary point of Ω, Δ is the parameter of blurring.

Note, that the operator A is a self-adjoint operator, that is $A^* = A$. If a complex vector space is finite-dimensional, this is equivalent to the condition that the matrix of A is Hermitian, i.e. equal to its conjugate transpose A*.

The function $rect(t)$ is defined as follows:

$$rect(t) = \begin{cases} 1, & -\frac{1}{2} \leq t \leq \frac{1}{2} \\ 0, & \text{otherwise} \end{cases}. \tag{10}$$

Suppose that the function $v(t)$ is defined in the interval as follows:

$$v(t) = \begin{cases} h_v, & -\frac{T_v}{2} \leq t \leq \frac{T_v}{2} \\ 0, & \text{otherwise} \end{cases}. \tag{11}$$

The functions $v(t)$ and $rect(t)$ are related by the expression below:

$$v(t) = h_v \cdot rect(\frac{t}{T_v}). \tag{12}$$

We also consider the following function $a(t)$:

$$a(t) = \begin{cases} \frac{1}{2\Delta}, & -\Delta \leq t \leq \Delta \\ 0, & \text{otherwise} \end{cases}. \tag{13}$$

The functions $a(t)$ and $rect(t)$ are related by

$$a(t) = \frac{1}{2\Delta} \cdot rect(\frac{t}{2\Delta}). \tag{14}$$

By applying the operator A to the function v, we obtain the convolution between $v(t)$ and $a(t)$,

$$u_0(t) = v(t) * a(t) = \int_{-\infty}^{+\infty} v(\tau) \cdot a(t-\tau)d\tau = \frac{1}{2\Delta} \int_{-\infty}^{+\infty} v(\tau) \cdot rect\left(\frac{t}{2\Delta} - \tau\right)d\tau$$

$$= \frac{1}{2\Delta} \int_{-\infty}^{+\infty} v(\tau) \cdot rect\left(\frac{t}{2\Delta} - \tau\right)d\tau = \frac{1}{2\Delta} \int_{t-\Delta}^{t+\Delta} v(\tau) \cdot 1 d\tau. \tag{15}$$

3 Frequency Analysis of Gradient Descent Method

A common way for solving the problem in Eq. (8) is to use the gradient descent method [11]. Iterations of the gradient descent method are given as

$$u_{k+1} = u_k - \alpha_{k+1} \cdot H'(u_k). \tag{16}$$

Here $H'(u)$ is the gradient of the functional $H(Au)$. In the functional space L_2 the gradient is given by

$$H'(u_k) = 2A^*(Au - u_0), \tag{17}$$

where A^* is an adjoin operator, α_k is a parameter of the gradient method. This parameter is selected in such a way to satisfy the following condition:

$$H(u_{k+1}) < H(u_k). \tag{18}$$

Recently, it has been shown [11] that the function u_{k+1}, obtained at the iteration $k + 1$ of the gradient descent method can be expressed as follows:

$$u_{k+1} = \left(\left(1 - 2\alpha A^2\right)^{k+1} A + \left(\left(1 - 2\alpha A^2\right)^k + \cdots + \left(1 - 2\alpha A^2\right)^1 + 1 \right) 2\alpha A^2 \right) (\tilde{u}). \tag{19}$$

The Fourier transformations of the functions u_{k+1} and $\tilde{u}$ are connected by the following equation:

$$\mathfrak{F}(u_{k+1}) = \left(1 - (1 - sinc(2 \cdot f)) \left(1 - 2\alpha sinc^2(2 \cdot f) \right)^{k+1} \right) \cdot \mathfrak{F}(\tilde{u}). \tag{20}$$

It is interesting to note that at the points $f = \frac{\pi n}{2\Delta}$ the function $\mathfrak{F}(u_k) = 0$ for any step k of the gradient descent method. Thus, at these points the Fourier transformation of the restored function takes zeroes regardless of the values of the Fourier transformation of the original function computed at these points.

4   Discrete Realization

The discrete function $rect(k)$ is defined for $k = 0, \ldots, n - 1$ as follows

$$rect(k) = \begin{cases} 1, & k \in [0; \Delta] \cup [n - \Delta; n - 1] \\ 0, & k \notin [\Delta + 1; n - \Delta - 1] \end{cases}. \tag{21}$$

The function $sincd(k)$ is the Fourier transformation of the discrete function $rect(k)$,

$$sincd(k) = \begin{cases} \frac{sin(\frac{\pi k(2\Delta + 1)}{n})}{(2\Delta + 1)sin(\frac{\pi k}{n})}, & k = 1, \ldots, n - 1 \\ 1, & k = 0 \end{cases}. \tag{22}$$

Let $v_k, k = 0, \ldots, n - 1$ denote the original discrete function. Let A be a centered uniform blur with cyclic boundary conditions as follows:

$$(Av)_k = \frac{1}{2\Delta + 1} \sum_{i=k-\Delta}^{k+\Delta} v_{i \, mod \, n}, \tag{23}$$

where is $k = 0, \ldots, n - 1$, Δ is the duration of the blur. In discrete case the function $a(k)$ has the following form:

$$a(k) = \frac{1}{2\Delta + 1} \cdot rect(k), \ k = 0, \ldots, n - 1. \tag{24}$$

The action of the linear operator A with cyclic boundary conditions is equivalent to the circular convolution,

$$Av = a * v. \tag{25}$$

Therefore, Eq. (20) can be rewritten as

$$\mathfrak{F}(u_{k+1})_l = \left(1 - (1 - sincd(l))\left(1 - 2\alpha sincd^2(l)\right)^{k+1}\right) \cdot \mathfrak{F}(v)_l, \ l = 0, \ldots, n - 1. \tag{26}$$

Note that if for some frequencies l the functions $\mathfrak{F}(v)_l \neq 0$ and $sincd(l) = 0$, then the method is unable to restore the signal because the spectrum of the convolution will contain zeroes.

5 Kernel of Linear Operator and Asymptotic Behavior of Gradient Descent Method

The set $Ker\mathfrak{A} = \{x \in X | \mathfrak{A}x = 0\}$ is called the kernel of a linear operator $\mathfrak{A} : X \to Y$, X and Y are Banach spaces. Suppose that the function $\tilde{u}$ is a solution of the problem in Eq. (8). Then the function $\tilde{u} + u_{ker}$ is also a solution of the problem. Are there solutions of the problem that could be obtained by the gradient descent method and which differ from functions of the form $\tilde{u} + u_{ker}$?

Proposition. The gradient descent method asymptotically converges to functions of the form $\tilde{u} + u_{ker}$.

Proof Equation (26) for the discrete frequency l of the Fourier transformation $\mathfrak{F}(u_k)_l$ of function u_k and for the k-th step of gradient descent method takes the following form:

$$\mathfrak{F}(u_k)_l = \left(1 - (1 - sincd(l))\left(1 - 2\alpha sincd^2(l)\right)^k\right) \cdot \mathfrak{F}(v)_l, \ l = 0, \ldots, n - 1. \tag{27}$$

Considering the action of the operator A to the difference $v - u_k$ we get

$$A(v - u_k) = a * (v - u_k). \tag{28}$$

Using the Fourier transformation Eq. (28) can be represented as

$$A(v - u_k) = a * (v - u_k) = \mathfrak{F}^{-1}(\mathfrak{F}(a) \cdot (\mathfrak{F}(v) - \mathfrak{F}(u_k))). \tag{29}$$

For a given discrete frequency l and taking into account Eq. (29) we obtain

$$
\begin{aligned}
\mathfrak{F}(a) \cdot (\mathfrak{F}(v) - \mathfrak{F}(u_k))_l &= \mathfrak{F}(a) \cdot \left(\mathfrak{F}(v) - \left(1 - (1 - sincd(l))(1 - 2\alpha sincd^2(l))^k\right) \cdot \mathfrak{F}(v) \right)_l \\
&= \mathfrak{F}(a) \cdot \left(\mathfrak{F}(v) - \mathfrak{F}(v) + \mathfrak{F}(v)(1 - sincd(l))(1 - 2\alpha sincd^2(l))^k \right)_l \\
&= \mathfrak{F}(a) \cdot \mathfrak{F}(v) \cdot (1 - sincd(l)) \cdot \left(1 - 2\alpha sincd^2(l)\right)_l^k \\
&= sincd(l) \cdot \mathfrak{F}(v) \cdot (1 - sincd(l)) \cdot \left(1 - 2\alpha sincd^2(l)\right)_l^k .
\end{aligned}
\tag{30}
$$

Note that $sincd(l) \cdot \mathfrak{F}(v) \cdot (1 - sincd(l)) \cdot (1 - 2\alpha sincd^2(l))_l^k$ is equal to 0 when $l = 0$, because $1 - sincd(0) = 0$. For such l that $sincd(l) = 0$ the expression also equals 0. For other values of l the expression $(1 - 2\alpha sincd^2(l))^k \to 0$ when $k \to \infty$ for sufficiently small α. Therefore, for any l Eq. (30) tends to zero, because the function $\mathfrak{F}(a) \cdot (\mathfrak{F}(v) - \mathfrak{F}(u_k))$ tends to zero in the frequency domain. It means that the function $\mathfrak{F}^{-1}(\mathfrak{F}(a) \cdot (\mathfrak{F}(v) - \mathfrak{F}(u_k)))$ also tends to zero in the time domain.

Thus the gradient descent method converges to functions of the form $v + u_{kl}$. If a kernel is not trivial then the considered example shows that u_{kl} can be a non-zero kernel element.

6 Linear Variation

Let $\Phi_u(t)$ be the number of regular components of a level set t for a continuous function [20]. The first Kronrod's linear variation is defined as

$$
V(u) = \int_{-\infty}^{+\infty} \Phi_u(t) dt.
\tag{31}
$$

Let w be a binary discrete function $w = (w_{i,j})$, where $w_{i,j} \in \{0, 1\}$, for all pairs (i,j). A subset of such pairs (i,j) when $w_{i,j} = 1$ and all elements of the subset are connected by the 8-connectivity, is called the connected component of the binary function w. For a number $k \in \mathbb{N}$ and a discrete function u we define the following indicator function χ:

$$
\chi_k(u_{ij}) = \begin{cases} 1, & u_{ij} \geq k \\ 0, & u_{ij} < k \end{cases}.
\tag{32}
$$

Definition. The number $V_k(u_{i,j})$ of connected components for each level k, $k \in \mathbb{N}$ of the discrete function u is called the number of connected components of the binary discrete function $\chi_k(u_{i,j})$.

Definition. The linear variation $V(u_{i,j})$ of a discrete function u is defined as follows:

$$
V(u_{ij}) = \sum_{k=0}^{+\infty} V_k(u_{ij}).
\tag{33}
$$

7 Smoothing of Oscillating Functions

In the preceding sections it was shown that the kernel function of a linear operator A is oscillating. Thus, to improve the accuracy of restoration by the gradient descent method one can use a smoothing operator applied to a restored image. The exact solutions of the variational problem in Eq. (6) for piecewise constant functions were obtained [18]. These results can be improved by a smoothing operator. We propose to utilize a method based on the use of linear variations [21, 22]. Figure 1 shows the two-dimensional discrete function w_k^1 constructed from the one-dimensional discrete function u_k. The size of the image is 200×40 pixels. Let w be the two-dimensional discrete function constructed from the one-dimensional discrete function v.

Figure 2 shows the discrete function w_k^2 restored by Eq. (28). This function is computed as follows: In Eq. (28) zeroes of the function $\mathfrak{F}(u_k)_l$ are replaced by values obtained with the help of a linear interpolation of neighboring non-zero pixels. In order to evaluate the restoration performance we use the distance between the functions f and g in the functional space L_2 denoted by (f, g). The following distances between the considered functions are obtained: $d(w, w_k^1) = 1306.7$; $d(w, w_k^2) = 61.8$.

Values of the linear variation for the considered functions are given as follows: $V(w) = 211$; $V(w_k^1) = 290$; $V(w_k^2) = 255$.

So, there is a correlation between the accuracy of restoration with respect to the metric L_2 and the linear variation values of the considered functions.

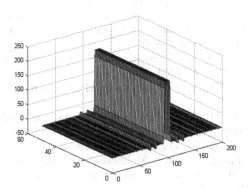

Fig. 2. The restored function w_k^2.

8 Conclusion

In this paper we have analyzed the performance of the gradient descent method in the frequency domain and showed that the method converges to the sum of the original undistorted function and the kernel of a linear distortion operator. For uniform linear degradation the kernel function is strongly oscillating. We have proposed a signal restoration method based on the use of topological characteristics of functions of two variables. The obtained results have been illustrated with the help of computer simulation.

Acknowledgments. The work was supported by the Ministry of Education and Science of Russian Federation, grant 2.1766.2014K and RFBR grant 13.01.00735.

References

1. Jain, A.K.: Fundamentals of Digital Image Processing. Prentice Hall, NY (1989)
2. Biemond, J., Lagendijk, R.L., Mersereau, R.M.: Iterative methods for image deblurring. Proc. IEEE **78**(5), 856–883 (1990)
3. Banham, M., Katsaggelos, A.: Digital image restoration. IEEE Sig. Process. Mag. **14**(2), 24–41 (1997)
4. Kundur, D., Hatzinakos, D.: Blind image deconvolution. IEEE Sig. Process. Mag. **13**(3), 43–64 (1996)
5. Sroubek, F., Flusser, J.: Multichannel blind iterative image restoration. IEEE Trans. Image Process. **12**(9), 1094–1106 (2003)
6. Chan, T., Wong, C.: Total variation blind deconvolution. IEEE Trans. Image Process. **7**(3), 370–375 (1998)
7. You, Y.L., Kaveh, M.: Blind image restoration by anisotropic regularization. IEEE Trans. Image Process. **8**(3), 396–407 (1999)
8. Zhuk, P.P.: Asymptotic behavior of the s-step method of steepest descent for eigenvalue problems in Hilbert space. Russ. Acad. Sci. Sb. Math. **80**(2), 467–495 (1995)
9. Cominetti, R.: Asymptotic convergence of the steepest descent method for the exponential penalty in linear programming. J. Convex Anal. **2**(1–2), 145–152 (1995)
10. Álvarez, C., Cabot, A.: On the asymptotic behavior of a system of steepest descent equations coupled by a vanishing mutual repulsion. In: Seeger, A. (ed.) Recent Advances in Optimization. LNEMS, vol. 563, pp. 3–17. Springer, Heidelberg (2006)
11. Bovik, A.C.: Handbook of Image and Video Processing. Academic Press, Orlando (2005)
12. Gonzalez, R.C., Woods, R.E., Eddins, S.L.: Digital Image Processing using Matlab, p. 209. Gatesmark Publishing, Knoxville (2009)
13. Kober, V., Ovseevich, I.A.: Image restoration with sliding sinusoidal transforms. Pattern Recogn. Image Anal. **18**(4), 650–654 (2008)
14. Rudin, L.I., Osher, S.: Nonlinear total variation based noise removal algorithms. Phys. D **60**, 259–268 (1992)
15. Chambolle, A., Lions, P.L.: Image recovery via total variational minimization and related problems. Numer. Math. **76**, 167–188 (1997)
16. Osher, S., Burger, M., Goldfarb, D., Xu, J., Yin, W.: An iterative regularization method for total variation based image restoration. Multiscale Model. Simul. **4**, 460–489 (2005)
17. Chambolle, A.: An algorithm for total variation minimization and applications. J. Math. Imaging Vis. **20**, 89–97 (2004)
18. Strong, D.M., Chan, T.F.: Exact solutions to total variation regularization problems. UCLA CAM Report (1996)
19. Snyman, J.A.: Practical Mathematical Optimization: An Introduction to Basic Optimization Theory and Classical and New Gradient-Based Algorithms. Springer, New York (2005)
20. Kronrod, A.: On functions of two variables. Uspehi Mat. Nauk. **1**(35), 24–134 (1950)
21. Makovetskii, A., Kober, V.: Modified gradient descent method for image restoration. In: Proceedings of SPIE Applications of Digital Image Processing XXXVI, vol. 8856, p. 885608-1 (2013)
22. Makovetskii, A., Kober, V.: Image restoration based on topological properties of functions of two variables. In: Proceedings of SPIE Applications of Digital Image Processing XXXV, vol. 8499, p. 84990A (2012)

Shape Matching Based on Skeletonization and Alignment of Primitive Chains

Olesia Kushnir$^{(\boxtimes)}$ and Oleg Seredin

Tula State University, Tula, Russia
kushnir-olesya@rambler.ru, oseredin@yandex.ru

Abstract. We introduce a new shape matching approach based on skeletonization and alignment of primitive chains. At the first stage the skeleton of a binary image is traversed counterclockwise in order to encode it by chain of primitives. A primitive describes topological properties of the correlated edge and consists of a pair of numbers: the length of some edge and the angle between this and the next edges. We offer to expand a primitive by the information about the radial function of the skeleton rib. To get the compact width description we interpolate radial function by Legendre polynomials and find the vector of Legendre coefficients. Thus the resulting shape representation by the chain of primitives includes not only topological properties but also the contour ones. Then we suggest the dynamic programming procedure of the alignment of two primitive chains in order to match correspondent shapes. Based on the optimal alignment we propose the pair-wise dissimilarity function which is evaluated on artificial image dataset and the Flavia leaf dataset.

Keywords: Binary image · Shape matching · Skeleton · Chain of primitives · Skeleton radial function · Legendre polynomials · Pair-wise alignment

1 Introduction

Shape recognition and shape analysis play important role in computer vision. A lot of shape comparison methods have been already proposed but the problem can't be considered as completely solved yet. Each method is founded on certain shape description technique, but generally all techniques could be classified on two main approaches – skeleton-based and boundary-based ones. A skeleton is constructed as a locus of centers of maximal circles inscribed into the shape [5]. The main problem of skeleton-based approach is the direct comparison of two arbitrary skeletons. Existing procedures have advantages and drawbacks. We propose a novel approach to featureless pair-wise comparison of skeletons. It takes into account not only topological properties of describing shape but also its thickness which characterized by skeleton radial function. Thus, the method combines skeleton-based and contour-based binary image descriptions and leads to comprehensive shape representation.

© Springer International Publishing Switzerland 2015
M.Y. Khachay et al. (Eds.): AIST 2015, CCIS 542, pp. 123–136, 2015.
DOI: 10.1007/978-3-319-26123-2_12

The paper is organised as follows. The next section provides a summary of related works. In Sect. 3 we present the idea of encoding the skeleton (a specific list of nodes and edges) as a chain of primitives. Each primitive contains information about topological characteristics of the corresponding edge of skeleton. Also we offer to take into consideration the object width using parametric description of skeleton radial function by Legendre polynomials. As it will be shown, the proposed skeleton description is invariant under translation, rotation and scaling. In order to compare chains of primitives we build the procedure of pair-wise alignment as it is traditionally used for amino-acids chains in bioinformatics (by the analogy with Levenshtein edit distance). Experiments with developed comparison function on the artificial Stars dataset and real-world Flavia leaf dataset will be presented in Sects. 4 and 5 respectively.

2   Related Work

Shape Comparison. Considering the method of shape representation, the binary images comparison techniques can be divided into two groups. The first group is based on skeletons and the second one – on boundary contours. The main approaches in the first group are (1) morphological shape comparison built on mathematical morphology of Serra [31] and morphological spectrums of the shape proposed by Maragos [19]. Vizilter suggests the method of computing the distance between two images based on their spectra. He utilizes shape medial axis for the fast spectrum calculation [35], (2) matching skeletons of objects via tree edit distance [13], (3) path similarity which depends on the shortest distance between all skeleton nodes [7], (4) path similarity that stands on the shortest distance between skeleton end nodes [2], (5) plenty of methods using features calculated on 2D or 3D skeleton (the number of edges, the aspect ratio between the length and the width of the skeleton, etc.) for solving certain classification problem [3,37]. The review of (2), (3) and (4) methods is given in [2].

In a group of techniques founded on shape contours three main approaches can be pointed out. The first one is formed on the Procrustes distance between ordered sets of objects boundary points [11]. The second approach is built on the comparison of boundary polygonal figures: the edges of the each figure are approximated by straight line segments of fixed length, the sequence of angles between the segments is used as string representation. Such representations are then compared by edit distance [24]. The third method is the mechanical model to compare the external outlines of the images. It is assumed that the contours are made of wire. The transformation of the contour is done by deforming the wire and characterized by mechanical work. The minimal work is regarded as a measure of distinction between images [27,30].

All the methods have strengths and weaknesses depending on the application task. The skeletal methods are more appropriate, for example, for the datasets including images with sophisticated well-pronounced shape and shapes with occlusions [2,29]. If objects differ just by their width or have rather simple convex and concave form, the methods based on contours can be more fruitful.

According to the latest research [33] there are only few methods that trying to combine skeletons and contours for the shape recognition purposes. The approach that we offer in this paper aims to join two representations utilizing the radial skeleton function. Finally we get the informative and discriminative but nevertheless compact shape description.

Skeletons. We will use the concept of skeleton-based binary image analysis. Mathematical framework of a skeleton has been formulated initially for continuous objects [5]. There are two ways to construct the skeleton of binary image: discrete and continuous ones. A discrete skeleton is usually defined as a binary image derived by a certain transformation of the initial image. It consists of pixel-wide lines and all of these lines are approximately equidistant from the boundary of the initial object.

The topological thinning methods are based on consequence recoloring of black border pixels into white from four sides while the object on the picture will become the connected pixel-wide set of lines. This set is a skeleton of the initial shape. The method of Distance Map uses the weights for the pixels which shows how far each pixel is from the border. Then the pixels with local maximal weights are considered as skeleton [28]. Comprehensive description of thinning methods is provided in [15].

It should be noted that the discrete methods of skeletonization are rather simple to realize. However, discrete skeletons have essential disadvantages in comparison with their continuous analogues. In methods of topological thinning non-Euclidean metrics are used for skeleton building (City block metric or Chessboard (Tchebychev) metric). Hence, discrete skeleton is not an exact medial axis of figure. Skeletonization methods by Distance Map may cause loss of skeleton connectivity. It is also hard to implement image affine transformations and shapes comparison using discrete skeletons because they are represented as binary images.

Continuous methods are free of disadvantages of discrete ones. Nevertheless they are more complicated because the resulting skeleton is computed as mathematical model (not an image). The process of computation is based on the Voronoi diagram and analyzed in [17,25]. In our research the algorithm of continuous skeletonization proposed in [20] is utilized because it is computationally efficient and appropriate for the real-time applications. The example of binary image and its skeleton is in Figs. 1a and b.

Pruning. Skeletons of both types — discrete and continuous — have the problem of "noisy" branches: small irregularities in figure boundary lead to occurrence of skeleton branches, unessential for shape analysis. The solution to the problem is pruning of non-significant branches depending on empirical threshold [1,20]. The example of pruned skeleton is in Fig. 1c.

Interesting approach was posed by Ogniewicz in [26]. He suggests to use the Voronoi diagram of the shape in order to prune noisy edges of the skeleton without human supervision and manually selected thresholds. Ogniewicz utilizes residual functions for pruning: each edge in skeleton is attributed by the distance between "anchor" points of Voronoi diagram, then the potential function of the

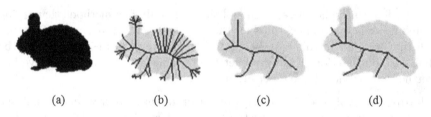

      (a)   (b)   (c)   (d)

Fig. 1. (a) – source image; (b) – skeleton; (c) – pruned and (d) – approximated skeletons

edge is calculated and the hierarchy of the edges (so called pyramid) is built. It allows to distinguish edges of different importance (or order) automatically.

Approximation. Often the skeleton regularized by pruning is not appropriate for the recognition tasks because it contains branches which consist of several (sometimes a lot of) consecutive edges. Such complicated presentation is not needed and we want to change each sequence of edges to one or several simple ribs. This process we call approximation. The idea of approximation is shown in Fig. 1d and can be described as follows. The distance between an inner skeletal vertex of complicated skeleton branch and the straight line between the ends of this skeleton branch is compared with the approximation threshold. If this distance is smaller than threshold, the branch is changed by one straight edge. The approximation threshold is computed as a portion of diameter of minimal circle circumscribed about the skeleton.

In this work we implement pruning and approximation with manually selected thresholds for each class of shapes, nevertheless we plan to realize some methods for automatic skeleton regularization ([32] for example). It will help us to test our comparison approach on various datasets.

3 Comparison of Skeletons Based on Primitive Chains

Skeleton is a good model for binary image description. It is constructed as a locus of centers of maximal circles inscribed into the shape. As far as it was already mentioned in previous Section we use the algorithm of continuous skeletonization proposed in [20].

The general way we propose is to describe the skeleton as follows: traversing skeleton from some initial node counterclockwise we regard current skeleton edge as a primitive. Each primitive can be expressed by some values. In the simplest case the length of the current edge and the angle between the current and the next edges are used [6]. To make description invariant under scale all values have to be normalized. Scaling unit for the length is the diameter of minimal circle circumscribed about the skeleton. The angle is normalized to 2π. The traversing process stops when the starting node is attained. The number of written length/angle combinations is $2M$, where M is the number of edges in skeleton. In Fig. 2 there is an example of description based on primitive chain

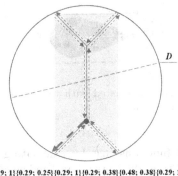

{0.29; 1}{0.29; 0.25}{0.29; 1}{0.29; 0.38}{0.48; 0.38}{0.29; 1}
{0.29; 0.25}{0.29; 1}{0.29; 0.38}{0.48; 0.38}

Fig. 2. Image, skeleton, diameter of circumcircle, traversing and the chain of primitives

Fig. 3. Shapes with equal skeleton topology (just one edge) and different radial function

for the skeleton of rectangle. It is clear that this representation depends on the starting point. So, we can obtain a set of different primitive chains for the same skeleton. However, they can be transformed one into the other by cyclic shift. Thus, the length normalization and cyclic shift transformation give us the skeleton description invariant under translation, rotation and scaling.

As shown in [14] it is valuable for many shape comparison applications to take into account the information about the width of the object. It can be extracted from the skeleton radial function that includes all radii of maximal circles inscribed into the shape with centers in the skeleton edges. Figure 3 shows shapes with the equal skeleton topology and different radial function.

To get the compact width description, in [14] we offer to interpolate radial function by Legendre polynomials and find the vector of Legendre coefficients for the each skeleton rib. Let us notice some key points of this process. As it was mentioned in Sect. 2, the continuous skeleton after pruning process usually contains branches, i.e. consecutive small edges. The values of the radii of maximal inscribed circles with centers in vertices of these edges (see Fig. 3) can be used for interpolation. Legendre polynomials are based on Legendre functions which are orthogonal to each other in the interval [-1, 1]. So, after interpolation we get some coefficients which are parametric descriptors of the shape local width. Practically it is enough to calculate 4 – 7 first coefficients. Thus for the each skeleton rib we obtain the vector of fixed length $\mathbf{p} = \{p_0, ..., p_n\}$, where n – the proper order of interpolating polynomial for all skeleton ribs. Then we can

Fig. 4. One skeleton branch is associated with two vectors of Legendre coefficients because of the direction of traversing

compare different radial functions just by calculating the squared Euclidean distance between them. The coefficients are utilized as coordinates in orthogonal space of Legendre functions: $f(\mathbf{p}', \mathbf{p}'') = \sum_{i=0}^{n} (p'_i - p''_i)^2$.

Is worth noting that while traversing the skeleton we come across each branch twice. So consideration must be given to the order of start and end points of the branch. It leads to one branch will be associated with two vectors of Legendre coefficients (see Fig. 4), which differs each other by signs of even coefficients. This way the width description is invariant under rotation of initial binary image.

To resume all mentioned above let us denote a primitive as a feature vector: $\omega = \{l, \alpha, \mathbf{p}\}$. The first component is the scaled length of the current edge, the second one is normalized angle between the current and the next edges, the third one – the vector of Legendre coefficients. Let Ω is a set of all possible primitives.

The idea of comparison of chains is originated in bioinformatics. It uses the symbolic sequences alignment for studying evolution similarity of proteins. The dissimilarity between two elements is specified by the substitution matrix. In case of primitive chains it is possible to calculate dissimilarity between primitives directly without any cost matrix because a primitive itself is a set of numbers.

The main concept of skeleton comparison is an alignment of two primitive chains. The mechanism of alignment is the following. There are two chains of different (in general) lengths N and K. The first one is called base chain $\mathbb{B}$ and the second one is the reference chain $\mathbb{R}$:

$$\mathbb{B} : b_1, ..., b_N = \{l_1, \alpha_1, \mathbf{p}_1\}\{l_2, \alpha_2, \mathbf{p}_2\}...\{l_N, \alpha_N, \mathbf{p}_N\} \in \Omega,$$
$$\mathbb{R} : r_1, ..., r_K = \{l_1, \alpha_1, \mathbf{p}_1\}\{l_2, \alpha_2, \mathbf{p}_2\}...\{l_K, \alpha_K, \mathbf{p}_K\} \in \Omega.$$

To indicate deletions and insertions of primitives in aligned chains, it is needed to extend reference chain by special elements called gaps ($g = \{-\}$, $g \in \Omega$).The extended reference chain $\mathbb{E}$ with length $\bar{K}$ is:

$$\mathbb{E} : g, r_1, g, r_2, g, ..., g, r_K, g = e_1, ..., e_{\bar{K}}$$
$$= \{-\}\{l_1, \alpha_1, \mathbf{p}_1\}\{-\}\{l_2, \alpha_2, \mathbf{p}_2\}\{-\}...\{-\}\{l_K, \alpha_K, \mathbf{p}_K\}\{-\} \in \Omega.$$

The alignment will be determined by the reference vector $\mathbf{z} = \{z_t\}$, $t = 1, ..., N$, where N – the number of elements in $\mathbb{B}$. The value of $z_t \in \{1, ..., \bar{K}\}$, where $\bar{K}$ – the number of elements in $\mathbb{E}$, determines the order number of the element in $\mathbb{E}$ which is referenced by the t–th element in $\mathbb{B}$. In the alignment each element in $\mathbb{B}$ has the reference to one certain element in $\mathbb{E}$ and several elements in $\mathbb{B}$ can have references to the same element in $\mathbb{E}$. Certain variant of the alignment produces fixed reference vector $\mathbf{z}$.

It is obvious that the goal is to find the optimal alignment (vector $\hat{\mathbf{z}}$) which shows our understanding of the comparison model. So, we have to minimize some criterion $J(\mathbf{z})$ relative to parameter $\mathbf{z}$. Our criterion is a sum of two parts. The first part we call node function and the second one we call edge function [21]:

$$J(\mathbf{z}) = \sum_{t=1}^{N} \psi_t(z_t) + \sum_{t=2}^{N} \gamma_t(z_{t-1}, z_t). \qquad (1)$$

Node function reflects the difference between two corresponding primitives in $\mathbb{B}$ and $\mathbb{E}$. It quadratically penalizes this difference: $\psi_t(z_t) = \rho(\omega'_t, \omega''_{z_t}), t \in \{1, ..., N\}, z_t \in \{1, ..., \bar{K}\}$ where

$$\rho(\omega', \omega'') = (l' - l'')^2 + (\alpha' - \alpha'')^2 + \sum_{i=0}^{n} (p'_i - p''_i)^2. \qquad (2)$$

Here we consider all differences equally but hypothesize that each of them can be punished with own weight depending on the application task.

The edge function (3) supplies the mutual order in pairwise alignment and defines the set of reasonable constrains for combination of references. Cross-reference is prohibited, that is the condition $z_t \geq z_{t-1}$ must be hold in all cases. Equality $z_{t-1} = z_t$ can take place in case of the references to the same gap. This situation is penalized by positive real parameter c. Otherwise it isn't allowed. Each "no-gap" in $\mathbb{E}$ indexed in range between two references (z_{t-1}, z_t) is also punished by c because it means that gaps appear in $\mathbb{B}$ as a result of alignment. In addition, every reference to the gap is fined by c. Varying the c parameter we can get rather differing classification results. Thus we have to tune this parameter depending on the application task. The prohibited cases in combination of references are penalized by ∞ value:

$$\gamma_t(z_{t-1}, z_t) = \begin{cases} \infty, & z_t < z_{t-1}, \\ \infty, & z_t = z_{t-1} \wedge e_{z_t} \neq g, \\ c, & z_t = z_{t-1} \wedge e_{z_t} = g, \\ \sum_{j=z_{t-1}+1}^{z_t-1} c \cdot I(e_j \neq g), \\ c, & e_{z_t} = g, \end{cases} \quad t \in \{2, ..., N\}. \qquad (3)$$

So, we have to minimize criterion (1). This task is solved by means of dynamic programming. Eventually the measure of dissimilarity between two chains (i.e. skeletons) is calculated as follows:

$$D(\mathbb{B}, \mathbb{R}) = \sqrt{\frac{1}{N+K} J(\hat{\mathbf{z}})}. \tag{4}$$

In Fig. 5 there is a simple examples of compared shapes and the values of dissimilarity measures between them according to (4) ($c = 0.2$).

Fig. 5. Examples of compared shapes and the values of dissimilarity measures

4   Model Example of Shapes Dissimilarity Calculation

For initial experiments with developed comparison function we used the artificial dataset with three classes: 1 – four-pointed, 2 — five-pointed and 3 — six-pointed stars. There are 23 objects in each class, total number of objects is 69. The archive of image database is available at http://lda.tsu.tula.ru/papers/starsDB. zip. Examples from each class with their skeletons are in Fig. 6.

We built the pair-wise dissimilarity matrix for 12 images (see Fig. 6). As dissimilarity we use the value of function (4). Values of dissimilarities for the same class are filled with gray color.

Generally dissimilarity measure is smaller for the images from the same class then for ones from the different classes. This fact allows us to suppose that we can solve the task of objects classification utilizing proposed function. We computed the pair-wise dissimilarity matrix for the whole dataset and applied *3*-nearest neighbors (*NN*) method to it. Results of classification are following: 1-th class (four-pointed stars) — 100 % right recognized; 2-th class (five-pointed stars) — 100 %; 3-th class (six-pointed stars) — 91.3 %.

5   Experiments on Real-World Flavia Leaf DataSet

To evaluate the suggested dissimilarity function we use Flavia leaf dataset [39]. The task of plant leaves classification is well-known in the literature [4,8,12,38]. There are different leaf datasets which can be used for testing purposes, for example "Leaves from Swedish Trees" [34], "ImageCLEF" [9], "Middle European Woods (MEW) 2010" [16], "ICL" [10].

In Flavia leaf dataset there are 32 types of species, the total number of objects is 1907. So the average number of leaves in each class is near sixty. The smallest

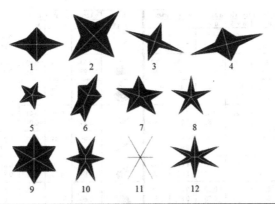

№	1	2	3	4	5	6	7	8	9	10	11	12
1	0.000	0.193	0.001	0.049	0.260	0.295	0.292	0.261	0.348	0.345	0.327	0.344
2	0.193	0.000	0.193	0.210	0.171	0.242	0.226	0.172	0.295	0.295	0.267	0.296
3	0.001	0.193	0.000	0.050	0.260	0.295	0.292	0.261	0.348	0.345	0.326	0.344
4	0.049	0.210	0.050	0.000	0.277	0.296	0.301	0.279	0.358	0.351	0.338	0.350
5	0.260	0.171	0.260	0.277	0.000	0.194	0.161	0.005	0.247	0.251	0.208	0.251
6	0.295	0.242	0.295	0.296	0.194	0.000	0.100	0.195	0.221	0.209	0.184	0.209
7	0.292	0.226	0.292	0.301	0.161	0.100	0.000	0.161	0.198	0.196	0.149	0.197
8	0.261	0.172	0.261	0.279	0.005	0.195	0.161	0.000	0.247	0.251	0.208	0.251
9	0.348	0.295	0.348	0.358	0.247	0.221	0.198	0.247	0.000	0.035	0.197	0.036
10	0.345	0.295	0.345	0.351	0.251	0.209	0.196	0.251	0.035	0.000	0.202	0.005
11	0.327	0.267	0.326	0.338	0.208	0.184	0.149	0.208	0.197	0.202	0.000	0.202
12	0.344	0.296	0.344	0.350	0.251	0.209	0.197	0.251	0.036	0.005	0.202	0.000

Fig. 6. Examples of three classes in each row from "Stars" dataset and correspondent pair-wise dissimilarity matrix with $c = 0.3$

class has got 50 objects and the largest one — 77 objects. We have to mention that the images of leaves were binarized and preprocessed via noise filtering. Examples from each class are in the "Class/Image" column of Fig. 7.

Let us notice that the dataset is rather difficult for classification: it contains large amount of classes, there are some classes which are very similar to each other and there are some classes with very different objects inside. While using the dissimilarity function without taking into account width properties, we get rather poor total classification rate (37 %). For example, it is impossible to find the difference among 1,2,3,9,17,... classes because of their indistinguishable skeletons (usually just one edge).

We use the k-NN classifier relying on the dissimilarity function (4). Classification rate for each class achieved using different numbers of Legendre coefficients (from 4 till 8) and 3- and 5-NN is shown in Fig. 7.

For more detailed analysis of misclassification it makes sense to examine the non-symmetric contingency table, also known as confusion matrix (see Fig. 8). The left column and the top row of the table show number of class. On the main diagonal the number of true positive cases is located. The bottom row shows how many objects each class contains, that is the sum of true positives and false positives. The right column points out how many objects were assigned to

Class/ Image	Accuracy with n Legendre Coefficients					Class/ Image	Accuracy with n Legendre Coefficients				
	n = 4	n = 5	n = 6	n = 7	n = 8		n = 4	n = 5	n = 6	n = 7	n = 8
1	0.81	0.85	0.92	0.86	0.86	17	0.96	0.96	0.96	0.96	0.96
	0.80	0.85	0.92	0.86	0.85		0.96	0.96	0.96	0.96	0.96
2	0.54	0.62	0.84	0.87	0.87	18	0.92	0.94	0.92	0.94	0.92
	0.44	0.56	0.86	0.89	0.86		0.95	0.95	0.95	0.95	0.95
3	0.29	0.40	0.43	0.43	0.45	19	0.69	0.74	0.75	0.74	0.69
	0.22	0.42	0.45	0.45	0.43		0.74	0.77	0.80	0.77	0.74
4	0.81	0.81	0.82	0.81	0.81	20	0.94	0.95	0.97	0.97	0.97
	0.82	0.82	0.82	0.83	0.83		0.94	0.95	0.97	0.95	0.97
5	0.86	0.93	0.93	0.93	0.93	21	0.95	0.93	0.97	0.97	0.98
	0.82	0.93	0.93	0.93	0.93		0.98	0.95	0.97	0.97	0.95
6	1.00	1.00	1.00	1.00	1.00	22	0.87	0.89	0.87	0.85	0.89
	1.00	1.00	1.00	1.00	1.00		0.84	0.85	0.85	0.85	0.87
7	0.82	0.73	0.82	0.87	0.85	23	0.55	0.73	0.65	0.67	0.65
	0.76	0.73	0.76	0.87	0.87		0.60	0.76	0.69	0.65	0.65
8	0.98	0.98	0.98	0.98	0.98	24	0.77	0.80	0.78	0.80	0.78
	0.98	0.98	0.98	0.98	0.98		0.77	0.77	0.77	0.78	0.77
9	0.20	0.13	0.24	0.25	0.27	25	0.52	0.52	0.57	0.61	0.63
	0.18	0.20	0.25	0.25	0.24		0.56	0.57	0.65	0.63	0.70
10	0.98	0.98	0.98	0.98	0.98	26	0.77	0.88	0.85	0.85	0.85
	0.98	0.98	0.98	0.98	0.98		0.79	0.83	0.81	0.87	0.83
11	0.60	0.70	0.64	0.62	0.62	27	0.98	0.98	0.96	0.96	0.96
	0.58	0.64	0.62	0.62	0.62		0.96	0.98	0.98	0.98	0.98
12	0.87	0.92	0.92	0.90	0.94	28	0.98	0.98	0.98	0.98	0.98
	0.89	0.92	0.90	0.92	0.92		0.96	0.96	0.96	0.96	0.96
13	0.50	0.54	0.58	0.56	0.60	29	0.58	0.68	0.70	0.79	0.81
	0.46	0.52	0.58	0.58	0.60		0.67	0.68	0.74	0.81	0.79
14	0.69	0.72	0.69	0.71	0.74	30	0.92	0.92	0.92	0.92	0.92
	0.69	0.69	0.69	0.74	0.72		0.92	0.92	0.92	0.94	0.92
15	0.57	0.57	0.73	0.83	0.80	31	1.00	1.00	1.00	1.00	1.00
	0.57	0.60	0.75	0.82	0.83		1.00	1.00	1.00	1.00	1.00
16	0.63	0.66	0.61	0.57	0.55	32	0.79	0.77	0.70	0.86	0.84
	0.61	0.57	0.59	0.54	0.54		0.79	0.73	0.71	0.82	0.82
						Total Accuracy	0.76	0.79	0.80	0.81	0.82
							0.76	0.78	0.81	0.82	0.81

Fig. 7. Classification rate for 32 species (top number in each cell means classification rate with *3-NN*, bottom one – classification rate with *5-NN*)

the i-th class, that is the sum of true positives and false negatives. The integer numbers in cells (i, j) mean how many objects of real class j were assigned to the class i. From this table we can evaluate different parameters of classifier quality: for example, the interclass dissimilarity, the intraclass variance, precision, recall and F1-measure. In particular, the average precision for the whole dataset is equal to 82 %, average recall is 83 % and average F1-measure is 82 %.

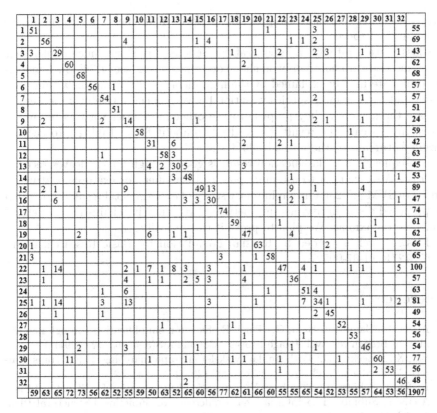

	1	2	3	4	5	6	7	8	9	10	11	12	13	14	15	16	17	18	19	20	21	22	23	24	25	26	27	28	29	30	31	32	
1	51																		1						3								55
2		56						4							1	4						1	1	2									69
3	3		29														1		1		2			2	3			1			1		43
4				60															2														62
5					68																												68
6						56	1																										57
7							54																	2				1					57
8								51																									51
9	2					2			14				1		1									2	1			1					24
10										58																	1						59
11											31		6						2			2	1										42
12						1						58	3															1					63
13												4	2	30	5				3									1					45
14														3	48									1							1		53
15	2	1		1					9							49	13							9	1			4					89
16		6												3	3	30					1	2	1								1		47
17																	74																74
18																		59				1								1			61
19			2								6		1	1				47				4						1					62
20	1																		63					2									66
21	3																3		1	58													65
22		1	14						2	1	7	1	8	3		3		1			47		4	1			1	1				5	100
23		1							4		1	1		2	5	3			4			36											57
24					1				6										1				51	4									63
25	1	1	14					3		13						3			1				7	34	1			1			2		81
26		1			1						1													2	45								49
27									1							1										52							54
28			1																1				1				53						56
29			2					3						1								1		1				46					54
30			11								1			1			1	1			1				1				60				77
31																						1								2	53		56
32													2																			46	48
	59	63	65	72	73	56	62	52	55	59	50	63	52	65	60	56	77	62	61	66	60	55	65	54	52	53	55	57	64	53	56		1907

Fig. 8. Contingency table for classification results with 7 Legendre coefficients, *5–NN*

6 Conclusion

We have proposed a new way to compare shapes of objects based on skeletoniza-tion and chain description. In contrast to different feature-based methods our approach allows to match arbitrary objects without searching for some special attributes of shape. Comparing to works of Bai, Klein, Bystrov, Vizilter and others it gives us much richer description of the skeleton representation. Besides the length of the skeleton edge and angle between edges we have introduced Legendre polynomials in order to describe the width of the objects. This gives us the representation of the skeleton as a chain of primitives; each primitive is a vector of length, angle and set of Legendre coefficients. Thus the resulting shape representation by the chain of primitives includes not only topological proper-ties as length and angle but also the contour ones. The proposed description invariant under translation, rotation and scaling. Pair-wise dissimilarity func-tion constructed on the optimal alignment of primitive chains gives promising results in real-world classification experiments. We achieve an average accuracy of about 82 % on Flavia leaf dataset. The average time of comparison of two images is 20 ms. We plan to apply some efforts in further work to improving computational speed.

Classification rate for different leaf datasets usually is not much higher than 73 % if methods are based only on shape description, e.g. see [4, 36]. Such moderate results are mostly explained by the high intra-species variation within some classes and the low inter-species variation between species with ovate leaves. To increase the accuracy they propose to use other modalities such as texture of leaves. In this case the result is raised up to about 80 %. Taking into account margin features rises the result to 96 % [18].

Consequently our classification result is good enough for the leaves recognition task. Also we plan to use more powerful classifier, for example, featureless version of SVM [22,23]. Our preliminary study of experimental results shows that the proposed dissimilarity measure between two primitive chains (i.e. skeletons) fulfills metric axioms. We plan to prove it theoretically.

This research is funded by Russian Foundation for Basic Research, grants 14-07-31271, 14-07-00527.

References

1. Attali, D., Sanniti di Baja, G., Thiel, E.: Skeleton simplification through non significant branch removal. Image Process. Commun. **3**(3–4), 63–72 (1997)
2. Bai, X., Latecki, L.J.: Path similarity skeleton graph matching. IEEE Trans. Pattern Anal. Mach. Intell. **30**(7), 1282–1292 (2008)
3. Balfer, J., Schöler, F., Steinhage V.: Semantic Skeletonization for Structural Plant Analysis. Submitted to International Conference on Functional-Structural Plant Models (2013)
4. Beghin, T., Cope, J.S., Remagnino, P., Barman, S.: Shape and texture based plant leaf classification. In: Blanc-Talon, J., Bone, D., Philips, W., Popescu, D., Scheunders, P. (eds.) ACIVS 2010, Part II. LNCS, vol. 6475, pp. 345–353. Springer, Heidelberg (2010)
5. Blum, H.: A transformation for extracting new descriptors of shape. Models Percept. Speech Vis. Form **19**(5), 362–380 (1967)
6. Bystrov, M.Y.: Structural approach application for recognition of binary image skeleton. Proc. Petrozavodsk State Univ. **2**(115), 76–80 (2011). (in Russian)
7. Demirci, M.F., Shokoufandeh, A., Keselman, Y., Bretzner, L., Dickinson, S.: Object recognition as many-to-many feature matching. Int. J. Comput. Vision **69**(2), 203–222 (2006)
8. Du, J.X., Huang, D.S., Wang, X.F., Gu, X.: Computer-aided plant species identification (CAPSI) based on leaf shape matching technique. Trans. Inst. Measur. Control **28**(3), 275–285 (2006)
9. ImageCLEF, Plant Identification (2012). http://imageclef.org/2012/plant
10. Intelligent Computing Laboratory, Chinese Academy of Sciences Homepage. http://www.intelengine.cn/English/dataset
11. Jänichen, S., Perner, P.: Aligning concave and convex shapes. In: Yeung, D.-Y., Kwok, J.T., Fred, A., Roli, F., de Ridder, D. (eds.) SSPR 2006 and SPR 2006. LNCS, vol. 4109, pp. 243–251. Springer, Heidelberg (2006)
12. Kadir, A., Nugroho, L.E., Susanto, A., Santosa, P.I.: A comparative experiment of several shape methods in recognizing plants (2011). arXiv preprint arXiv:1110.1509

13. Klein, P., Tirthapura, S., Sharvit, D., Kimia, B.: A tree-edit-distance algorithm for comparing simple, closed shapes. In: Proceedings of the eleventh annual ACM-SIAM Symposium on Discrete Algorithms. Society for Industrial and Applied Mathematics, pp. 696–704 (2000)
14. Kushnir, O., Seredin, O.: Parametric description of skeleton radial function by legendre polynomials for binary images comparison. In: Elmoataz, A., Lezoray, O., Nouboud, F., Mammass, D. (eds.) ICISP 2014. LNCS, vol. 8509, pp. 520–530. Springer, Heidelberg (2014)
15. Lam, L., Lee, S.-W., Suen, C.Y.: Thinning methodologies - a comprehensive survey. IEEE Trans. Pattern Anal. Mach. Intell. **14**(9), 869–885 (1992)
16. LEAF - Tree Leaf Database, Inst. of Information Theory and Automation ASCR, Prague, Czech Republic. http://zoi.utia.cas.cz/tree_leaves
17. Lee, D.: Medial axis transformation of a planar shape. IEEE Trans. Pat. Anal. Mach. Int. PAMI **4**(4), 363–369 (1982)
18. Mallah, C., Cope, J., Orwell, J.: Plant leaf classification using probabilistic integration of shape, texture and margin features. Computer Graphics and Imaging/798: Signal Processing, Pattern Recognition and Applications (CGIM2013), Acta Press (2013). doi:10.2316/P.2013.798-098
19. Maragos, P.: Pattern spectrum and multiscale shape representation. IEEE Trans. Pattern Anal. Mach. Intell. **11**(7), 701–716 (1989)
20. Mestetskiy, L., Semenov, A.: Binary image skeleton - continuous approach. VIS-APP **1**, 251–258 (2008)
21. Mottl, V.V., Blinov, A.B., Kopylov, A.V., Kostin, A.A.: Optimization techniques on pixel neighborhood graphs for image processing. In: Jolion, J.-M., Kropatsch, W.G. (eds.) Graph-Based Representations in Pattern Recognition. Computing Supplement, vol. 12, pp. 135–145. Springer, Wien (1998)
22. Mottl, V., Seredin, O., Dvoenko, S., Kulikowski, C., Muchnik, I.: Featureless pattern recognition in an imaginary Hilbert space. In: Proceedings of 16th International Conference on Pattern Recognition, vol. 2, pp. 88–912 (2002)
23. Mottl, V., Krasotkina, O., Seredin, O., Muchnik, I.: Kernel fusion and feature selection in machine learning. In: Proceedings of the Eighth IASTED International Conference on Intelligent Systems and Control, Cambridge, USA, pp. 477–482 (2005)
24. Neuhaus, M., Bunke, H.: Edit distance-based kernel functions for structural pattern classification. Pattern Recogn. **39**(10), 1852–1863 (2006)
25. Ogniewicz, R., Kubler, O.: Hierarchic voronoi skeletons. Pattern Recogn. **28**(3), 343–359 (1995)
26. Ogniewicz, R.: Automatic medial axis pruning by mapping characteristics of boundaries evolving under the euclidean geometric heat flow onto Voronoi skeletons. Harvard Robotics Laboratory Technical report, pp. 95–114 (1995)
27. Reier, I.A.: Plane figure recognition based on contour homeomorphism. Pattern Recogn. Image Anal. **11**(1), 242–245 (2001)
28. Sanniti di Baja, G., Thiel, E.: Computing and comparing distance-driven skeletons. In: Aspects of Visual Form Processing, pp. 465–486 (1994)
29. Sebastian, T.B., Kimia, B.: Curves vs. skeletons in object recognition. Sig. Process. **85**(2), 247–263 (2005)
30. Sederberg, T.W., Greenwood, E.: A physically based approach to 2-D shape blending. Comput. Graph. **26**(2), 25–34 (1992)
31. Serra, J.: Image Analysis and Mathematical Morphology. Acad. Press, London (1982)

32. Shen, W., Bai, X., Yang, X., Latecki, L.J.: Skeleton pruning as trade-off between skeleton simplicity and reconstruction error. Sci. China Inf. Sci. **56**(4), 1–14 (2013)

33. Shen, W., Wang, X., Yao, C., Bai, X.: Shape recognition by combining contour and skeleton into a mid-level representation. In: Li, S., Liu, C., Wang, Y. (eds.) CCPR 2014, Part I. CCIS, vol. 483, pp. 391–400. Springer, Heidelberg (2014)

34. Söderkvist, O.: Computer Vision Classification of Leaves from Swedish Trees. Diss, Linköping (2001)

35. Vizilter, Y.V., Sidyakin, S.V., Rubis, A.Y., Gorbatsevich, V.S.: Morphological shape comparison based on skeleton representations. Pattern Recogn. Image Anal. **22**(3), 412–418 (2012)

36. Wang, B., Brown, D., Gao, Y., La Salle, J.: MARCH: Multiscale-arch-height description for mobile retrieval of leaf images. Information Sciences (2014). http:// dx.doi.org/10.1016/j.ins.2014.07.028

37. Wang, C., Gui, C.-P., Liu, H.-K., Zhang, D., Mosig, A.: An image skeletonization based tool for pollen tube morphology analysis and phenotyping. J. Integr. Plant Biol. **55**(2), 131–141 (2013)

38. Wang, Z., Chi, Z., Feng, D.: Shape based leaf image retrieval. Vis. Image Sig. Process. IEE Proc. **150**(1), 34–43 (2003)

39. Wu, S.G., Bao, F.S., Xu, E.Y., Wang, Y.-X., Chang, Y.-F., Xiang, Q.-L.: A leaf recognition algorithm for plant classification using probabilistic neural network. In: 2007 IEEE International Symposium on Signal Processing and Information Technology, pp. 11–16 (2007)

Color Image Restoration with Fuzzy Gaussian Mixture Model Driven Nonlocal Filter

V.B. Surya Prasath[1]([✉]) and Radhakrishnan Delhibabu[2]

[1] University of Missouri-Columbia, Columbia, MO 65211, USA
prasaths@missouri.edu
http://web.missouri.edu/~prasaths
[2] Knowledge Based System Group, Higher Institute for Information Technology
and Information Systems,
Kazan Federal University, Kazan, Russia

Abstract. Color image denoising is one of the classical image processing problem and various techniques have been explored over the years. Recently, nonlocal means (NLM) filter is proven to obtain good results for denoising Gaussian noise corrupted digital images using weighted mean among similar patches. In this paper, we consider fuzzy Gaussian mixture model (GMM) based NLM method for removing mixed Gaussian and impulse noise. By computing an automatic homogeneity map we identify impulse noise locations and utilize an adaptive patch size. Experimental results on mixed noise affected color images show that our scheme performs better than NLM, anisotropic diffusion and GMM-NLM over different noise levels. Comparison with respect to structural similarity, color image difference, and peak signal to noise ratio error metrics are undertaken and our scheme performs well overall without generating color artifacts.

Keywords: Gaussian mixture · Type-2 fuzzy sets · Nonlocal means · Image denoising · Mixed noise

1 Introduction

Image smoothing filters are widely used in image processing and computer vision as a pre-processing/denoising step. Variational regularization [1], anisotropic diffusion [2–4] and neighborhood filters [5,6] are widely used and literature is rich with different variations of these approaches. One of the main requirement in image smoothing is the preservation of salient edges and can be posed as a feature preservation [7]. Nonlocal means which is based on weighted mean of similar patches is proven to be a good candidate in removing noise effectively using information via self similarity.

The aforementioned techniques though provide good results in removing additive Gaussian noise, the mixed noise case of Gaussian and impulse require further improved modeling and experimentation. Recently, a GMM based NLM for mixed noise suppression is proposed by [8]. Motivated by this approach,

© Springer International Publishing Switzerland 2015
M.Y. Khachay et al. (Eds.): AIST 2015, CCIS 542, pp. 137–145, 2015.
DOI: 10.1007/978-3-319-26123-2_13

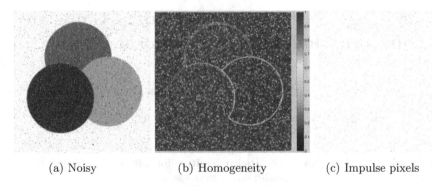

(a) Noisy   (b) Homogeneity   (c) Impulse pixels

Fig. 1. Synthetic color image *Shapes* used in our experiment for obtaining the homogeneity map threshold. (a) Input image obtained by adding Gaussian (std $\sigma = 10$) and impulse (probability $p = 0.1$) noises (b) Homogeneity map (in range $[0, 1]$) computed using a gray-scale version of the input noisy color image. Note that edges and impulse noise regions are of different values. (c) Identified impulse pixels using the homogeneity map (see Sect. 2.2). Better viewed online and zoomed in (Color figure online).

we consider a homogeneity map for determining impulse pixels and adaptively change the patch sizes and use nonlocal means (NLM) filter for removing noise. In this work, we use type-2 fuzzy [9] GMM model for obtaining similarity between different patches. Experimental results on noisy synthetic and real color images indicate we obtain better results when compared with GMM-NLM, anisotropic diffusion methods.

Rest of the paper is organized as follows. Section 2 introduces the fuzzy GMM and homogeneity map driven NLM approach for color image restoration. Section 3 illustrates the advantages of our method on various noise color images and detailed comparison with related schemes is undertaken. Finally, Sect. 4 concludes the paper.

2 Fuzzy GMM Driven Nonlocal Means

2.1 Nonlocal Means

Nonlocal means (NLM) proposed by Buades et al. [6] works well for denoising Gaussian noise corrupted images by utilizing similarity of local patches. By using not only the pixel neighborhood but from all over the image, NLM provides good denoising results. For a given noisy (gray scale) image $u_0 : \Omega \to [0, 255]$, Ω image domain of size $m \times n$, the NLM obtains the following resultant image

$$u(i, j) = \frac{\sum_{k,l} W_{i,j}(k, l) u_0(k, l)}{\sum_{k,l} W_{k,l}}, \quad \forall (i, j) \in \Omega \tag{1}$$

The weights are given as

$$W_{i,j}(k, l) = \exp \frac{-\|u_0(P_{i,j} - P_{k,l})\|_\sigma^2}{\rho^2} \tag{2}$$

where $\|\cdot\|_\sigma^2$ is Gaussian weighted Euclidean distance (σ standard deviation), $P_{i,j}$ is a patch (of size $r \times r$, centered at pixel (i,j)). The final denoising result depends on both the similarity of patches (on the patch space $\mathcal{P}$) and the size of the patches involved. We modify the patch size using an homogeneity map to obtain best possible results under impulse noise in this paper and we describe the process next.

2.2 Homogeneity Based Patch Size Estimation

For denoising mixed noise case a better modeling is required to make use of the NLM filter. For this purpose, we use a homogeneity map based on smoothed gradients to decide the patch sizes used in NLM, which we describe next.

1. For each pixel (i,j) compute a homogeneity map: $\mathcal{H}(i,j) := 1 - |G_s \star \nabla u_0|^2$, where G_s is a Gaussian kernel of variance $s > 0$. Note that for color images we compute the map using the gray-scale version[1]. Figure 1(b) shows an example for the computed homogeneity map on a synthetic noisy color image corrupted by mixed noise.
2. Check if $\mathcal{H}(i,j) < Th$ then use the following decision rule decide if (i,j) correspond to edge or impulse pixel:
   - $\mathcal{H}(i,j) > median(P_{i,j} \setminus (i,j)) \implies$ impulse pixel, enlarge patch size from $r \times r$ to $r' \times r'$ where $r' = round(\xi \cdot ratio \cdot |G_s \star \nabla u_0|^2 + 3)$ with $\xi = \min(m,n)$, scale $ratio = 5\%$.
   - $\mathcal{H}(i,j) > median(P_{i,j} \setminus (i,j)) \implies$ edge pixel, do not enlarge patch size.

We apply NLM (1) using this decision rule for handling detected impulse noise (Fig. 1(c)) and the smoothing parameter for the Gaussian is chosen as $\sigma = 2$ for Gaussian noise std $\sigma = 10$ to 30. Note that step 1 takes care of the Gaussian noise by using isotropic smoothing of gradient map and step 2 computes impulse noise pixels from a given image using the obtained gradient map.

2.3 Fuzzy GMM for Patch Similarity via Histograms

We next utilize the type-2 fuzzy Gaussian mixture model (GMM) parameters for determining similarity of local patches instead of the Euclidean distance used in Eq. (2). To do this, we first utilize the CIELa*b* color space and construct histogram in a-b chromaticity space [8], see Fig. 2(a) which shows color cloud for the synthetic image *Shapes*. Then we use fuzzy GMM to model color histogram via expectation-maximization (EM) for parameter estimation. The final similarity between two patches $P_{i,j}$ and $P_{k,l}$ is computed using earth mover's distance between the computed fuzzy GMM parameters of the patches. Note that a similar approach taken in [8] uses traditional GMM to estimate the histogram distributions of patches and is prone to noisy data as exemplified by

---

[1] Color image converted using MATLAB's `rgb2gray` which uses the following formula $I_g = 0.2989 * R + 0.5870 * G + 0.1140 * B$, where R, G, B are the red, green, and blue channels of the given color image. Images are normalized to $[0,1]$.

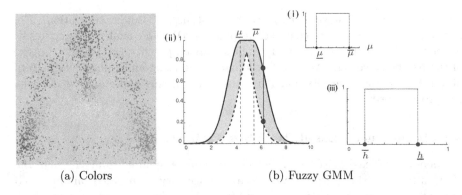

(a) Colors   (b) Fuzzy GMM

Fig. 2. (a) Color cloud of synthetic *Shapes* color image (Fig. 1(a)) used within our histogram. (b) Type-2 fuzzy Gaussian mixture model involves Gaussian primary membership functions (i). (ii) Mean boundaries (iii) Primary membership function as an interval. Image adapted from [9] with permission (Color figure online)

experiments, see Sect. 3.2. In contrast, fuzzy GMM is proven to obtain good or better results under a footprint of uncertainty [9]. It uses M mixture components of multivariate Gaussians and is given for a d dimensional observation vector $\mathbf{x} = (x_1, \ldots, x_d)$,

$$p(\mathbf{x}|\Theta) = \sum_{i=1}^{M} \alpha_i \, \mathcal{N}(\mathbf{x}|\mu_i, \Sigma_i) \tag{3}$$

where $\Theta = (\alpha_1, \ldots, \alpha_M, \mu_1, \ldots, \mu_M, \Sigma_1, \ldots, \Sigma_M)$, $\sum_{i=1}^{M} \alpha_i = 1$, and

$$\mathcal{N}(\mathbf{x}|\mu_i, \Sigma_i) = \frac{1}{\sqrt{(2\pi)^d |\Sigma_i|}} \exp\left[-\frac{1}{2} (\mathbf{x} - \mu_i)' \Sigma_i^{-1} (\mathbf{x} - \mu_i) \right] \tag{4}$$

with mean $\mu_i \in [\underline{\mu}_i, \overline{\mu}_i]$, $\forall i$, we refer to [9] for more details. Each component is the Gaussian primary membership function h with uncertain mean, see Fig. 2(b). Without loss of generality, we assume diagonal covariance matrix Σ^2. Initially the number of components are set 20 and decreased if α_m is less than 0.01.

3   Experimental Results

3.1   Setup and Parameters

We have implemented the fuzzy GMM using EM algorithm with NETLAB toolbox of MATLAB [10] and the NLM implementation (with default parameters) outlined in [6]. NLM filter is computationally expensive as the patch space needs

---

[2] For a non-diagonal symmetric matrix Σ there exists orthogonal matrix $\mathcal{O}$ and diagonal matrix Λ such that $\Sigma^{-1} = \mathcal{O}\Lambda\mathcal{O}$.

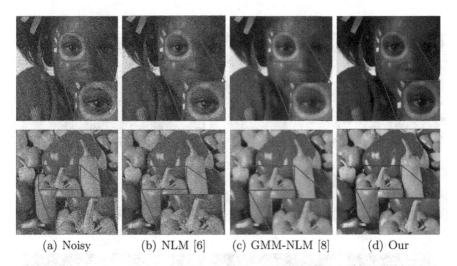

(a) Noisy   (b) NLM [6]   (c) GMM-NLM [8]   (d) Our

Fig. 3. Comparison between GMM versus FGMM driven NLM filtering results on noisy *Girl*, *Peppers* color images. (a) Noisy images, obtained by adding Gaussian (std $\sigma = 20$) and impulse (probability $p = 0.2$) noises Restoration results from: (b) NLM [6], (b) GMM-NLM [8], (c) Proposed approach. Better viewed online and zoomed in (Color figure online).

to be expanded according to the homogeneity map. Our current implementation in MATLAB on a Mac Laptop with 2.3 GHz Intel Core i7, 8 GB RAM takes a minute to obtain final denoising result for an image of size $256 \times 256 \times 3$. To fix the thresholding parameter $Th = 0.7$ in the homogeneity map (see Sect. 2.2 and Fig. 1(b)), we used the synthetic image (see Fig. 1(a)) and added different noise levels of both Gaussian and impulse noises and compared the final restoration accuracy. This is a tuning parameter of our scheme and can be fixed for different imaging modalities by considering a similar simulation on synthetically generated image. To make comparisons with GMM-NLM we used the same parameter set in [8] for the NLM filter.

3.2 Comparison Results

Figure 3 shows results on noisy color images for NLM [6], GMM-NLM [8], anisotropic diffusion [2] and our scheme. Note that nonlocal means assumes Gaussian noise and hence can not handle impulse noise (Fig. 3(b)) without a modification in the modeling of the filter. In contrast to GMM-NLM (Fig. 3(c)) our scheme (Fig. 3(d)) obtains better preservation colors and no blurring can be observed near edges.

To compare different denoising results quantitatively we utilize the mean absolute error (MAE), peak signal to noise ratio (PSNR) and mean structural similarity (MSSIM) as error metrics. Table 1 provides comparison between GMM-NLM [8] with our method for different noise levels on two color images

(a) Noisy

(b) GMM-NLM

(c) Our

Fig. 4. (a) Input color image (*Khram na krovi*). Denoised by schemes: (b) GMM-NLM [8] and (c) Our method. Better viewed online and zoomed in (Color figure online).

Table 1. Comparison of PSNR (dB), MAE and MSSIM error metric values on different noisy (Gaussian std σ, Impulse probability p) color *Girl* and *Peppers* images for GMM-NLM [8]/our proposed scheme.

Girl

p-σ	PSNR			MAE			MSSIM		
	10	20	30	10	20	30	10	20	30
0.1	25.31	24.33	22.46	8.66	9.73	12.61	0.7342	0.7111	0.6824
	27.89	25.90	23.77	7.81	8.86	11.39	0.8233	0.8190	0.7394
0.2	25.89	24.10	22.15	9.49	10.59	14.05	0.7004	0.6734	0.6282
	26.61	25.84	23.44	8.40	9.39	12.79	0.7961	0.7527	0.7110
0.3	24.44	23.57	21.36	10.43	11.94	16.30	0.6490	0.6284	0.5954
	25.55	24.47	22.65	9.24	10.20	15.42	0.7343	0.6988	0.6583

Peppers

p-σ	PSNR			MAE			MSSIM		
	10	20	30	10	20	30	10	20	30
0.1	24.76	23.67	21.82	8.32	9.67	13.06	0.7298	0.7174	0.6733
	26.99	24.92	22.75	7.07	8.39	12.21	0.8105	0.8003	0.7567
0.2	24.22	23.42	21.40	9.15	10.57	14.45	0.6983	0.6645	0.6188
	25.48	24.23	23.02	8.50	9.03	13.82	0.7411	0.7243	0.6920
0.3	24.44	23.57	21.36	10.43	11.94	16.30	0.5989	0.6478	0.6109
	25.84	24.15	22.09	9.83	11.02	15.76	0.6764	0.6591	0.6340

Girl and *Peppers* given in Fig. 3. Note that the performance of NLM is omitted as it can not remove impulse noise effectively. These results show that fuzzy version of GMM can improve the NLM filter for handling mixed Gaussian and impulse noise across various levels. Next, Table 2 shows comparison results with NLM [6], GMM-NLM [8], anisotropic diffusion (AD) [2] with our scheme for

Table 2. Comparison of PSNR (dB), MAE and MSSIM error metric values on different noisy (Gaussian std σ, Impulse probability p) color *Peppers* image for NLM [6], GMM-NLM [8], AD [2], and our proposed scheme.

(p/σ)	PSNR				MAE				MSSIM			
	GMM	NLM	AD	Our	GMM	NLM	AD	Our	GMM	NLM	AD	Our
(0.1/10)	24.76	18.54	24.33	**25.55**	8.32	10.49	10.69	**7.56**	0.7298	0.6552	0.6234	**0.8105**
(0.2/20)	23.42	19.48	22.52	**24.37**	10.57	13.21	14.22	**10.06**	0.6645	0.5785	0.5622	**0.7243**
(0.3/30)	20.77	20.84	20.83	**21.75**	16.37	16.59	18.06	**15.63**	0.6109	0.5302	0.5114	**0.6340**

Peppers color image. Our scheme outperforms all the other schemes and we also obtain better mean structural similarity (MSSIM) values across different noise levels indicating salient structures are preserved.

Finally in Fig. 4 shows a real color image which is corrupted by high ISO noise and a comparison of GMM-NLM and our method. As can be seen by comparing close-up of the images, our scheme obtains better structure preservation and noise in homogenous regions are completely removed.

4   Conclusions

In this paper, we considered a fuzzy GMM driven nonlocal means for color image restoration for mixed noise case. By using FGMM for determining similarity of patches used in NLM we have derived an efficient scheme which can handle different noise levels. Experimental results on a variety of color images indicate we obtain better performance than NLM and other traditional denoising methods. Comparison results in terms of various error metrics indicate we obtain good denoising results overall and without generating color artifacts noticed in other schemes. Extensive sweeping of parameters in NLM method (initial patch size $r \times r$, weight normalization ρ) and adaptive homogeneity (enlargement $r' \times r'$, scale *ratio*) map needs to be done for further improving the results and defines our future work in this direction. Another area of exploration is the speed-up of fuzzy GMM-NLM method computations and extending it to multi-spectral images [11] by using coupling of different channels.

Acknowledgments. This work was done while the first author was visiting the Center for Scientific Computation and Mathematical Modeling (CSCAMM), University of Maryland, College Park, MD, USA in the summer of 2014. The first author thanks the CSCAMM institute for their great hospitality and support during the visit.

References

1. Prasath, V.B.S., Singh, A.: A hybrid convex variational model for image restoration. Appl. Math. Comput. **215**(10), 3655–3664 (2010)
2. Perona, P., Malik, J.: Scale-space and edge detection using anisotropic diffusion. IEEE Trans. Pattern Anal. Mach. Intell. **12**(7), 629–639 (1990)
3. Prasath, V.B.S., Vorotnikov, D.: On a system of adaptive coupled PDEs for image restoration. J. Math. Imaging Vis. **48**(1), 35–52 (2014)
4. Prasath, V.B.S., Delhibabu, R.: Automatic contrast parameter estimation in anisotropic diffusion for image restoration. In: Ignatov, D.I., Khachay, M.Y., Panchenko, A., Konstantinova, N., Yavorsky, R.E. (eds.) AIST 2014. CCIS, vol. 436, pp. 198–206. Springer, Heidelberg (2014)
5. Tomasi, C., Manduchi, R.: Bilateral filtering for gray and color images. In: IEEE International Conference on Computer Vision, pp. 59–66 (1998)
6. Buades, A., Coll, B., Morel, J.M.: A review of image denoising methods, with a new one. Multiscale Model. Simul. **4**(2), 490–530 (2005)

7. Prasath, V.B.S., Moreno, J.C.: Feature preserving anisotropic diffusion for image restoration. In: Fourth National Conference on Computer Vision, Pattern Recognition, Image Processing and Graphics (NCVPRIPG 2013), India, pp. 1–4, December 2013

8. Luszczkiewicz-Piatek, M.: Gaussian mixture model based non-local means technique for mixed noise suppression in color images. In: Choraś, R.S. (ed.) Image Processing & Communications Challenges. Advances in Intelligent Systems and Computing, vol. 313, pp. 75–83. Springer, Switzerland (2015)

9. Zeng, J., Liu, Z.Q.: Type-2 Fuzzy Graphical Models for Pattern Recognition. Studies in Computational Intelligence, vol. 666. Springer, Berlin (2015)

10. Nabney, I.: NETLAB: Algorithms for Pattern Recognitions. Springer, London (2002)

11. Prasath, V.B.S., Singh, A.: Multispectral image denoising by well-posed anisotropic diffusion scheme with channel coupling. Int. J. Remote Sens. 31(8), 2091–2099 (2010)

A Phase Unwrapping Algorithm
for Interferometric Phase Images

Andrey Sosnovsky[✉]

Ural Federal University, pr. Mira, 19, Yekaterinburg 620002, Russian Federation
sav83@e1.ru

Abstract. Phase unwrapping is the most complicated and unreliable stage of interferometric data processing, which is often used in remote sensing techniques. For real radar scenes, the phase unwrapping problem doesn't have a unique solution due to phase discontinuity caused by the phase noise and aliasing. A phase unwrapping algorithm for interferometric phase images based on 3d-phase function branch merging and cutting is proposed. It improves reliability of the unwrapped phase used for digital elevation models generation.

Keywords: Synthetic aperture radar images · InSAR systems · Phase unwrapping algorithms

1 Introduction

Nowadays, the digital elevation models (DEM) generation and the relief motion detection become the main problems, which are solved on the basis of contemporary synthetic aperture radar (SAR) remote sensing systems. Such systems are so-called synthetic aperture radar interferometers or InSAR [1–3].

One of the key problems of interferometric radar data processing is the problem of phase unwrapping, *i.e.* conversion of a relative phase which is wrapped into $[-\pi, \pi]$-interval to an absolute (unwrapped) phase. Unsatisfactory solution of this problem can undermine all advantages of the interferometric method of the remote sensing of the Earth [3–6]. This work is devoted to developing of the new algorithm for the phase unwrapping. The algorithm improves the accuracy of the absolute phase by reducing the influence of phase residues, which cause the phase gaps and other artifacts after unwrapping.

The main approach to the phase unwrapping problem is integration of the phase gradient [3,4]. However, due to presence of the phase gaps caused by topography features, the phase noise and aliasing the integration results depend on the path of integration. So, it leads to occurrence of the unwrapping artifacts and to the absolute (restored) phase corruption.

2 An Unwrapping Algorithm in 3D-space

To eliminate influence of the result from the path of integration, the following method of phase unwrapping for SAR interferograms may be proposed.

© Springer International Publishing Switzerland 2015
M.Y. Khachay et al. (Eds.): AIST 2015, CCIS 542, pp. 146–150, 2015.
DOI: 10.1007/978-3-319-26123-2_14

Now construct a 3d-phase image by the following law: $\Phi_{m,n,k} = \psi_{m,n} \cdot e^{2\pi k/N}$, where $\Phi_{m,n,k}$ is a three-dimensional relative phase interferogram, N is the number of phase samples per one phase turnaround, $k = \overline{0..Nr}$, r is the minimal absolute phase range (number of turnaround), $\Phi_{m,n,k}$ is a three-dimensional discrete function (3d-image) that is periodic along the coordinate k with period N. Let's select an isosurface of this function by any value ϕ_0 from the interval $[-\pi, \pi]$ ($\phi_0 = 0$, for example). This isosurface may be interpreted as a binary periodic function $\tilde{\Phi}_{m,n,k}$ (binary image) and, at the same time, as a multi-valued discrete function of two variables $\Psi_{m,n}^{(L)}$ where L is the index for its unique branch $\Psi_{m,n}^{(L)} = \Phi_{m,n,k}|_{\phi_0=0}$ (see Fig. 1a). Under the phase gaps absence of in the interferogram, $\Psi_{m,n}^{(L)}$ branches wouldn't have self-intersections, and any of them, $\Psi_{m,n}^{(0)}$ for example, coincides with the desired absolute phase $\psi_{m,n}$. The effect of quantization is easily removed by adding the differential phase

$$\phi_{m,n}^{\delta} = arg\{e^{j[\Psi_{m,n}^{(0)} - \phi_{m,n}]}\}, \tag{1}$$

to quantized phase $\Psi_{m,n}^{(0)}$. So:

$$\hat{\psi}_{m,n} = \Psi_{m,n}^{(0)} + \phi_{m,n}^{\delta}, \tag{2}$$

where $\hat{\psi}_{m,n}$ is an interferogram absolute phase estimate.

Another way to generate $\Psi_{m,n}^{(0)}$ is to create a point cloud where the point coordinates (m, n, p) are determined as follows:

$$\{m, n, p\} = \{m, n, \phi_{m,n} + 2\pi k\}, \tag{3}$$

where $k = \overline{0..r}$ (Fig. 1b). Both methods are equivalent, but the post-processing algorithms may differ.

Phase gaps lead to self-intersections of $\Psi_{m,n}^{(L)}$ resulting in appearance of the "jumpers" between adjacent unambiguous branches, while the periodicity of the $\Psi_{m,n}^{(L)}$ is completely retained. Presence of the jumpers expectantly complicates allocation of unambiguous $\Psi_{m,n}^{(L)}$ branches.

At least three approaches to elimination of the jumpers and, therefore, unambiguous branches allocation may be offered:

1. Manual jumpers cutting, which can be realized in any 3d-editor. Unlikely 2d-unwrapping (using Goldstein residue cut algorithm, for example), a 3d-image is usually more human understandable.
2. Installation of the protection zones around the gaps, which sizes are chosen such that the gaps were fully covered by these zones. Also, exclusion of the areas with high phase spatial gradient may be useful here.
3. Creation of a separating surface (hyperplane) between branches using pattern recognition algorithms or the network ones (for the point cloud).

It should be noted that the problem of unambiguous branches allocation for the $\Psi_{m,n}^{(L)}$ not always takes place even if the gaps occur. If residues are closely allocated and in some other cases, the $\Psi_{m,n}^{(L)}$ branches may initially occur unambiguous.

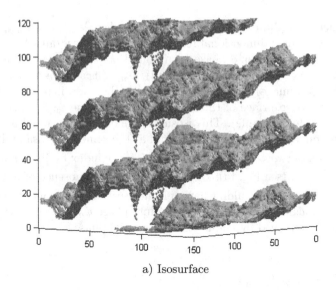

a) Isosurface

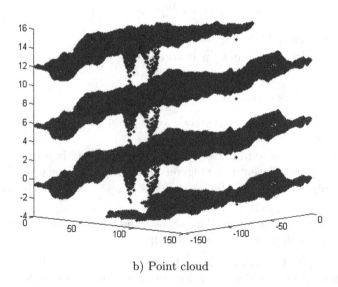

b) Point cloud

Fig. 1. A 3D phase image of 2D SAR interferogram

3   Experimental Results

Now test the method on the space-based SAR interferogram (Fig. 2a) obtained in the L-band (ALOS PALSAR, 1.2 GHz, the spatial resolution is about 7 m). An area under interferogram has significant relief difference (an open-pit in the center of the scene) and large forest areas. Unwrapping with the traditional Region Growing algorithm, which is one of the most suitable unwrapping algorithms for

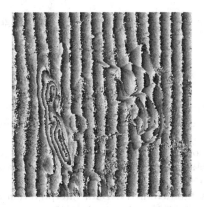

a) An interferogram

b) Region Growing algorithm   c) proposed algorithm

Fig. 2. An interferogram (a) and the interferogram processing results; Region Growing algorithm (b) and the proposed one (c)

different interferogram types [4,5], leaves significant artifacts easily detectable visually (Fig. 2b). Processing with the proposed algorithm doesn't leave visible artifacts (Fig. 2c). Thus, the proposed algorithm is better coped with the task. A quantitative accuracy assessment for the scene is not available because of poor reference DEM for this territory.

4   Conclusion

The phase unwrapping method for SAR interferogams is proposed which assumes conversion of two-dimensional interferograms into three-dimensional space with following analysis of the isosurface or point cloud. The algorithm is tested on the L-band SAR interferogram and showed better results compared with ones of the traditional phase unwrapping algorithm "Region Growing".

Acknowledgment. The work was supported by the RFBR grants Nos. 13-07-12168, 13-07-00785.

References

1. Neronskij, L.B., Kobernichenko, V.G., Zraenko, S.M.: Digital representation of the land surface radar images for the Almaz-1 SAR. Issledovanie Zemli iz Kosmosa **4**, 33–43 (1993)
2. Elizavetin, I.V., Ksenofontov, E.A.: Investigation results of precision evaluation Earth surface relief possibility using SAR interferometry data. Issledovanie Zemli iz Kosmosa **1**, 75–90 (1996)
3. Hanssen, R.F.: Radar Interferometry: Data Interpretation and Error Analysis, p. 328. Kluwer Academic Publishers, Dordrechrt (2001)
4. Shuvalov, R.I.: Matematicheskoe modelirovanie fazovogo gradienta dlja zadachi razvjortki fazy v kosmicheskoj radiolokacionnoj topograficheskoj interferometrii: dis.kand. tekhn. nauk [Mathematical modeling of phase gradient for the task of phase unwrapping in space-basrd radar topographic interferometry], p. 207. Moscow (2011). (In Russian)
5. Kobernichenko, V.G., Sosnovsky, A.V.: Particular qualities of digital elevation maps generation in interferometric SAR technology. SPIIRAS Proc. **5**(28), 194–208 (2013)
6. Sosnovsky, A.V.: A phase unwrapping algorithm for InSAR data processing. In: 2014 24th International Crimean Conference Microwave and Telecommunication Technology (CriMiCo-2014), Sevastopol, Crimea, Russia, 7–13 September, vol. 24, pp. 1155–1156 (2014)

Robust Image Watermarking on Triangle Grid of Feature Points

Alexander Verichev[1,2] and Victor Fedoseev[1,2(✉)]

[1] Samara State Aerospace University, Samara, Russia
{alexanderverichev, vicanfed}@gmail.com
[2] Image Processing Systems Institute, RAS, Samara, Russia

Abstract. The paper presents a digital image watermarking technique robust to geometric distortions. This technique is based on a novel procedure of building a set of primitives for embedding using Delaunay triangulation on a set of feature points, and uses additive embedding method with linear correlation detector. Much attention is paid to the problem of choosing the most appropriate feature points detector. Conducted experiments demonstrate robustness of the proposed method to a range of geometric distortions.

Keywords: Digital watermarking · Robust watermarking · Geometric distortions · Delaunay triangulation · Feature points · Harris detector · SIFT

1 Introduction

One of the most advanced techniques for copyright protection of digital images is robust watermarking. The major issue concerning this technique is achieving robustness to various geometrical distortions of an image being protected. There are two commonly used methods to solve this problem. The first approach suggests embedding a watermark into coefficients of an invariant to geometrical distortions transform. The most widely used is Fourier-Mellin transform [1, 2], since it is invariant to the most common distortions: rotation, scale and translation, usually abbreviated as RST. Such methods are described in [3–5]. The second approach uses so-called *feature*, or *interest points* robust to geometrical transformations, which define regions for embedding in spatial domain. In this paper we call these regions *primitives for embedding*. The second approach potentially allows constructing a watermarking system robust to a wider range of geometrical distortions than RST.

Bas et al. in [6] proposed a rather promising method which uses feature points. In their algorithm, a set of feature points is used to build a Delaunay triangulation and a watermark is embedded in each of the triangles independently. It is possible to identify locations of the watermark primitives at the extraction stage with high accuracy, provided the image didn't undergo any non-affine transformation and a large number of the source feature points were detected.

However, the suggested system has several drawbacks. First and foremost, embedding in triangles demands usage of a triangular-shaped reference watermark pattern, forcing a stair-step border of the reference pattern due to discrete nature of

M.Y. Khachay et al. (Eds.): AIST 2015, CCIS 542, pp. 151–159, 2015.
DOI: 10.1007/978-3-319-26123-2_15

digital images as well as leading to potential problems with the mapping of the pattern to a particular triangle in the image. Moreover, Harris corner detector used in [6] to find a set of feature points might not be the most suitable for the task.

This paper describes a digital watermarking system that achieves robustness to geometrical distortions by means of feature points. Following the same lines of Bas' method, we only left the triangulation step unchanged and enhanced all the others. In Sect. 2 we describe the procedure of choosing the most appropriate feature points detector and present the corresponding results. In Sect. 3 we propose a method of building a set of primitives for embedding as well as a procedure of forming a signal to be embedded. Section 4 describes embedding and detection procedures. Finally, in Sect. 5 we show some results of experimental research of the developed system.

2   Feature Points Extraction

In our search for the most suitable detector for the watermarking system we investigated three feature point detectors: Achard-Rouquet et al. [7], Harris and Stephens [8] and Scale Invariant Feature Transform (SIFT) [9]. To do this we calculated the performance quality measure for each of them using the following steps:

1. Searching for a set of feature points P_{orig} on the source image.
2. Altering the image by i^{th} transformation (listed below).
3. Searching for a set of feature points $P_{dist,i}$ on the altered image.
4. Building Delaunay triangulation $Tri_{dist,i}$ using the set $P_{dist,i}$.
5. Calculating $S_{dist,i}$:

$$S_{dist,i} = \frac{TP_i}{N_i},\qquad(1)$$

where TP_i denotes the number of triangles in $Tri_{dist,i}$ such that all of their vertices belong to the set P_{orig}, N_i is the number of triangles in $Tri_{dist,i}$.
6. Obtaining the overall quality measure S:

$$S = \frac{1}{M}\sum_i S_{dist,i},\qquad(2)$$

where M is the number of applied distortions.

The distortions used in the experiment:

- Rotation (angles ranging from 20° to 220° with step 20°) followed by the subsequent inverse rotation by the same angle.
- Scaling (factors ranging from 0.9 down to 0.5 with step 0.1).
- JPEG compression (ratios ranging from 90 down to 50 with step 10).
- Noise addition (a uniform noise with absolute peak values ranging from 50 down to 10 with step 10 for 256 grayscale image levels).

For the present and all the following experiments we used seven images from the well-known Waterloo test set [10] (cf. Fig. 1). The results are presented in Fig. 2. It is clear that SIFT detector in most cases shows superior performance compared to the other detectors on all test images, which suggests that the most appropriate detector is SIFT. Consequently, we decided to choose it for our system.

Fig. 1. Test images: barb, boat, bridge, goldhill, lena, mandrill, peppers

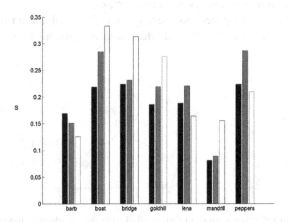

Fig. 2. Values of quality measure S for three feature point detectors on test images. Grey bar denote Harris detector, black – Achard-Rouquet, white – SIFT

3   Building a Set of Primitives for Embedding

Having found a set of highly repeatable feature points, we can build areas for embedding a watermark pattern, called *primitives for embedding*. First, a Delaunay triangulation is built over the set of the feature points. It has several valuable properties: it is uniquely defined for a given set of points, it maximizes the sum of the minimum angles of triangles hence it consists of triangles that are close to equilateral [11]. In addition, there exist efficient algorithms for building Delaunay triangulation.

We could use the triangles as possible primitives but, unfortunately, there are problems associated with the mapping of a triangular-shaped reference pattern to a particular triangle. Specifically, consider one possible approach suggested in [6]. A reference mark has the shape of a right-angled triangle; the mapping is done based on the angles of the two triangles such that the right angle of the reference pattern is mapped to the largest angle of the extracted triangle (as illustrated in Fig. 3). However, such an approach may lead to numerous errors while working with triangles which are close to equilateral (recall that Delaunay triangulation tends to build such triangles). In addition, we have to deal with the already mentioned problem of approximating a stair-step border of a reference triangle.

To alleviate these issues we decided to use quadrangles as primitives for embedding. To obtain the new primitives every triangle is split into three quadrangles by the point of intersection of the medians (as illustrated in Fig. 4). In that case the reference pattern is a matrix (square or rectangular) of values ±1. To map it to an arbitrary quadrangle, the quadrangle is first split into regular grid with the number of rows and columns matching the number of rows and columns of the reference matrix. Then each element of the reference matrix is mapped to the corresponding cell of the grid (cf. Fig. 5). This way two of the four vertices can always be identified as being a feature point and the point of intersection of the medians (points A and D in Fig. 4, respectively), so the mapping can always be defined uniquely.

Mapping the reference pattern to every quadrangle we obtain an image $w(n_1, n_2)$ that consists of the values ±1 or 0 and which size is equal to the size of the image

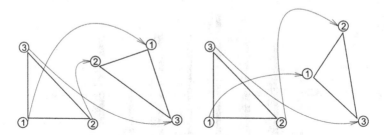

Fig. 3. Mapping of reference pattern (a right-angled triangle) to arbitrary triangle, as suggested in [6]

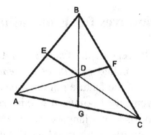

Fig. 4. Splitting of triangle ABC into three quadrangles: AEDG, BEDF, CGDF

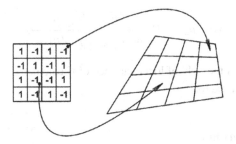

Fig. 5. Splitting of a quadrangle into regular grid and mapping reference pattern (a matrix of ±1) to the cells of the grid

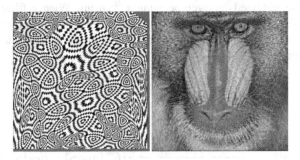

Fig. 6. Example of embedded signal $w(n_1, n_2)$ and result of embedding in mandrill test image. Black, grey and white pixels denote -1, 0 and 1

being processed. The left of the Fig. 6 illustrates an example of $w(n_1, n_2)$ for a reference pattern of the size 8×8. We call $w(n_1, n_2)$ *the embedded signal* since it is the entity that we actually embed in the image.

4 Embedding and Detection Algorithms

4.1 Embedding Algorithm

The embedding begins with forming an embedded signal $w(n_1, n_2)$. A reference pattern is mapped to every quadrangle; therefore the procedure is redundant which results in an increase of the system's robustness. The embedding of the obtained embedded signal is done according to the additive model:

$$C^W(n_1, n_2) = C(n_1, n_2) + \alpha(n_1, n_2) * w(n_1, n_2), \tag{3}$$

where $C(n_1, n_2)$ is the source image, $C^W(n_1, n_2)$ is the watermarked image, $w(n_1, n_2)$ is the embedded signal, and $\alpha(n_1, n_2)$ denotes coefficients that shape the embedded signal according to the characteristics of the human visual system. In our research we used the following mask proposed by Piva et al. [12]:

$$\alpha(n_1, n_2) = D \cdot \frac{\sigma^2_{C,R \times R}(n_1, n_2)}{\max_{n_1, n_2} \sigma^2_{C,R \times R}(n_1, n_2)}, \tag{4}$$

where $\sigma^2_{C,R \times R}(n_1, n_2)$ is the field of local variance of C calculated in the square window $R \times R$; and $D > 0$ is a factor.

4.2   Detection Algorithm

Our goal is to find out whether or not a given image C^W has a watermark W embedded. First of all, we find feature points, build a grid of triangles and split each of them into quadrangles Q_i. Then, using the set of Q_i the embedded signal $w(n_1, n_2)$ is formed. Better results are obtained by preprocessing C^W using Wiener filter. To do this, masking coefficients $\hat{\alpha}(n_1, n_2)$ are estimated using the image C^W in the same way the coefficients $\alpha(n_1, n_2)$ were estimated using the source image C. This allows estimating the modulated embedded signal:

$$\hat{w}(n_1, n_2) = \hat{\alpha}(n_1, n_2) * w(n_1, n_2). \tag{5}$$

Now we can apply Wiener filter as follows:

$$\tilde{C}^W(n_1, n_2) = \frac{\sigma^2_{\hat{w}}(n_1, n_2)}{\sigma^2_{\hat{w}}(n_1, n_2) + \sigma^2_{C^W}(n_1, n_2)} \left(C^W(n_1, n_2) - E_{C^W}(n_1, n_2) \right), \tag{6}$$

where $\sigma^2_{\hat{w}}(n_1, n_2)$ and $\sigma^2_{C^W}(n_1, n_2)$ are the local variances of $\hat{w}$ and C^W, respectively, $E_{C^W}(n_1, n_2)$ is the local mean of the image C^W. All these quantities are calculated in the 9×9 local neighborhood of each point (n_1, n_2). The filtered image $\tilde{C}^W(n_1, n_2)$ is used to calculate correlation values z_i:

$$z_i = \frac{1}{\mu(Q_i)} \sum_{n_1, n_2 \in Q_i} \tilde{C}^W(n_1, n_2) * w_i(n_1, n_2), \tag{7}$$

where $\mu(Q_i)$ is the area of Q_i, and $w_i(n_1, n_2)$ is defined as

$$w_i(n_1, n_2) = \begin{cases} w(n_1, n_2), & (n_1, n_2) \in Q_i, \\ 0 & otherwise, \end{cases} \tag{8}$$

i.e. its elements are equal to the corresponding elements of the embedded signal for pixels that belong to Q_i and are zero elsewhere.

Finally, we decide that the watermark W was embedded in C^W if

$$Z = \frac{1}{N_Q} \Sigma_{i=1}^{N_Q} z_i > \tau, \tag{9}$$

where N_Q is the number of the primitives and τ is an adaptive threshold found as follows. Using C^W we perform detection of a set of different from W watermarks $\{W_j\}$ and applying (8) obtain the correlation values Z_j. Assuming Z_j are normally distributed (evidence of which are given in [6]) we calculate the sufficient statistics m and s^2. Since an event $\{Z_j > \tau\}$ results in wrong detection of the watermark W_j it is effectively a false alarm; choosing a particular acceptable value of $Pr(Z_j > \tau)$ we can find τ from this equation:

$$Pr(Z_i > \tau) = \int_\tau^\infty \frac{1}{\sqrt{2\pi s^2}} e^{-\frac{(x-m)^2}{2s^2}} dx. \tag{10}$$

5 Experimental Research

5.1 Robustness to Misdetection

To assess the robustness to misdetection (when the system can't tell the true watermark from some other and detects the latter instead of the former) we conducted an experiment. First, 100 test watermark patterns were generated and the 40th was embedded into the *barb* test image. Then we calculated the correlation values Z of this image and each of the test patterns using (8). As can be seen in Fig. 7, the correlation value for the true pattern is significantly greater than for any other, so it is possible to effectively distinguish between different watermarks.

5.2 Robustness to Geometrical Distortions

To investigate the robustness of the proposed system to rotation and scaling transformations we conducted another experiment. A watermark was embedded in every test

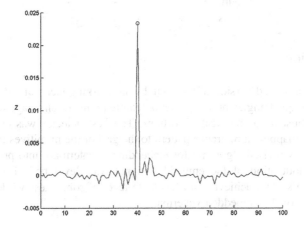

Fig. 7. Correlation values of the extracted watermark with 100 watermark patterns (*barb* image)

image and each of the following transformations were applied: rotation of the water-marked image by the angles ranging from 5° to 40° with step 5°, and scaling with scale factor ranging from 0.5 to 1.7 with step 0.2. Then we performed detection of the embedded watermark.

The results are presented in Tables 1 and 2. They lead us to the conclusion that the proposed system is robust to rotations by rather large angles – up to 20°–25°. As for robustness to scaling operations, it can be considered acceptable – generally a water-mark withstands scaling with factors ranging from 0.9 to 1.5; it worth noting that the distortions introduced by scaling with factors less than one affect the embedded watermark more, since such transformations result in the loss of some information.

Table 1. Results of watermark detection after rotation; + indicates success

	5	10	15	20	25	30	35	40
barb	+	+	+	+	−	−	+	−
boat	+	+	+	+	+	+	+	+
bridge	+	+	+	+	+	+	+	−
goldhill	+	+	+	+	+	+	+	+
lena	+	+	+	+	−	+	−	−
mandrill	+	+	+	+	+	+	+	−
peppers	+	+	+	+	+	−	+	−

Table 2. Results of watermark detection after scaling; + indicates success

	0,5	0,7	0,9	1,1	1,3	1,5	1,7
barb	−	−	+	+	+	+	−
boat	+	+	+	+	+	+	+
bridge	+	+	+	+	+	+	+
goldhill	+	+	+	+	+	−	+
lena	−	−	+	+	+	+	+
mandrill	−	+	+	+	+	+	+
peppers	−	+	+	+	+	+	−

6   Conclusion

In this paper a modified version of the digital watermarking technique [6] is proposed. It uses a grid of quadrangles built on a set of feature points of an image for embedding watermarks robust to geometrical distortions. The best detector was chosen and the new method of mapping a reference pattern to the grid of the primitives for embedding was suggested. A revised algorithm for embedding a watermark into primitives aided by applying modulation of the embedded signal was developed. The experiments showed that the system achieves an acceptable level of robustness while maintaining imperceptibility of the embedded watermark.

Acknowledgements. This work was supported by RFBR (Project 13-01-12080), by the Russian President grant program (Project MK-4506.2015.9), by the state job № 2014/198, and by the RF Ministry of education and science in the framework of the implementation of the Program of increasing the competitiveness of SSAU among the world's leading scientific and educational centers for 2013–2020 years.

References

1. Sheng, Y., Duvernoy, J.: Circular-Fourier–radial-Mellin transform descriptors for pattern recognition. JOSA A **3**, 885–888 (1986)
2. Glumov, N.I., Kuznetsov, A.V.: Detection of local artificial changes in images. Optoelectron. Instrum. Data Process. **47**(3), 207–214 (2011)
3. Zheng, D., Zhao, J., Saddik, A.E.: RST-invariant digital image watermarking based on log-polar mapping and phase correlation. IEEE Trans. Circuits Syst. Video Technol. **13**(8), 753–765 (2003)
4. Glumov, N.I., Mitekin, V.A.: The new blockwise algorithm for large-scale images robust watermarking. In: Proceedings of ICPR-2010, pp. 1453–1456 (2010)
5. Glumov, N.I., Mitekin, V.A.: The algorithm for large-scale images robust watermarking using blockwise processing. Comput. Opt. **35**(3), 368–372 (2011)
6. Bas, P., Chassery, J.-M., Macq, B.: Geometrically invariant watermarking using feature points. IEEE Trans. Image Proc. **11**(9), 1014–1028 (2002)
7. Achard-Rouquet, C., Bigorgne, E., Devars, J.: Un détecteur de points caractéristiques sur des images multispectrales, extension vers un détecteur sub-pixellique. In: Proceedings of GRETSI 1999, pp. 627–630 (1999)
8. Harris, C., Stephens, M.: A combined corner and edge detector. In: Proceedings of Fourth Alvey Vision Conference, pp. 147–151 (1988)
9. Lowe, D.G.: Distinctive image features from scale-invariant keypoints. Int. J. Comput. Vis. **60**(2), 91–110 (2004)
10. Waterloo Grey Set: University of Waterloo Fractal coding and analysis group, Mayer Gregory Image Repository (2009). http://links.uwaterloo.ca/Repository.html
11. De Berg, M., Cheong, O., Van Kreveld, M., Overmars, M.: Computational Geometry: Algorithms and Applications. Springer, Berlin (2008)
12. Piva, A., Barni, M., Bartolini, F., Capellini, V.: DCT-based watermark recovering without resorting to the uncorrupted original image. In: Proceedings of ICIP-1997, vol. 1, pp. 520–523 (1997)

Pattern Recognition and Machine Learning

Traffic Flow Forecasting Algorithm Based on Combination of Adaptive Elementary Predictors

Anton Agafonov[1,2(✉)] and Vladislav Myasnikov[1,2]

[1] Samara State Aerospace University (SSAU), Samara, Russia
ant.agafonov@gmail.com, vmyas@rambler.ru
[2] Image Processing Systems Institute of the Russian Academy of Sciences
(IPSI RAS), Samara, Russia

Abstract. In this paper the problem of traffic flow prediction in the transport network of a large city is considered. For fast calculation of predictions, partition of a transport graph into a certain number of subgraphs based on the territorial principle is proposed. Next, we use a dimension reduction method based on principal components analysis to describe the spatio-temporal distribution of traffic flow condition in subgraphs. A short-term (up to 1 h) traffic flow prediction in each subgraph is calculated by an adaptive linear combination of elementary predictions. In this paper, the elementary predictions are Box-Jenkins time-series models, support vector regression, and the method of potential functions. The proposed traffic prediction algorithm is implemented and tested against the actual travel times over a large road network in Samara, Russia.

Keywords: Transport network · Traffic flow · Traffic flow prediction · Algorithms combination · Potential functions method · Box-Jenkins model · SVR

1 Introduction

Development and wide usage of communication systems, GPS, computer vision systems, active and passive sensors of different types and purposes has led to the possibility of solving complex problems, such as creating "smart cities" [1] and the Intelligent Transportation Systems (ITS) problems [2]. In this paper we consider a short-term road traffic prediction problem, one of many subproblems that must be solved for a full and effective solution of these problems. At present, for example, various online services and/or mobile applications for the road traffic prediction allow road users to analyze traffic situation in the city area and plan their travel.

There are many works dedicated to the short-term traffic flow prediction problem. Some detailed surveys and classifications can be found in the review papers [3–5]. The basic approaches to solving this problem are:

- regression models;
- time series models;

M.Y. Khachay et al. (Eds.): AIST 2015, CCIS 542, pp. 163–174, 2015.
DOI: 10.1007/978-3-319-26123-2_16

- neural networks;
- support vector regression.

Despite a huge number of works in the study area, the proposed solutions are not currently complete and have serious limitations that do not allow using them for the road traffic prediction in a large city directly. The main part of existing works is dedicated to the traffic prediction on separate segments of the road network based on a small neighborhood, as a rule, the intersection. Obviously, this approach ignores the significant information about the state of the road network as a whole. In addition, the segmental approach has high computational and technological complexity for usage in large cities/metropolitan areas.

Additionally, most of the existing models and algorithms do not account for the spatio-temporal data redundancy. Direct confirmation of the spatial redundancy is given in [6], where it is noted that Origin-Destination flows can be accurately modeled with 10 % of independent components or dimensions. The spatial data redundancy has been explicitly or indirectly used in a number of short-term forecasting works [7, 8]. However, any such usage of temporal redundancy is not known to authors.

This work is devoted to the development of the method (mathematical model and its fitting algorithm) for the short-term traffic flow parameters estimation for a road network of a large city. The method is developed in such a way as to eliminate the above-mentioned limitations and to be able to use different algorithms of predicting in its composition.

The next section presents the method and the network description used by the method. Section 3 provides examples on the transport network in Samara. Finally, we present our conclusion and recommendations for further work.

2 Methodology

The proposed method consists of the following steps:

1. Partition of a road network graph into overlapping subgraphs on a territorial basis and formation of a feature vector (vector of traffic flow parameter values) for each subgraph.
2. Reduction of the feature vector dimension for each subgraph by eliminating the space-time dependence of the flow parameter values.
3. Calculation of a set of elementary predictions for each subgraph. The proposed method uses three types of elementary predictions. In the first case, elementary predictions are based on the method of potential functions with a measure of closeness between feature vectors introduced by analogy with the bilateral filtering. It is possible that predictions will not be formed due to the finiteness of the selected kernels. This situation is used on the next step of the proposed method as additional (control) information. In the second case, the elementary prediction based on support vector machines is used. Finally, in the third case the elementary prediction based on the classical scalar and vector time-series models is used. The result of each elementary prediction is the vector of the forecast traffic flow parameters or (for the first case) an indication, that prediction forming is impossible.

4. Aggregation of elementary predictions constructed for each road network subgraph by using an adaptive linear combination of elementary predictions is used. Adaptability is introduced by taking into account additional (control) information arising in the impossibility of forming elementary predictions by the potential functions method. In this case a linear combination of elementary predictions is formed for the reduced set of forecasts.
5. Calculation of the final values of predictive traffic flows throughout the road network as a linear combination of forecasts for all subgraphs. In this case, only those parameters that correspond to the network segments (edges of the graph), which are part of several subgraphs are averaged.

Next subsections present the main steps of the proposed method.

2.1 Notation and Basic Relations

The road network is considered as a directed graph, the edges $w \in \mathbf{W}$ of which correspond to the road links and nodes represent road intersections. The geometric coordinates of the start and end points of the road link $w \in \mathbf{W}$ are denoted as $\left(x_0^w(0), x_1^w(0)\right)$ and $\left(x_0^w(1), x_1^w(1)\right)$. The edge direction determines the direction of the vehicle movement at the appropriate network link. The traffic flow parameter is defined as a function $v : \mathbf{W} \times \mathbf{T} \to \mathbf{R}$ that at a certain time $t \in \mathbf{T}$ for a certain edge $w \in \mathbf{W}$ determines its value $v(w, t)$. There are three main variables to visualize a traffic stream:

- mean travel time;
- density;
- flow.

Further, under the traffic flow parameter we will mean any of these variables. In the experiments presented in the final section we will use the mean travel time.

It is assumed that the input actual and historically available information is presented as a set of parameter values for all road network links:

$$v(w, t), \quad w \in \mathbf{W}, \quad t = t^* - n\Delta \ (n = 0, 1, \ldots),$$

where t^* is the current time.

The output information of the method contains short-term forecast traffic flow parameters values for all road links:

$$v(w, t), \quad w \in \mathbf{W}, \quad t = t^* + n\Delta \ \left(n = \overline{1, N}\right).$$

2.2 Representation of the Network Graph by Feature Vector of Its Subgraphs

We suppose that the method of partition of a road network graph into subgraphs must satisfy a number of difficult to formalize requirements that can be conventionally formulated as follows:

- the method should be regular;
- the road network subgraphs should be connected;
- the subgraphs should be approximately of the same size (by number of edges and the size of the covered area);
- the edges of each subgraphs should be compactly located (i.e. corresponding road links located geographically close);
- the number of edges in the subgraph should not be small, i.e. for real road network graphs it should be from a few dozen to several hundred;
- it should be easy to define a closeness measure between a pair of subgraphs (used in the following section).

Taking into account the above-mentioned requirements, we propose the following method of the graph representation using subgraphs.

We select the number of formed subgraphs $K_0 \cdot K_1$. Subgraph $\mathbf{W}_k$ with the number $k = k_0 K_1 + k_1$ $(k_0 = \overline{0, K_0 - 1}; k_1 = \overline{0, K_1 - 1})$ contains such edges from the set $\mathbf{W}$, the coordinates of at least one vertex of which are in the corresponding rectangular area Π_{k_0,k_1}:

$$\mathbf{W}_{k_0 K_1 + k_1} \equiv \left\{ w \in \mathbf{W} : \quad \bar{x}^w(0) \in \Pi_{k_0,k_1} \vee \bar{x}^w(1) \in \Pi_{k_0,k_1} \right\}$$

where

$$\Pi_{k_0,k_1} \equiv \left[x_0^{min} + \frac{k_0}{K_0} \left(x_0^{max} - x_0^{min} \right), x_0^{min} + \frac{k_0 + 1}{K_0} \left(x_0^{max} - x_0^{min} \right) \right]$$
$$\times \left[x_1^{min} + \frac{k_1}{K1} \left(x_1^{max} - x_1^{min} \right), x_1^{min} + \frac{k_1 + 1}{K_1} \left(x_1^{max} - x_1^{min} \right) \right].$$

and $x_s^{min} = \min\limits_{\substack{\zeta=0,1; \\ w \in \mathbf{W}}} x_s^w(\zeta)$, $x_s^{max} = \max\limits_{\substack{\zeta=0,1; \\ w \in \mathbf{W}}} x_s^w(\zeta)$, $s = 0, 1$.

The number of subgraphs disposed vertically and horizontally K_0, K_1 is selected empirically, but in such a way as to satisfy the number of edges requirement.

For further processing each subgraph $\mathbf{W}_k$ is represented by its description, i.e. a feature vector that describes the traffic flow in the subgraph at a particular time. The feature vector is generated as follows.

Let $\{w_s^k\}_{s=0}^{S^k-1}$ be an edges set of the subgraph $\mathbf{W}_k$ ordered in a certain way (method of ordering is not important), number of edges is S^k. Then the feature vector is used as a descriptor of this subgraph and has the following form:

$$\bar{v}_M^k(t) = (v(w_0, t), \dots, v(w_0, t - M\Delta), \dots, v(w_{S^k-1}, t - M\Delta))^T \tag{1}$$

where $M > 1$ is the number of the historical traffic flow parameter values used in the feature vector for each road link. For convenience, it can be written as the projection operation:

$$\bar{v}_M^k(t) = \bar{v}_M(t)|_{\mathbf{W}_k}, \quad k = \overline{0, K - 1}.$$

2.3 Dimension Reduction of the Subgraph Description with the Use of Spatial and Temporal Redundancy of the Traffic Data

The representation of a road network subgraph by the feature vector (1) has the essential informational redundancy. The proposed method for dimensionality reduction is to transform the original representation (1) to a compact representation with a small number of components obtained using the principal component analysis procedure.

In the context of this work, PCA method consists of the following steps (performed for each subgraph $\mathbf{W}_k$ separately):

1. Calculate the estimate of the covariance matrix of feature vectors $\vec{v}_M^k(t)$ (vectors implementation corresponds $\widetilde{N}$ time points in the past):

$$C^k = \frac{1}{\widetilde{N}} \sum_{n=0}^{\widetilde{N}-1} \left(\vec{v}_M^k(t - \Delta n) - \vec{v}^k\right)\left(\vec{v}_M^k(t - \Delta n) - \vec{v}^k\right)^T$$

where mean vector $\vec{v}^k = \frac{1}{\widetilde{N}} \sum_{n=0}^{\widetilde{N}-1} \vec{v}_M^k(t - \Delta n)$.

2. Calculate the eigenvalues $\lambda_1^k, \ldots, \lambda_{MS^k}^k$ and the eigenvectors of the covariance matrix C^k. The ordering is produced such that $\lambda_1^k \geq \lambda_2^k \geq \ldots \geq \lambda_{MS^k}^k \geq 0$.
3. Calculate the relative error square of the vector representation (1) by a given number of principal components as the ratio of the residual variance for the sample variance:

$$\delta_{k,r}^2 = \frac{\sum_{\ell=r+1}^{MS^k} \lambda_\ell^k}{\sum_{\ell=1}^{MS^k} \lambda_\ell^k}$$

4. The number of principal components R is selected by relative error $\delta_{k,r}$ to provide the required accuracy of representation $\delta_{threshold}^2$.
5. Form a principal components matrix M^k the size $MS^k \times R$ of the covariance matrix of eigenvectors corresponding to first R eigenvalues.
6. Calculate the new feature vector of the reduced dimension as projection of the original feature vector on principal components:

$$\vec{\vartheta}_M^k(t) = \left(M^k\right)^T \vec{v}_M^k(t)$$

The result of applying the dimensionality reduction method is the vector $\vec{\vartheta}_M^k(t)$ of $R(R < MS^k)$ components that describes the current state (with M sampling time in past) of the subgraph $\mathbf{W}_k$ at the specific time moment t.

The resulting description vector $\vec{\vartheta}_M^k(t)$ and the original feature vector $\vec{v}_M^k(t)$ are used to build a set of elementary predictions.

Next subsections describe elementary algorithms for short-term traffic flow prediction.

2.4   Traffic Flow Prediction with the Use of Time-Series Models

Traffic flow prediction using time-series models is usually done by one of the following methods:

- Autoregressive integrated moving average model (ARIMA) and its extension, seasonal ARIMA;
- Vector autoregression moving-average model (VARMA);
- Space-time ARMA (ST-ARMA).

In this paper, for the elementary predictive calculations and the study of their effectiveness we used the vector model VARMA. Prediction result, which is obtained by using the time-series method for all the segments in the given k-th subgraph, is denoted as follows:

$$\bar{v}_0^k(t + n\Delta) = TS(k, M, t, n\Delta)$$

Description of the time-series model order choice and model fitting is presented, for example, in [9].

2.5   Traffic Flow Prediction with the Use of Machine Learning Methods

Among machine learning methods for short-term traffic flow prediction are typically used the following methods and algorithms:

- linear regression;
- support vector regression (SVR);
- potential functions method;
- nearest neighbor algorithm;
- neural networks.

In this paper for making elementary predictions and their effectiveness studies we use the SVR method and the potential functions method. The prediction result obtained by using the SVR method for all the segments in a given k-th subgraph will be denoted as follows:

$$\bar{v}_0^k(t + n\Delta) = SVR(k, M, t, n\Delta)$$

For identifying parameters in SVM, grid-search is used to pick up optimal parameter values.

The potential functions method allows predicting traffic flow parameter values using closeness of descriptive vectors $\bar{\vartheta}_M^k(t)$ at different times. The main relation characterizing the computational procedure for obtaining predictive values for our problem has the following form (the "$\varnothing$" value in the formula given below corresponds to the fact that the result of the prediction is not defined/lacking):

$$\bar{v}_0^k(t+n\Delta) = PF_\sigma(k, M, t, n\Delta)$$

$$\equiv \begin{cases} \dfrac{\sum\limits_{m=0}^{\tilde{N}-1} \bar{v}_0^k(t-m\Delta+n\Delta)R_\sigma\left(\bar{\vartheta}_M^k(t), \bar{\vartheta}_M^k(t-m\Delta)\right)}{\sum\limits_{m=0}^{\tilde{N}-1} R_\sigma\left(\bar{\vartheta}_M^k(t), \bar{\vartheta}_M^k(t-m\Delta)\right)}, & \sum\limits_{m=0}^{\tilde{N}-1} R_\sigma\left(\bar{\vartheta}_M^k(t), \bar{\vartheta}_M^k(t-m\Delta)\right) > ; 0, \\[4ex] \varnothing, & \sum\limits_{m=0}^{\tilde{N}-1} R_\sigma\left(\bar{\vartheta}_M^k(t), \bar{\vartheta}_M^k(t-m\Delta)\right) = 0. \end{cases}$$

$$(2)$$

where $R_\sigma\left(\bar{\vartheta}_M^k(t), \bar{\vartheta}_M^k(t-m\Delta)\right)$ is the kernel of the formed value, which decreases monotonically with increasing divergence between vectors $\bar{\vartheta}_M^k(t), \bar{\vartheta}_M^k(t-m\Delta)$. In this paper we propose to use the following function, which takes into account closeness of descriptive vectors not only in the same subgraph, in which we predict traffic flow parameters, but also in "adjacent" to the k-th H subgraphs:

$$R_\sigma\left(\bar{\vartheta}_M^k(t), \bar{\vartheta}_M^k(t-m\Delta)\right) = \begin{cases} \frac{1}{\sigma\sqrt{2\pi}} \exp\left(-\frac{\rho^2}{2\sigma^2}\right), & \rho \leq 4\sigma, \\ 0, & \rho > 4\sigma. \end{cases} \quad (3)$$

where

$$\rho = \left\| \bar{\vartheta}_M^k(t) - \bar{\vartheta}_M^k(t-m\Delta) \right\|^2 + \frac{1}{H} \sum_{h=0}^{H-1} \left\| \bar{\vartheta}_M^h(t) - \bar{\vartheta}_M^h(t-m\Delta) \right\|^2.$$

The σ parameter determines the maximum distance between descriptive vectors that will be used in the forecast. By varying this parameter, we obtain different predictive values of traffic flow features.

In the chosen research variant the potential functions method with the finite kernel, see (2) and (3), has its specificity: for small values σ parameter a formally obtained result can be marked as "$\varnothing$", i.e. result is not defined. This situation occurs if the current state of the network does not have any close prototype in the historical data. In the proposed solution this situation is used to construct an adaptive computational procedure of elementary predictions aggregation, which is described in the next section.

2.6 Adaptive Linear Combination of Elementary Predictions

An algorithm of elementary predictions as adaptive linear combination is used to calculate the final prediction of traffic flow features for each input subgraph. Adaptability of combination is introduced by analyzing facts of occurrence undefined prediction values ("Ø") in a set of prediction algorithms using the potential functions method considered above. The proposed adaptive combination method is described more formally below.

Let $\{\sigma_q\}_{q=0}^{Q-1}$ $(\sigma_q \in \mathbf{R}_+)$ be a monotonically decreasing numbers sequence: $\sigma_q > \sigma_{q+1}$ $(q = \overline{0, Q-2})$.

Given the form of the relations, see (2) and (3), the following assertions are obvious:

$$R_{\sigma_q}\left(\bar{\vartheta}_M^k(t), \bar{\vartheta}_M^k(t - m\Delta)\right) = 0 \quad \Rightarrow \quad R_{\sigma_{q+1}}\left(\bar{\vartheta}_M^k(t), \bar{\vartheta}_M^k(t - m\Delta)\right) = 0,$$
$$PF_{\sigma_q}(k, M, t, n\Delta) = \emptyset \quad \Rightarrow \quad PF_{\sigma_{q+1}}(k, M, t, n\Delta) = \emptyset.$$

Then, for Q kernels chosen for this method of potential functions with different $\{\sigma_q\}_{q=0}^{Q-1}$ parameters, there is $(Q+1)$ different situation possible when certain predictions have defined/undefined values.

The idea of adaptive linear combination is to construct independent linear combinations of elementary predictions (i.e. linear regression of elementary predictions) for each of $(Q+1)$ possible situations.

The proposed adaptive combination can be formally presented as follows:

$$v_0^k(t+n\Delta) = \begin{cases} \alpha_0^0 TS(k, M, t, n\Delta) + \alpha_1^0 SVR(k, M, t, n\Delta), \\ \qquad\qquad PF_{\sigma_0}(k, M, t, n\Delta) = \emptyset, \\ \alpha_0^{\tilde{q}} TS(k, M, t, n\Delta) + \alpha_1^{\tilde{q}} SVR(k, M, t, n\Delta) \\ \qquad + \sum_{q=0}^{\tilde{q}-1} \alpha_{q+2}^{\tilde{q}} PF_{\sigma_q}(k, M, t, n\Delta), \\ \qquad \tilde{q} = 1 + \arg \max_{q=0, Q-1}\left(PF_{\sigma_q}(k, M, t, n\Delta) \neq \emptyset\right). \end{cases} \quad (4)$$

In the presented adaptive linear combination, the setting is required for the next set of real coefficients $\left\{\alpha_q^{\tilde{q}}\right\} \begin{array}{l} \tilde{q} = \overline{0, Q}; \\ q = \overline{0, \tilde{q}+1} \end{array}$. Fitting these coefficients is performed independently for each of $(Q+1)$ identification problems by least squares method.

2.7 A Computational Procedure for Calculating Predictions of Traffic Flow Parameters

The final step of the proposed method is the calculation of traffic flow parameters' predictions in the current graph for the entire road network as a linear combination of predictions (4) for subgraphs. This operation turns out to be necessary because some road links may be included in several subgraphs simultaneously and, therefore, have several prediction values. The required prediction value $\hat{v}(w, t)$ for the edge w at the time t is proposed to be calculated by the following formula:

$$\hat{v}(w, t) = \frac{\sum\limits_{k=0}^{K-1} \bar{v}_0^k(t)\big|_w}{\sum\limits_{k=0}^{K-1} \bar{v}_0^k(t)\big\|_w},$$

where $\bar{v}_0^k(t)\big|_w = \begin{cases} v(w_r, t), & w = w_r \in \mathbf{W}_k; \\ 0, & \text{otherwise.} \end{cases}$ is the prediction value for the edge w, obtained for the k-th subgraph:

$\bar{v}_0^k(t)\big\|_w = \begin{cases} 1, & w = w_r \in \mathbf{W}_k; \\ 0, & \text{otherwise.} \end{cases}$ indicates whether an edge w is in the subgraph k.

3 Test Results and Analysis

The proposed traffic prediction algorithm is implemented and tested against the actual travel times over a large road network in Samara, Russia. Travel times are calculated by GPS-data as described in [10].

The road network consists of 3,387 segments. The road network graph is partitioned into subgraphs by area size of 1 square kilometer. On average each subgraph contains 50 edges. The size of feature vectors $M = 6$, the time interval $\Delta = 10$ minutes, i.e. the feature vector contains historical data for the last hour. The prediction of new feature values up to 1 h ahead was issued every 10 min. An experimental investigation has been made on a sample of traffic flow parameters for 12 weekdays.

3.1 Accuracy Measurement

To evaluate the performance of the forecasting method, the differences between the predicted travel times by the adaptive combination method, by elementary algorithms, and the actual travel time are compared. The prediction accuracy is evaluated by computing the mean absolute error (MAE) and the mean absolute percentage error (MAPE) depending on the prediction horizon. The elementary algorithms, described previously, feature support vector regression (SVR), VARMA, the potential functions method with the largest value $\sigma(PF(0))$, defining predictive travel times for all road links, and the potential functions method with the lowest value $\sigma(PF(4))$, defining predictive travel times for about 20 % of road links.

The tests were conducted by cross-validation method; the validation set contains traffic flow parameters in one day. Figure 1 shows the mean absolute error and mean absolute percentage error for different prediction horizon values on a control set.

Our experimental test shows, that the adaptive combination model gives better results in terms of accuracy compared with elementary algorithms for almost the whole range of prediction horizon values. The more detailed accuracy analysis has shown that the biggest error is reached at congested road links. This fact makes relevant the

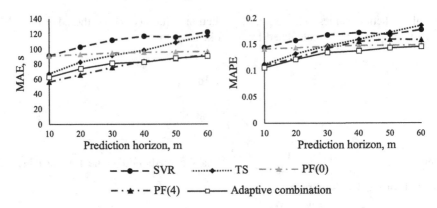

Fig. 1. MAE and MAPE for different prediction horizon values

development of data pre-filtering methods for training and performance estimation, as well as the development of congestion detection methods with a subsequent change in the prediction model.

3.2   Computation Measurement

Since the prediction method must be run continuously, the computation time of the method is a critical consideration. This experimental test shows computation time of the proposed adaptive combination algorithm for a different number of used principal components, providing the necessary accuracy of the original feature vector (Fig. 2). Traffic flow parameters has been predicted for a subgraph with 438 edges and prediction horizon up to 1 h, using laptop computer (Intel Core i5-3740 3.20 GHz, 8 GB RAM).

The detailed analysis of the computation time of particular algorithms shows that most of the time is taken by the traffic flow prediction using the support vector regression method.

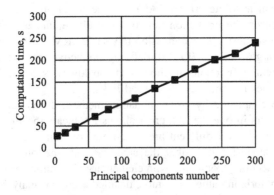

Fig. 2. Computation time for different number of principal components

4   Conclusion and Future Work

In this paper we have proposed a new original method of short-term traffic flow parameters prediction for a large city road network, based on the model of adaptive combination of elementary prediction algorithms.

The proposed method has the following properties that are not inherent jointly to any paper on the topic known to authors:

- It allows to calculate predictions for all road links of the large city network;
- It takes into account the spatial and temporal redundancy of the analyzed traffic data;
- It has adaptability to the analyzed situation in road network. Adaptability means that the parameters of the constructed combination depend of the presence or absence of prototypes for prediction.

In the experimental tests the proposed prediction algorithm has shown the best result compared with elementary algorithms used for short-term traffic predictions: time-series VARMA model, SVR, and the potential functions method.

Direction for further study includes:

- studies related to selection of data pre-filtering methods for training and performance estimation of forecasting methods;
- studies related to development of congestion detection methods with a subsequent change in the prediction algorithm.
- studies related to the analysis of efficiency of certain prediction algorithms not discussed in this paper.

Acknowledgements. This work was supported by the Russian Foundation for Basic Research (RFBR) grant №13-07-12103-ofi-m, grant №13-01-12080-ofi-m, grant №12-07-0021-a and by the Ministry of Education and Science of the Russian Federation in the framework of the implementation of the Program of increasing the competitiveness of SSAU among the world-leading scientific and educational centers for 2013–2020 years.

References

1. Batty, M., Axhausen, K.W., Gianotti, F., Pozdnoukhov, A., Bazzani, M., Wachowicz, M., Ouzounis, G., Portugali, Y.: Smart cities of the future. Eur. Phys. J. Spec. Top. **214**(1), 481–518 (2012)
2. Hall, R.: Handbook of Transportation Science, p. 737. Kluwer Academic Publishers, Dordrecht (2003)
3. Vlahogianni, E.I., Karlaftis, M.G., Golias, J.C.: Short-term traffic forecasting: where we are and where we're going. Transp. Res. Part C Emerg. Technol. **43, Part 1**, 3–19 (2014)
4. Bolshinsky, E., Freidman R.: Traffic flow forecast. Israel Institute of Technology. Technical Report, 15 p. (2012)
5. Faouzi, N.E., Leung, H., Kurian, A.: Data fusion in intelligent transportation systems: progress and challenges. A survey. Inf. Fusion **12**(1), 4–10 (2011)

6. Lakhina, A., Papagiannaki, K., Crovella, M., Diot, C., Kolaczyk, E.D., Taft, N.: Structural analysis of network traffic flows. ACM SIGMETRICS Perform. Eval. Rev. **32**(1), 61–72 (2004)
7. Guorong, G., Yanping, L.: Traffic flow forecasting based on PCA and wavelet neural network. Inf. Sci. Manag. Eng. (ISME). **1**, 158–161 (2010)
8. Jin, X., Zhang, Y., Yao, D.: Simultaneously prediction of network traffic flow based on PCA-SVR. In: Liu, D., Fei, S., Hou, Z., Zhang, H., Sun, C. (eds.) ISNN 2007, Part II. LNCS, vol. 4492, pp. 1022–1031. Springer, Heidelberg (2007)
9. Box, G.E., Jenkins, G.M., Reinsel, G.C.: Time Series Analysis: Forecasting and Control (4th edn.), p. 784. Wiley, New York (2008)
10. Agafonov, A.A., Myasnikov, V.V.: An algorithm for traffic flow parameters estimation and prediction using composition of machine learning methods and time series models. Computer Optics **38**(3), 539–549 (2014)

Analysis of the Adaptive Nature of Collaborative Filtering Techniques in Dynamic Environment

Khaleda Akhter and Sheikh Muhammad Sarwar$^{(\boxtimes)}$

Institute of Information Technology, University of Dhaka, Dhaka, Bangladesh
bit0232@iit.du.ac.bd, smsarwar@du.ac.bd

Abstract. Collaborative filtering (CF) has been an active area of research for a long time. However, most of the works available in the literature either focuses on handling cold start problems (when CF fails to make acceptable prediction due to the lack of ratings) or emphasizes on improving CF performance in terms of some evaluation statistics. Very few of them addressed the problem and issues of updating from a cold start affected initial stage to a steady one. To cope with this progressive nature of CF, we propose to model the entire life cycle of Recommender System (RS). Specifically, we suggest a combination of two neural network based CF techniques for the implementation of a complete RS framework. We propose to adopt the cold start based algorithm proposed by Bobadilla *et al.* for the initial stage. For the later stage we propose a new algorithm based on neural network. We suggest to adopt these two algorithms in different stages of CF to ensure better performance and uniformity throughout the RS life cycle.

Keywords: Collaborative filtering · Recommender system · Similarity measures · Performance

1 Introduction

As one of the most successful approaches of building RS, Collaborative filtering (CF) has been utilized by the on-line community for a long period of time. The basic assumption of CF is that, if two on-line users rate a set of items similarly, or have likewise behaviors (e.g. buying, watching, listening), then they will do the same in the future. The ratings can either be explicit indications, for example a value on a scale from 1 (disliked) to 5 (liked), or implicit indications, such as purchases or click-through rates [1].

On the basis of how CF determines the similarity among users, CF methods can be categorized as either model-based or memory-based algorithms [2]. According to Breese *et al.* [3], memory-based algorithms approaches the collaborative filtering problem by identifying a neighborhood of similar users to make predictions. Our work here focuses more on the model-based technique, which uses the pure rating data to build a statistical or machine learning based "model". This model is later used to make recommendations and thus reduces the efforts to analyze the complete data-set every time.

© Springer International Publishing Switzerland 2015
M.Y. Khachay et al. (Eds.): AIST 2015, CCIS 542, pp. 175–186, 2015.
DOI: 10.1007/978-3-319-26123-2_17

In this competitive world of E-Business it is very important for industry to strength RS to provide fast and accurate recommendations for attracting more customers. But in order to produce high quality predictions, CF has to overcome many real life challenges like cold start problem, scalability issue, dynamic update etc. [4]. Cold start is one of the most common and crucial problems in the real-world recommendation tasks. It describes a situation in which a general CF fails to make acceptable prediction due to the lack of ratings [5]. A new user may have either no or very few ratings. Still such users expect to get recommendation for prospective preferred items. Since CF makes recommendation based on co-rated items [6], it is very difficult to make recommendation to new users. This may cause business to lose new users. As a consequence, there are several studies on cold start problems. Nonetheless, these algorithms are very critical and time consuming, since they have to make recommendation using very few ratings [5].

The entire process of recommendation is a dynamic phenomenon. The more time users spend on the system the more ratings they make [7]. That results in a continuously evolving database. Therefore, after a considerable amount of time, there are enough ratings in the system database to make better recommendation. However, if the system was initially designed to handle cold start situation, now it has to go through an unnecessary complex process to make recommendation. This could lead to slow and inappropriate predictions. When a regular user does not get fast and accurate recommendation from their trusted website, they could easily move on to another one. However, updating an entire system is a complex and costly process. Besides, in future new community may again join the system creating further cold start situations.

Motivated by this problem, we propose to adopt two collaborative filtering techniques in two stages of recommendation system. Our assumption states that designing a recommendation system as a complete cycle from the initial stage would ensure overall better performance. In our work, we suggest a complete recommendation life cycle by combining two separate algorithms to be implemented at two different stages of the RS. These algorithms are chosen because of the resemblance in their methodology and computation process. The algorithm that we suggest to adopt in the cold start situation was proposed by Bobadilla et al. [2] and it is known as the MJD similarity measure (Mean-Jaccard-Differences), which uses the learning ability of neural network to optimize their similarity computation matrix. For the second algorithm we proposed to use a modified version of another neural network based similarity computation method proposed by Mannan et al. [8] called the ANN method. We suggest making the modification in the second algorithm to make it more similar to MJD suitable in the cold start setting. Our assumption is that, their similarity will ensure reducing the update costs we mentioned above.

2   Preliminary Concept

2.1   Problem Specification

In Collaborative filtering system we are given a set of users $U = \{u_1, \ldots, u_m\}$ and a set of items $I = \{i_1, \ldots, i_n\}$. An $m \times n$ matrix is used to represent the ratings given to the item set by the users. If a user $u \in U$ rates an item $i \in I$, then the rating is expressed as $r_{(u,i)}$. For our work, we use a rating scale of $[1,5]$. If $r_{(u,i)} = 0$ then it represents that user i has not rated item j. The average rating habit of any user in the system is an important factor for many CF techniques. We represent the average rating of any user $u_t \in U$ as $\overline{r_t}$.

2.2   Rating Prediction

For any target user t (for whom we make recommendation) we determine a neighborhood N_t, i.e. a group of people who are similar to the target user. In collaborative filtering, to predict the rating $r_{(t,j)}$ for the target user on any item j, the following equation is used:

$$r_{(t,j)} = \overline{r_t} + \frac{\sum_{n \in N_t} sim(n,t) \times \left(r_{(n,j)} - \overline{r_n}\right)}{\sum_{n \in N_t} sim(n,t)}, \text{ where} \tag{1}$$

$\overline{r_t}$ and $\overline{r_n}$ respectively are the averages of ratings made by the given user t and another user n with whom target user have some similarity, $i.e.$ a neighbor of t. $r_{(n,j)}$ is the rating of the neighbor n for item j. In every collaborative filtering techniques the main problem is to define the similarity function $sim(n,t)$ between two users.

2.3   Mean Absolute Error

Mean Absolute Error (MAE) is used to measure the errors made in rating predictions in order to evaluate the performance of CF techniques [8]. Therefore, lower MAE values indicate better performance. MAE of the RS for the set of training users U_{tr} is obtained by the following equation:

$$MAE = \frac{1}{|U_{tr}|} \sum_{u \in U} \frac{\sum_{i \in I_u} |Pr_{(u,i)} - Or_{(u,i)}|}{|I_u|} \tag{2}$$

Here I_u represent the training items rated by the user u. $Pr_{(u,i)}$ and $Or_{(u,i)}$ respectively means the predicted rating and original rating made by user u for item i.

2.4   Coverage

According to [9] coverage can be defined as the measure of the domain of items over which the algorithm can make recommendations. If I is the set of available items and I_p is the set of items for which a prediction can be made, then according to [10] a basic expression for prediction coverage can be given by:

$$Coverage = \frac{|I_p|}{|I|}. \tag{3}$$

3   Background Study

The architecture of collaborative filtering has been an active area of research, for a rather long period of time. However, all these available works are either on handling cold start situations or on improving CF performance in regular situations. In case of cold start situations, the authors of papers [11–13] proposed several notable methods to solve the new user problem. However none of them investigated their methods updating ability from a cold start stage to a normal one. On the contrary, [3,14] are some of the regular algorithms who focused solely on improving the CF performance, but only in the given circumstance, when enough data is available.

The dynamic updating issue of CF was addressed by Yu *et al.* [4]. They proposed to model the user-item matrix as a bipartite graph and constructed a random walk on this graph to calculate the user similarity. They suggested applying Monte Carlo algorithm on random walk to compute the similarity between existing users of the system. They named their random walk process as *preference* propagation. Their algorithm showed better performance in extreme cold start situation and can handle dynamic update, but only for a small number of new users. However, in real life applications new users are constantly added to the system and so performance degradation is likely to happen if their approach is applied.

Paolo *et al.* was the first to introduce the concept of a complete recommendation cycle [7]. They suggested implementing two algorithms: an item-based CF instead of a cold start algorithm technique at the initial stage and an SVD (Singular Value Decomposition) based algorithm for the steady state of the recommender system. But according to [9], item-based algorithms cannot give proper prediction and ignores unpopular items especially in situations where data is sparse. Moreover, they also conducted their experiments on a dataset with binary values to prove their hypothesis. Thus, it is difficult to extrapolate the method's performance in case of systems that use different rating scales like those with a rating range of 1–5 or 1–100.

4   Proposed Methodology

4.1   Hypothesis

CF assumes that two users who have rated items similarly in the past, would do the same in the future [9]. As a result, a good recommendation can be made with very few information. Table 1 gives an example of user-item rating matrix. Here users u_1 and u_2 have somewhat similar rating habits. Hence, recommendation could be made for u_2 on items i_2 and i_5. However, to make it possible, it is necessary that u_1 and u_2 co-rated a significant amount of common items. Therefore, if most users (such as, u_3 and u_4) don't have commonly rated items, then it is difficult for CF to give recommendation [12]. In the literature, this situation is known as a "cold start new user" problem [5].

Table 1. User item rating matrix

User/Item	i_1	i_2	i_3	i_4	i_5	i_6
u_1	5	2	–	–	3	–
u_2	5	–	–	2	–	–
u_3	–	–	–	–	–	3
u_4	–	–	3	–	–	–

Usually most CF systems provide rating opportunity to a small subset of the users before adding a new user to the system. As a result, when new users enter the system they have no or very few ratings and it is very hard to make recommendations for them. In such cases, cold start algorithms are adopted to make reasonable recommendations. However, cold start algorithms are very critical and time consuming since they have to make recommendation using very few ratings. Moreover, recommender system is a dynamic system with a consistently evolving database. The more time users spend on the system, the more ratings they make. Thus, if a system decides to keep using a cold start algorithm with normal database, it could mean unnecessary calculation and low performance. However, updating an entire model is a costly and time consuming process. Since recommendation is an evolving process there might arise a situation when a new community joins the system in future. This may create a new user cold start problem in a steady environment.

Our hypothesis is that the recommender system should be modeled as a life cycle by combining a cold start and a performance-oriented algorithm for ensuring overall better performance. In our work we propose a recommender system life cycle by combining two stages. In the initial stage new users are added to the community. We name that stage as the **Frozen Stage**. After a considerable amount of time, when the database have enough data, we refer to that stage as the **Warm Stage**. For a RS database with approximately 1000 items we define all the above terms mathematically and they are shown below:

Definition 1. *Cold start users:* $U_{cold} = \{u \in U : 0 \le |I_u| \le 10\}$, *where* I_u *is a set of rated items by user* u.

Definition 2. *Normal users:* $U_{warm} = \{u \in U : |I_u| > 10\}$, *where* I_u *is a set of rated items by user* u.

Definition 3. *Frozen Stage: In the Frozen stage maximum, users in the system are new users, i.e.* $|U_{cold}| > |U_{warm}|$.

Definition 4. *Warm stage: In the Warm stage maximum, users in the system are normal users, i.e.* $|U_{warm}| > |U_{cold}|$.

If in the future more users are added to the community, the system could again shift back to its frozen stage to make appropriate recommendation, thus forming a proper life cycle. Figure 1 shows a complete demonstration of our proposed recommendation life cycle.

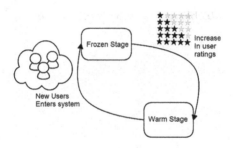

Fig. 1. Recommendation life cycle

4.2   Proposed Recommendation Life Cycle

In our work we propose a recommendation life cycle with two model-based algorithms. Model-based algorithms are the type of algorithms that approaches the collaborative filtering problems by using the database to extract information. Thus, reduces the effort to use the entire database to make recommendation each time. Besides, most of the model-based computations are done off-line, which ensures much faster recommendation. For our recommendation cycle we have chosen two neural network based algorithms.

Artificial neural networks (ANN) is a computational model that was inspired by the operation of a human brain. In our work we propose to choose two algorithms that use a matrix consisting of six individual similarity values to determine the resemblance between two users. We chose these two algorithms because of their similarity in structure and computation process. For the frozen stage we propose to select the process stated by Bobadilla *et al.* [2] for cold start users. For the warm stage we suggest to use the altered version of the algorithm introduced by Mannan *et al.* [8]. We propose to make some alteration in the later process to make it more similar to the frozen stage algorithm.

Set of Similarity Measures. Bobadilla *et al.* [2] in their work suggested that to obtain better recommendation it is not enough to consider only the numerical information of ratings available in CF technique. They suggested using the numerical information that represents similar rating habits of two users and also the non-numerical information based on distribution and number of votes cast by each pair of users. Based on their hypothesis Bobadilla *et al.* [2] suggested to model recommender system using a linear combination of six components and tried to optimize MAE. The first four components V^0, V^1, V^3, V^4 denotes the cases that represent both the similar and opposite rating habit of two users. The fifth component sim_{msd} is one of the traditional similarity measure techniques, which includes the arithmetic average of the squared differences between two users' votes. The last component $sim_{jaccard}$ utilizes the information based on distribution and total number of items voted by both users. The definitions of the components are illustrated below:

$$\{V^0, V^1, V^3, V^4, sim_{msd}, sim_{jaccard}\} \tag{4}$$

Here V^0 is the number of items with the same value in all the pairwise ratings of users x and y, V^1, V^3 and V^4 is the number of items with a difference of ratings 1, 3 and 4 respectively in all the pairwise ratings of users x and y, sim_{msd} is similarity-based on Mean Squared Difference between x and y, $sim_{jaccard}$ is Jaccard similarity measure.

The normalized equation for the aforementioned components is given below:

Let us assume that, the difference of ratings made by two users on any co-rated item is d and $\forall_d \in [0, 4]$

$$I^d_{(x,y)} = \{i \in I | r_{(x,i)} \neq 0 \wedge r_{(y,i)} \neq 0 \wedge |r_{(x,i)} - r_{(y,i)}| = d\} \tag{5}$$

For defining the cases in which user x and user y have rated with a difference of d we get,

$$V^d_{(x,y)} = \frac{|I^d_{(x,y)}|}{\sum_{d=0}^{4} |I^d_{(x,y)}|} \tag{6}$$

Let $I_{(x,y)} = \{i \in I | r_{(x,i)} \neq 0 \wedge r_{(x,i)} \neq 0\}$ be a set of items rated simultaneously by both users. Now the equation for similarity based on Mean Squared Difference is given below:

$$sim_{msd}(x,y) = 1 - \frac{\sum_{i \in I(x,y)} (r_{(x,i)} - r_{(y,i)})^2}{|I_{(x,y)}|} \tag{7}$$

We assume that $I_u = \{i \in I | r_{(u,i)} \neq 0\}$ is the set of items rated by user u. Thus, the Jaccard similarity for user x and user y can be defined as below:

$$sim_{jaccard}(x,y) = \frac{|I_x \cap I_y|}{|I_x \cup I_y|}. \tag{8}$$

Algorithm for Frozen Stage. Bobadilla *et al.* [2] named their similarity measure matrix as MJD metric (Mean-Jaccard-Differences). The proposed metric is formulated as follows:

$$sim_{mjd}(x,y) = \frac{1}{6} w_1 v^0_{(x,y)} + \frac{1}{6} w_2 v^1_{(x,y)} + \frac{1}{6} w_3 v^3_{(x,y)} + \frac{1}{6} w_4 v^4_{(x,y)} \tag{9}$$
$$+ \frac{1}{6} w_5 sim_{msd}(x,y) + \frac{1}{6} w_6 sim_{jaccard}(x,y)$$

Here, each similarity measure is assigned by a weight (w_i) to identify their individual degree of importance. In order to optimize the result [2] proposed to use a model based on artificial neural network; more specifically it is an ADELINE network with the following learning rule:

$$W \leftarrow W + \eta(d - o)X \tag{10}$$

Here η is the learning rate, d is the desired (real number) output and o is the output of the neural network. This learning rule is also referred to as Widrow-Hoff learning rule. Bobadilla *et al.* [2] suggested using a linear activation function adopting this Widrow-Hoff [15] method to make the adjustment of the weights. Their proposed equation is given in Eq. 11:

$$w_i(t+1) = w_i(t) + \alpha \cdot x_i(MAE(t)_{mjd(t)} - MAE(t)_{jmsd(t)}), \text{ where} \quad (11)$$

$i \in [1, 6]$, $x_1 = V^0_{(x,y)}$, $x_2 = V^1_{(x,y)}$, $x_3 = V^3_{(x,y)}$, $x_4 = V^4_{(x,y)}$, $x_5 = sim_{msd(x,y)}$, and $x_6 = sim_{jaccard(x,y)}$.

For training phase neural network [2] uses a set of user pairs (u_c, u_n) as input. Such that, u_c represents a cold-start user and u_n represents any normal user. Here $MAE(t)_{jmsd(t)}$ is an upper bound, which is determined using the traditional similarity measure JMSD [1] but for all users (not only cold-start ones).

Algorithm for Warm Stage. In the warm stage of the RS we adopt the modified solution of Mannan *et al.* [8], who proposed a multilayer feed-forward neural network based solution using a novel similarity function. Their neural network design was composed of a traditional sigmoid function with one input layer, one hidden layer and one output layer. Their input layer required five input values and those can be easily computed using Eq. 6 by varying the value of d from 0 to 4. They also named these five input values as *vector values*. In order to model similarity using ANN, they proposed three hidden nodes in the hidden layer. The desired similarity is calculated using Eq. 12. In order to create an optimal neural network, the system uses error back-propagation algorithm until the expected similarity (minimum error function is set) is achieved from the output node. The design of the similarity function as a multilayer feed-forward neural network (ANN) is shown in Fig. 2a.

$$sim_w(u_x, u_n) = \frac{r_{(x,i)} - \overline{r_x}}{r_{(n,i)} - \overline{r_n}} \quad (12)$$

In order to keep uniformity with the frozen state algorithm, we make some changes in the input of the above stated system. Instead of five factor values, we

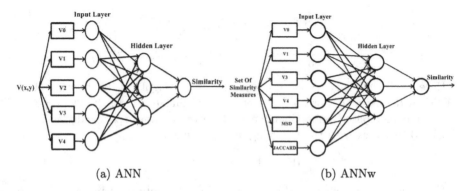

(a) ANN   (b) ANNw

Fig. 2. Design of (a) the original (ANN) and (b) the modified multilayer feed-forward neural network (ANNw)

propose to pass the set of similarity measures previously identified in Eq. 4. In our experimentation we show that changing the input value would not affect the performance of the system. According to [2] the proposed set of similarity values ensures that the system would provide better result. Based on their hypothesis we used their proposed set of similarity measures as the input model of the aforementioned neural network. The reason behind the proposed alteration was to ensure uniformity between the frozen stage and warm stage algorithm and better performance. The design of the improved similarity function as a multilayer feed-forward neural network (ANNw) is shown in Fig. 2b.

5 Experimental Results

5.1 Experimental Configuration

In order to test our hypothesis and evaluate the performance of our altered algorithm, we conduct a series of experiments on the dataset MovieLens. The MovieLens database was jointly developed by GroupLens (GroupLens) and Internet Movie Database (IMDB) Inc. The description of the dataset is listed in the Table 2. In our first experiment, we evaluate the performance of our proposed warm stage algorithm. To this end we divide our dataset with an approximate split ratio of 80:20, i.e. we perform five-fold cross validation. To test our hypothesis on recommendation life cycle we had to create a database with cold start users. However, MovielLens dataset has a substantially large data density. Therefore, to create a frozen stage we randomly removed 5–20 ratings of users who have 20–30 ratings similarly to the paper [2]. Then we analyze the behavior of both the algorithms in frozen and warm stages.

5.2 Experimental Result

In our first experiment, we compare our warm stage algorithm ANNw with the original ANN version [8]. Figure 3a indicates the MAE error obtained for Movielens using ANN, ANNw, Pearson correlation (COR), cosine (COS), Mean Squared Difference (MSD). Figure 3b shows their comparative performance analysis in terms of coverage. ANNw shows slightly better results than ANN both in terms of MAE and coverage for any value of K (the number of neighbors for each user).

Table 2. Movielens data set

Dataset name	Movielens
Users	6040
Movies	3952
Ratings	1,000,209
Min and max values	1–5
Average density	4.9 %

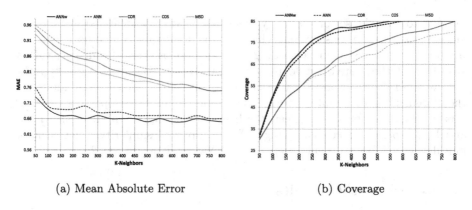

(a) Mean Absolute Error   (b) Coverage

Fig. 3. ANN, traditional metrics and modified ANNw's comparative performance analysis in terms of accuracy (a) MAE (Mean Absolute Error) and (b) coverage using Movielens dataset

Table 3. A comparative analysis of MJD and ANNw in terms of MAE

Stages of recommendation systems	Frozen state	Warm state
MJD	**0.83**	0.85
ANNw	1.04	**0.71**

Thus, it is proved that the proposed algorithm performs better than its ancestor. Next to prove our hypothesis, a comparative analysis of MJD and ANNw is shown in Table 3.

Table 3 shows that frozen stage algorithm MJD performs exceptionally better than the Warm Stage ANNw algorithm in cold start situation, but its performance somehow remains the same even with increased number of ratings. However, ANNw improves its performance drastically with more ratings. Bobadilla *et al.* [2] showed that MJD performs better than traditional CF algorithms like Pearson correlation (COR), cosine (COS) and JMSD in cold start situation using Movielens dataset. From Fig. 3 we can see and infer that ANNw outperforms ANN and many other traditional methods. Thus, it is easy to comprehend that although both the algorithm's performance degrades in other situations, they perform better in their respective domains. This proves our hypothesis that two algorithms should be adopted in two stages of the recommendation life cycle to ensure better overall performance.

6   Conclusion

In this paper we address the challenges and issues caused by the dynamic nature of a recommendation system. In our work, we introduce the concept of recommendation life cycle and define the stages through which the recommender system goes in dynamic setting. Specifically, we define the two stages of a recommender system (frozen stage and warm stage) and propose to adopt two separate algorithms for each of them; we suggested two neural network based CF algorithms for the two stages. For ensuring better performance and maintaining uniformity throughout the system we suggested some changes to be made in the best existing warm stage algorithm. Our results show that the proposed new version of the algorithm performs better than its ancestor. However, several questions still remain unanswered; among them are "When does a system switch from frozen stage to a warm stage?" or "How do we truly distinguish between a frozen stage and a warm stage?". We wish to address these problems by further analyzes of our hypothesis. Our future endeavors would possibly include pairing several memory-based, trust-based or other CF algorithms to form recommendation life cycle and study their behaviors. We also would like to experiment on several other datasets like Epinion, Yahoo! Music dataset etc. to further test our hypothesis.

References

1. Liu, H., Hu, Z., Mian, A., Tian, H., Zhu, X.: A new user similarity model to improve the accuracy of collaborative filtering. Knowl.-Based Syst. **56**, 156–166 (2014)
2. Bobadilla, J., Ortega, F., Hernando, A., Bernal, J.: A collaborative filtering approach to mitigate the new user cold start problem. Knowl.-Based Syst. **26**, 225–238 (2012)
3. Breese, J.S., Heckerman, D., Kadie, C.: Empirical analysis of predictive algorithms for collaborative filtering. In: Proceedings of the Fourteenth Conference on Uncertainty in Artificial Intelligence, pp. 43–52. Morgan Kaufmann Publishers Inc (1998)
4. Rong, Y., Wen, X., Cheng, H.: A monte carlo algorithm for cold start recommendation. In: Proceedings of the 23rd International Conference on World Wide Web, pp. 327–336. International World Wide Web Conferences Steering Committee (2014)
5. Schafer, J.B., Frankowski, D., Herlocker, J., Sen, S.: Collaborative filtering recommender systems. In: Brusilovsky, P., Kobsa, A., Nejdl, W. (eds.) Adaptive Web 2007. LNCS, vol. 4321, pp. 291–324. Springer, Heidelberg (2007)
6. Linden, G., Smith, B., York, J.: Amazon.com recommendations: item-to-item collaborative filtering. IEEE Internet Comput. **7**(1), 76–80 (2003)
7. Cremonesi, P., Turrin, R.: Analysis of cold-start recommendations in iptv systems. In: Proceedings of the Third ACM Conference on Recommender Systems, pp. 233–236. ACM (2009)
8. Mannan, N.B., Sarwar, S.M., Elahi, N.: A new user similarity computation method for collaborative filtering using artificial neural network. In: Mladenov, V., Jayne, C., Iliadis, L. (eds.) EANN 2014. CCIS, vol. 459, pp. 145–154. Springer, Heidelberg (2014)
9. Su, X., Khoshgoftaar, T.M.: A survey of collaborative filtering techniques. Adv. Artif. Intell. **2009**, 4 (2009)
10. Ge, M., Delgado-Battenfeld, C., Jannach, D.: Beyond accuracy: evaluating recommender systems by coverage and serendipity. In: Proceedings of the Fourth ACM Conference on Recommender Systems, pp. 257–260. ACM (2010)
11. Ahn, H.J.: A new similarity measure for collaborative filtering to alleviate the new user cold-starting problem. Inf. Sci. **178**(1), 37–51 (2008)

12. Said, A., Jain, B.J., Albayrak, S.: Analyzing weighting schemes in collaborative filtering: cold start, post cold start and power users. In: Proceedings of the 27th Annual ACM Symposium on Applied Computing, pp. 2035–2040. ACM (2012)
13. Schein, A.I., Popescul, A., Ungar, L.H., Pennock, D.M.: Methods and metrics for cold-start recommendations. In: Proceedings of the 25th Annual International ACM SIGIR Conference on Research and Development in Information Retrieval, pp. 253–260. ACM (2002)
14. Deng, Y., Wu, Z., Tang, C., Si, H., Xiong, H., Chen, Z.: A hybrid movie recommender based on ontology and neural networks. In: Proceedings of the 2010 IEEE/ACM International Conference on Green Computing and Communications and International Conference on Cyber, Physical and Social Computing, pp. 846–851. IEEE Computer Society (2010)
15. Widrow, B., Hoff, M.E.: Adaptive switching circuits. In: IRE WESCON Convention Record Part 4, pp. 96–104. IRE, New York (1960)

A Texture Fuzzy Classifier Based on the Training Set Clustering by a Self-Organizing Neural Network

Sergey Axyonov[1,2(✉)], Kirill Kostin[1], and Dmitry Lykom[1]

[1] National Research Tomsk Polytechnic University, Tomsk, Russia
{axyonov,kak,wedun}@tpu.ru
[2] Tomsk State University of Control Systems and Radio Electronics,
Tomsk, Russia

Abstract. The paper presents a fuzzy approach to the texture classification. According to the classifier the texture class is represented as a set of clusters in N-dimensional feature space that allows generating a cluster or clusters with an arbitrary shape and precisely reflecting any group of the vectors connected with the class. For each texture class it configures the self-organizing features map and estimates a degree of the overlap of the neighboring classes. Upon matching the maps each of them creates a set of fuzzy rules reflecting the feature value statistical distribution in its clusters. Advantages of the system are simplicity of the structure generation, functioning and performance. The suggested classification technique is universal and can be used not only as a texture analyzer but independently for many other real-world classification tasks.

Keywords: Texture classification · Self-organizing features map · Fuzzy inference

1 Introduction

There are some good approaches for image segmentation task based on many different color and texture features (TF) [1–3]. An energy approach based on the Law's texture features [4] showed the best results in a number of different classification problems [5]. According to the approach TFs are extracted by special two-dimensional filters detecting changes in the level, spots, waves, etc. The similarity in TF between image pixel neighborhoods says that they belong to the same texture area. In the given research we focus on the usage of energy approaches for a task of classification. Thus we are not going to obtain a set of regions based on their features but want to demonstrate what types of the texture covered the image. In research it is suggested to use a set of fuzzy sets associated with the different parts of a pattern type. And the first question was how to separate the pattern features to make their analysis simple and demonstrative. For each detected separate pattern features it is required to generate a reliable classifier based on the statistical distribution of TFs inside of the local area. To estimate the classifier quality we decided to use the textures from some bird's-eye images of three tree species (mangrove, pine, and spruce).

© Springer International Publishing Switzerland 2015
M.Y. Khachay et al. (Eds.): AIST 2015, CCIS 542, pp. 187–195, 2015.
DOI: 10.1007/978-3-319-26123-2_18

2   Pattern Representation as a Set of SOM Clusters

All analyzing (training and testing) textures were presented in the same scale and under mid-day light. For each pattern type training images provided 40 80 × 80 pixels patterns and each pattern was processed by 12 5 × 5 Law's filters. For our task we used 12 5 × 5 Law's features for the texture description. The list includes L5L5 (change in level on the ordinate and the abscissa), L5R5 (change in a level on the abscissa and change in ripple on the ordinate), W5L5 (change in a wave on the abscissa and change in level on the ordinate), W5W5 (change in a wave on the both of axis) TFs for three main (red, green and blue) colors [6].

Figure 1 illustrates some texture patterns and their TF distribution for Red L5L5 and Blue W5L5 coordinates. It's seen that a group of vectors for the pattern type isn't represented as a round- or ellipse- shaped cluster. For this reason an association of the pattern type with an ellipsoid shaped cluster is ineffective and allows to get the low classification ability [7, 8]. It is considered that a representation of TF cluster as a set of the interconnected nodes can reflect more accurately the cluster shape and therefore classify correctly an image.

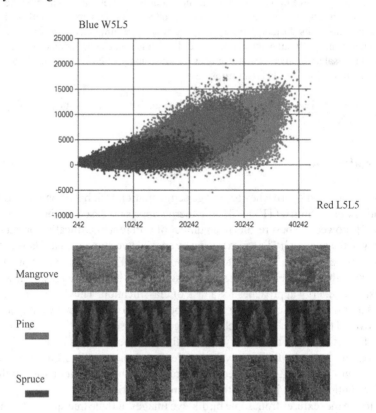

Fig. 1. Patterns and their Red L5L5-Blue W5L5 projections (Color figure online)

Figure 2 shows a classification problem for two classes possessing the objects with two normalized features X1 and X2. The vectors of the first class are shown as orange points and green polygons represent the second class vectors. The curve between polygons and points means possible boundary between classes and it can be obtained by well-known machine learning methods [8, 9, 10, 11] or in case of the space transformation of the initial set we can obtain the linear separable vector groups [12]. The suggested method uses operations with the local vector groups that belong to the same class. Instead of the only set we create some independent sets for each class and for each new set the self-organized features map (SOM) [8] is created. Figure 2 presents an example of the SOM mapping. Here the vectors of two classes were processed by 3 × 3 SOM. The advantage of the map is that the cluster center distribution in the map corresponds to their node physical location inside of the map [13]. Unfortunately not all nodes in SOM have their real clusters (they are known as dead neurons) and they have to be excluded from the further analysis.

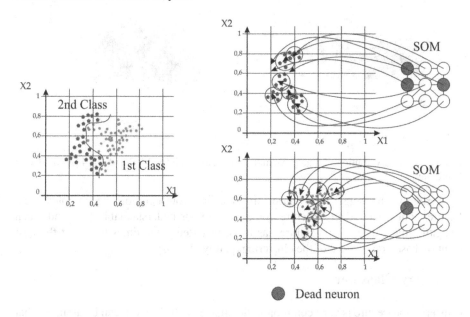

Fig. 2. SOM projecting on the training set containing vectors of two classes

The good partitioning will be found for an arbitrary shape of the vector group even if a group visually consists of some clusters located far from each other. More complex SOM structure (usage of the grid with bigger sizes and 3D grid) allows to reflect complex shape more accurately [14, 15]. In our research we used 3 × 3 and 4 × 4 SOMs to clusterize Law's TF of tree classes. Figure 3 shows Red L5L5-Blue W5L5 and Green L5L5-Blue L5R5 projections for the mangrove and spruce classes respectively. Each cluster has its color and their centers are imaged as black diamonds. Dead neurons were removed from the figure.

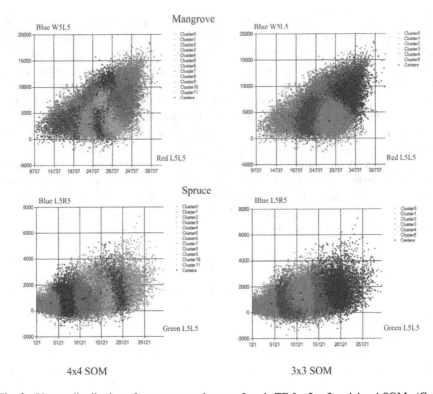

4x4 SOM   3x3 SOM

Fig. 3. Vector distribution of mangrove and spruce Law's TF for 3×3 and 4×4 SOMs (Color figure online)

Certainly, the bigger SOM grid can take into consideration the complex local features, but a redundant structure comes to be poor understandable [16] and when designing the clustering procedure has to be restarted some times to detect the best solution based on indexes of the clustering quality [17, 18].

3 Fuzzy Classifier

Clustering procedure is very common in the image analysis and it can be assumed that only the detected SOM centers can be used for the classification but in fact if the cluster possesses the smallest distance to an input vector it doesn't mean the vector is classified as a type associated with the cluster. Hence, the classification procedure has to be based on the statistical distribution of TFs inside of the clusters belonging to the class.

We suggest to use the fuzzy approach [19, 20] to classify TF of the pixel neighborhood.

In our model the fuzzy sets classifying a texture type are the sets with the normalized distribution function for the vectors belonging to the same cluster. So, for each cluster and for each TF it finds the histogram of TF distribution inside of the cluster. Then all bars in the histogram are normalized by the maximum bar value in the histogram. Now we've got the histograms that can be observed and used as membership functions (MFs) in the fuzzy rules.

Figure 4 illustrates MFs for the fuzzy sets associated for Red L5L5 and Blue W5L5 TF for the Mangrove pattern type.

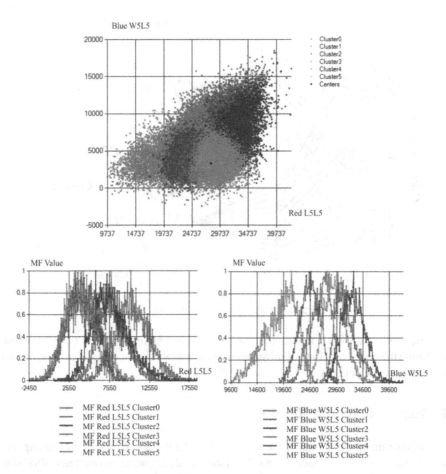

Fig. 4. Membership functions for the fuzzy sets of Red L5L5 and Blue W5L5 TFs for mangrove class (Color figure online)

The fuzzy model uses the product operation in the rule layer, and for aggregation the MAX operation. In case of product operation in the fuzzy rule the threshold detecting a success status of the classification procedure depends on a number of conditions in the rule [21]. So, if we take 12 MFs for each rule and correct class detecting the needs with the least MF value = 0.5 (for each consequent), then the threshold value is $(0.5)^{12}$. If the final output is bigger than the threshold it claims the inference system classifies the TF vector.

Figure 5 shows the view of the fuzzy classifier for a pattern type. The entire texture classifier model includes a set of the classifier for all classes. The output of the classifier is the class ID with the biggest outcome from its local fuzzy inference model.

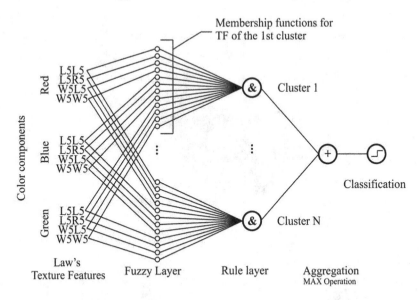

Fig. 5. Fuzzy inference system structure

The system allows to add iteratively the new classes (pattern types) without retraining the previous adjusted subsystems as each new inference model for a new class is computed independently.

4   Experiments and Results

To estimate the performance of the suggested model there were tested all training and testing patterns by fuzzy models. As the system analyzes separate pixel neighborhood after the classification procedure we obtain a 80 × 80 matrix with the class identifier for each neighborhood in the pattern. The class identifier for the entire pattern corresponds to the most occurring class ID in the matrix. Practically the most occurring class ID in the patterns takes more than 80 % from the entire identifier. All testing patterns for mangrove, pine and spruce patterns were recognized correctly.

The second test estimates the model accuracy by analyzing the natural images. Figure 6 shows the outcomes for three different texture types. Here red pixels in the classification mean the system detected texture neighborhood looks like a mangrove pattern, a green color is associated with the pine type and the violet corresponds to the spruce class.

The classification highlights the significant intolerance to the scale. The system can recognize correctly the close and far image parts.

Original Image   Classification

Mangrove forest

Pine wood

Spruce taiga

Fig. 6. Results of classification for some test images

5   Conclusions and Future Work

The obtained results show the high efficiency of the classifier model if the patterns have the close scale texture in the training set. The model doesn't analyze specially the changes in light and zoom in different image parts and these topics are the next stage of our research. In the approach we fix the SOM grid structure and its sizes are not based on the shapes of TF group belonging to the same cluster. The important question is a SOM grid complexity. The future algorithm has to consider the TF cluster shape and offers the best grid for the arbitrary shape description. The quality of classification is also based on boundary clusters between two classes. Probably a more accurate model requires an additional cluster analysis to rectify the boundary by splitting the boundary clusters.

Acknowledgements. The research was partially supported by grant 14-07-31090 MOL_A of the Russian Foundation for Basic Research. The work is performed within the framework of the Program for competitive growth of Tomsk Polytechnic University.

References

1. Acharya, T., Ray, A.K.: Image Processing: Principles and Application. Wiley, New York (2005)
2. Ahn, T.C., Roh, S.B., Yin, Z.L., Kim, Y.S.: Design radial basis function classifier based on polynomial neural network. In: Cho, Y.I., Kim, D., Matson, E.T. (eds.) Soft Computing in Artificial Intelligence, pp. 107–116. Springer, Berlin (2014)
3. Costello, D.P., Kenny, P.A.: Fat segmentation in magnetic resonance images. In: Dougherty, G. (ed.) Medical Image Processing. Techniques and Applications, pp. 89–114. Springer, Berlin (2011)
4. Laws, K.I.: Texture image segmentation. University of Southern California, Technical Report USCIPI No 940 (1979)
5. Palus, H.: Color image segmentation: selected techniques. In: Lukac, R., Plataniotis, K.N. (eds.) Color Image Processing. Methods and Applications, pp. 103–128. CRC Press, Boca Raton (2007)
6. Petrou, M., Sevilla, P.G.: Image Processing Dealing with Texture. Wiley, New York (2006)
7. Joellsen, S.R., Benediktsson, J.A., Sveinsson, J.R.: Random forest classification of remote sensing data. In: Chen, C.H. (ed.) Image Processing for Remote Sensing, pp. 61–78. CRC Press, Boca Raton (2008)
8. Nishii, R., Eguchi, S.: Supervised image classification of multi-spectral images based on statistical machine learning. In: Chen, C.H. (ed.) Image Processing for Remote Sensing, pp. 79–106. CRC Press, Boca Raton (2008)
9. Mitsa, T.: Temporal Data Mining. CRC Press, Boca Raton (2010)
10. Schutt, R., O'Neil, C.: Doing Data Science. O'Relly Media, California (2014)
11. Starck, J.-L., Murtagh, F., Bijaoui, A.: Image Processing and Data Analysis: The Multiscale Approach. Cambridge University Press, Cambridge (2000)
12. Yoshioka, T., Fujinaka, T., Omatu, S.: SAR image classification by support vector machine. In: Chen, C.H. (ed.) Image Processing for Remote Sensing, pp. 341–354. CRC Press, Boca Raton (2008)
13. Dasiopoulou, S., Spyrou, E., Kompatsiaris, Y., Avrithus, Y., Strintzis, M.G.: Semantic processing of color images. In: Lukac, R., Plataniotis, K.N. (eds.) Color Image Processing. Methods and Applications, pp. 259–284. CRC Press, Boca Raton (2007)
14. Gonzalez, R.C., Woods, R.E.: Digital Image Processing. Pearson Education, New York (2005)
15. Han, J., Kamber, M., Pei, J.: Data Mining. Concepts and Techniques, 3rd edn. Morgan Kaufmann, Los Altos (2010)
16. Palmer-Brown, D., Jayne, C.: Self-organization and modal learning: algorithms and applications. In: Bianchini, M., Maggini, M., Jain, L.C. (eds.) Handbook on Neural Information Processing, pp. 379–400. Springer, Berlin (2013)
17. Gevers, T., van de Weijer, J., Stokman, H.: Color feature detection. In: Lukac, R., Plataniotis, K.N. (eds.) Color Image Processing Methods and Applications, pp. 203–224. CRC Press, Boca Raton (2007)
18. Larose, D.T., Larose, C.D.: Discovering Knowledge in Data, 2nd edn. Wiley, New York (2014)

19. Alavala, C.R.: Fuzzy Logic and Neural Networks. New Age International Publishers, New Delhi (2010)
20. Castillejos, H., Ponomaryov, V.: Fuzzy image segmentation algorithms in wavelet domain. In: Dadios, E.P. (ed.) Fuzzy Logic – Algorithms, Techniques and Implementations, pp. 127–146. InTech Publishing, Rijeka (2012)
21. Hoeppner, F., Klawonn, F., Kruse, R., Runkler, T.: Fuzzy Cluster Analysis. Wiley, New York (2000)

Learning Representations in Directed Networks

Oleg U. Ivanov[1]([✉]) and Sergey O. Bartunov[2,3]

[1] Lomonosov Moscow State University, Moscow, Russia
tigvarts@gmail.com
[2] National Research University Higher School of Economics (HSE), Moscow, Russia
sbos@sbos.in
[3] Computational Center of the Russian Academy of Sciences, Moscow, Russia

Abstract. We propose a probabilistic model for learning continuous vector representations of nodes in directed networks. These representations could be used as high quality features describing nodes in a graph and implicitly encoding global network structure. The usefulness of the representations is demonstrated on link prediction and graph visualization tasks. Using representations learned by our method allows to obtain results comparable to state of the art methods on link prediction while requires much less computational resources. We develop an efficient online learning algorithm which makes it possible to learn representations from large and non-stationary graphs. It takes less than a day on a commodity computer to learn high quality vectors on Live-Journal friendship graph consisting of 4.8 million nodes and 68 million links and the reasonable quality of representations can be obtained much faster.

Keywords: Representation learning · Graph embedding · Link prediction · Graph visualization · Social network analysis

1 Introduction

Graph data are ubiquitous: online social networks, web pages connected by hyperlinks, peer-to-peer file sharing networks and even protein interactions are naturally represented as graphs. However, it may be difficult to make predictions about objects organized in a graph assuming that predictions about connected objects are dependent since this may require to analyze the whole network. While it is possible to solve such problems using the framework of probabilistic graphical models [1], it may be non-trivial to derive efficient learning and inference algorithms for a particular model and exact inference is often intractable.

At the same time, machine learning methods that do not consider *structured* (that is, dependent) input or output space such as support vector machines, decision trees or feedforward neural networks became widely used. The independence of predictions for a collection of objects is a common assumption allowing for fast and even parallel learning algorithms.

Since it may be hard to work with raw data, objects are usually represented by a set of real numbers called *features*. There are many ways in which raw data

M.Y. Khachay et al. (Eds.): AIST 2015, CCIS 542, pp. 196–207, 2015.
DOI: 10.1007/978-3-319-26123-2_19

could be transformed into their feature representation suitable for learning algorithm and the choice of representation is crucial for prediction performance. The
classic approach to machine learning involves hand-designed features prepared by
domain experts with prediction problem of interest in mind. This paper address
the problem of automatic feature extraction for nodes in a directed graph which
could be used for independent predictions on a graph in order to overcome the
limitations of non-structured learning algorithms.

In last decade there is growing interest in machine learning community to
representation learning methods which allow to automatically extract useful features from complex objects such as images [2] and sound data [3]. While state of
the art representation learning methods are usually intended to learn distributed
representations with deep neural architectures, recently two lightweight shallow
architectures were proposed for learning distributed word representations named
Skip-gram and Continuous bag of words [4]. Despite much more simple formulations these two models could be trained much faster than recurrent neural
network and outperform it in several tasks involving learned representations.

Motivated by success of these models we propose shallow bilinear model for
learning representations of nodes in a graph. We show that using representations
of relatively small dimensionality (about 30) it is possible to not just to recover
the original graph, but also to accurately predict future links. Since learned low-
dimensional vectors implicitly encode positions of nodes in the global structure
of the network they may serve as high-quality input features in various data
mining tasks.

2 Bilinear Link Model

Consider a directed network $G = (V, E)$, where V is the set of nodes and $E = \{(u_i, v_i)\}_{i=1}^{|E|} \subseteq V \times V$ is the set of links. We denote $d_+(u)$ as the number of
outgoing links for node u and $d_-(u)$ as the number of ingoing links for node u.

The most obvious way to numerically represent node u is to encode its local
connectivity information as binary vector r_u such that $r_{uv} = 1$ if $(u, v) \in E$
and $r_{uv} = 0$ otherwise. We may also consider a twice larger representation that
accounts for incoming links as well. While such naive representation allows to
reconstruct the network without loss of information, it has two important limitations which make it not very useful for a number of applications. First, it
accounts only for *local* information and ignores *global* structure of the network
which is also very important. For example, two nodes may not share any connections at all but still be considered close as belonging to the same structural
component of the network. The second limitation is that length of the representation depends linearly on the number of nodes V which may be not practical
for large networks with hundreds millions of nodes.

Below we describe our method for learning rich representations which over-
comes these limitations by simultaneously compressing the graph and learning global interactions between nodes. We define the probabilistic Bilinear Link

Model (BLM) which explains local connections in the network by latent representations of the nodes. Thus, we associate each node u in the graph with its *input* and *output* representations which we denote as $In_u \in \mathbb{R}^D$ and $Out_u \in \mathbb{R}^D$, where D is the dimensionality of the latent space. Further we denote the set of all representations as $\theta = \{(In_u, Out_u)\}_{u \in V}$.

We assign the probability to each link (u, v) with source node u fixed according to the following bilinear softmax model:

$$p(v|u, \theta) = \frac{\exp(In_u^T Out_v)}{\sum_{w \in V} \exp(In_u^T Out_w)} \tag{1}$$

We maintain two representations for each node to explicitly express the orientation of a link by using corresponding representations.

In many applications it is required to calculate joint link probability. We may express it as $p(u, v|\theta) = p(u)p(v|u, \theta)$ and use maximum likelihood principle to estimate θ and $p(u)$:

$$J(\theta) = \sum_{i=1}^{|E|} \log p(u_i, v_i|\theta) = \sum_{i=1}^{|E|} \log p(v_i|u_i, \theta) + \sum_{i=1}^{|E|} \log p(u_i) \to \max_{\theta, p(u)} \tag{2}$$

We can show that maximum is attained when $p(u) = \frac{d_+(u)}{\sum_{w \in V} d_+(w)} = \frac{d_+(u)}{|E|}$.

Unfortunately both evaluating $p(v|u, \theta)$ and computing the corresponding likelihood gradient requires normalizing over the entire network and thus (2) cannot be optimized efficiently.

The full gradient of (2) can be computed in $O(|V|^2 D)$ time and the $|V|^2$ term makes gradient descent method unacceptable for working with big data. Although we could employ stochastic gradient optimization and follow the direction of stochastic gradient estimated by a single randomly chosen link, each iteration will result into $O(|V|D)$ cost and the epoch (that is, pass over all links) requires $O(|E||V|D)$ time which is unacceptable for training on large graphs.

The most computationally expensive part is connected with normalization (denominator in (1)). Recently, noise contrastive estimation (NCE), a method for training unnormalized probabilistic models has been proposed [5]. The feature of this method is estimation a normalizing parameter $Z_u = \ln \sum_{w \in V} \exp(In_u^T Out_w)$ not any more as a function of θ but as an additional parameter of the model.

Further we show how to optimize the bilinear link model with NCE. We use an unnormalized model

$$p_{NCE}(v|u, \alpha) = \exp(In_u^T Out_v - Z_u)$$

as an estimation of $p(v|u, \theta)$, where $\alpha = \{(In_u, Out_u)\}_{u \in V}$, $Z_u \in \mathbb{R}$. Thus,

$$p_{NCE}(u, v|\alpha) = p(u)p_{NCE}(v|u, \alpha) = p(u) \exp(In_u^T Out_v - Z_u)$$

NCE optimizes the following logistic regression objective which separates objects from true data distribution $p_d(u, v)$ and samples from some *noise distribution* $p_n(u, v)$:

$$L_m(u, v, \alpha) = \ln \frac{p_{NCE}(v_i|u_i, \alpha)}{p_{NCE}(v_i|u_i, \alpha) + \nu p_n(\tilde{v}_i)}, L_n(u, v, \alpha) = \ln \frac{\nu p_n(\tilde{v}_i)}{p_{NCE}(\tilde{v}_i|\tilde{u}_i, \alpha) + \nu p_n(\tilde{v}_i)},$$

$$J_{NCE}(\alpha) = \frac{1}{|E|} \left(\sum_{i=1}^{|E|} L_m(u_i, v_i, \alpha) + \sum_{i=1}^{\nu|E|} L_n(\tilde{u}_i, \tilde{v}_i, \alpha) \right) \to \max_{\alpha} \qquad (3)$$

The choice of $p_n(u, v)$ is almost arbitrary, the most important requirement is that p_n should be nonzero whenever p_d is nonzero. We denote a set of edges generated from p_n as $\{(\tilde{u}_i, \tilde{v}_i)\}$, where $i \in \{1, \ldots, \nu|E|\}$, $\nu \in \mathbb{N}$, so there are ν times more noise samples than true data. In our study we used $p_n(u, v) = p(u)p_n(v)$, where $p_n(v)$ is a noise distribution for fixed source node, $p_n(v) \neq 0$ for all $v \in V$.

It was shown in [5] that optimization problem $J_{NCE}(\alpha) \to \max_{\alpha}$ leads to correct estimation of parameters α. In the limit (when amount of true data goes to infinity) the global maximum of J_{NCE} attains at such α^* that $p(u, v|\alpha^*) = p_d(u, v)$, where p_d is the true distribution (assuming that such α^* actually exists).

We can optimize $J_{NCE}(\alpha)$ (3) using stochastic gradient ascent. This allows us to handle online setting and the method is applicable for non-stationary graphs. On each iteration of the optimization we follow the direction of stochastic gradient estimated by single link. The computation complexity of each iteration is $O(|D|)$, so one epoch requires $O(|E|\nu D)$ time. This is much faster than $O(|E||V|D)$ for original model.

The model may tend to divergence. We can show that for $D = |V|$ optimal value of θ tends to infinity. We approach this problem by introducing regularizer $R(\alpha) = \sum_{u \in V} \left((\nu + 1)d_+(u)||In_u||_2^2 + (\nu + 1)d_-(u)||Out_u||_2^2 \right)$. In this work we use weighted L_2-regularizer, because it has the property of isotropy for In_u and Out_u.

$$J_{NCE,R}(\alpha) = J_{NCE}(\alpha) + \gamma R(\alpha) \qquad (4)$$

This kind of regularizer puts hub nodes closer to the center of latent space. So on one hand the distribution on a target for link with a source fixed in a hub is less peaked. On the other hand a hub becomes a more likely target for all nodes in the network.

Denoting $\{\tilde{v}_{i,j}\}$ as a set of nodes generated from $p_n(v)$. Since $p(\tilde{u}) = p(u)$ we optimize the following function instead of $J_{NCE,R}(\alpha)$ (4) to simplify the implementation.

$$\hat{J}_{NCE,R}(\alpha) = \frac{1}{|E|} \left(\sum_{i=1}^{|E|} \left[L_m(u_i, v_i, \alpha) + \sum_{j=1}^{\nu} L_n(u_i, \tilde{v}_{i,j}, \alpha) \right] \right) + \gamma R(\alpha) \qquad (5)$$

One can see that if $d_+(u) = 0$, then In_u would never be changed during optimization, while Out_v is learned for every v because $p_n(v) \neq 0$. Therefore, Out representations are informative for all nodes, unlike In representations. Thus output vectors are more suitable as node features.

3   Related Work

The classic approach to the problem of transforming nodes in a graph into real vectors involves projection on eigenvectors of graph Laplacian [6]. This is shown to minimize the squared euclidean distance between representations of adjacent nodes. Such technique is also known as graph PCA since it preserves maximum variance in terms of euclidean commute time distance [7]. Despite these interesting theoretical properties spectral projections as feature vectors have certain disadvantages as we show in our experiments on visualization. In particular, while dissimilar nodes have large distances between corresponding spectral projections similar ones are located very close to each other making it hard to distinguish between them.

The important distinction between graph PCA and BLM is that our method is intended to work on directed networks while most results about spectral properties of graph Laplacian were obtained for undirected graphs. There are several developments on analysis of directed graph Laplacians [8,9] but no straightforward way to obtain vector representations using it is available for our best knowledge.

Recently an alternative approach for learning node features based on factorization of adjacency matrix [10] was proposed. It is similar to our method in its bilinear formulation, but in contrast to BLM lacks for probabilistic formulation, i.e. no direct way to obtain link probability is provided. On the other hand, this approach could be straightforwardly adapted to the case of undirected graphs while BLM is not designed for this case.

4   Experiments

4.1   Implementation Details

To train models on huge networks, we use asynchronous parallel stochastic gradient ascent. The model was stored in the shared memory and optimized by several workers concurrently with no synchronization between them. Such approach was recently justified [11] by the fact that stochastic gradient estimates are usually sparse and concurrent updates rarely conflict. Moreover, any synchronization would affect the performance with almost no gain in test performance.

In our experiments we used constant learning rate 0.01, but using advanced learning rate schedules such as AdaGrad [12] should significantly increase speed of convergence and slightly increase the representation quality.

Representation vectors were initialized with small random numbers in a way that $\mathbb{E}In_u^T Out_v = 0$ and $\mathbb{E}Z_u = \ln|V|$. Regularization constant in (4) was set to $\gamma = 0.003$. We used noise distribution $p_n(v) = \frac{d_+(v)+d_-(v)}{2|E|}$ and ratio of noise links $\nu = 25$. We performed 200 epochs of stochastic gradient ascent on each dataset and as one can see in the experiments it was enough to converge.

For experiments we used C++ realization of the above algorithm with data representations stored in std::vectors and pthreads library for parallel computing.

4.2   Graph Visualization

In this section we demonstrate the efficiency of our method for graph visualization task. We use our algorithm for learning 2-dimensional vectors which could

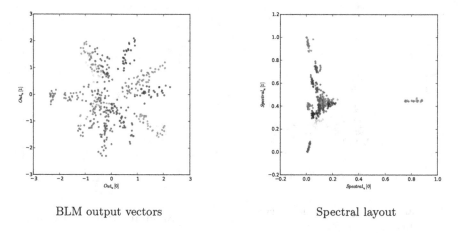

BLM output vectors   Spectral layout

Fig. 1. Visualization of randomly generated scale-free network

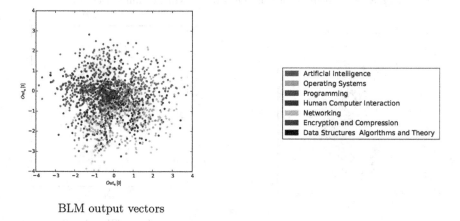

BLM output vectors

Fig. 2. Visualization of Cora citation network communities [13]

be used as node projections. As it was stated above, *Out* vectors are better at representing nodes so further we use them for visualization. We first conduct an experiment on randomly generated scale-free network. We used well known random graph generator [15] which is a standard tool for evaluation of various graph algorithms on synthetic data. We generated a 500 nodes graph with 19 non-intersecting a-priori known communities[1].

We visualized this graph on Fig. 1 using BLM with $D = 2$ and projection on eigenvectors of graph Laplacian (we further refer to this method as *spectral layout*). Spectral layout proceeds only undirected graphs, so directed graphs are considered as undirected graphs for visualization. One can see that our model

---

[1] In this benchmark we used flags "-N 500 -k 15 -maxk 45 -mu 0.2 -t1 2 -t2 1 -minc 5 -maxc 30 -on 0 -om 0".

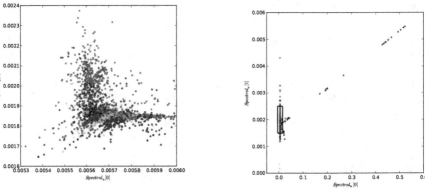

Spectral layout (zoom of the selected area on the right figure)

Spectral layout (all nodes)

Fig. 3. Visualization of Cora citation network communities [13]

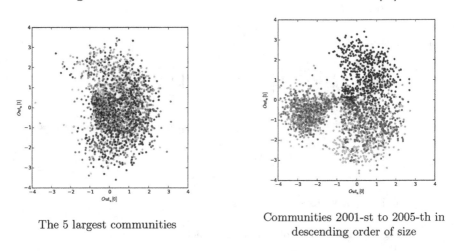

The 5 largest communities

Communities 2001-st to 2005-th in descending order of size

Fig. 4. BLM visualization (not more than 500 random nodes from each community) of LiveJournal [14] communities

is suitable for visualising graphs: on one hand the communities compactness property is satisfied, on the other hand nodes from one community are distant enough to see the difference between them.

On a Fig. 2 we can see our layout and on a Fig. 3 we can see spectral layout for Cora citation network. Each paper in Cora is associated with a topic. Topics themselves are organized in a tree hierarchy. For this experiment we selected the 20 most popular topics. On the Figs. 2 and 3 we joined intersected topics, such as Artificial Intelligence/Machine Learning/Neural Networks and Artificial Intelligence/Vision and Pattern Recognition in one label Artificial Intelligence and one color. For colors with number of nodes exceeding 500, we chose 500 random nodes

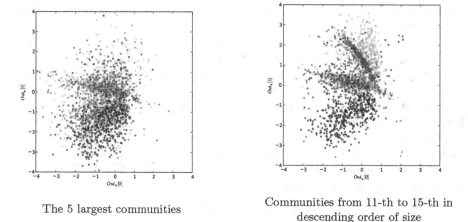

The 5 largest communities

Communities from 11-th to 15-th in descending order of size

Fig. 5. BLM visualization (not more than 500 random nodes from each community) of YouTube [14] communities

of this color to visualize. We can see that in BLM model each topic corresponds to a certain direction (or few directions). The disadvantage of spectral layout is that its vectors do not reasonable describe the distance between two nodes. On the Fig. 3 (note the scale of both axes!) we can see, that distances between pairs of black nodes are unevenly distributed and may be hundreds times more than distance between other topics and their diameters. Thus, spectral layout is hard to be used as reasonable feature vectors for graph data. However, spectral layout may be better for visual community detection if the communities are known.

On Figs. 4 and 5 we visualized the largest communities for LiveJournal and YouTube social networks. We can see, that the visualization of the largest communities is not very informative. The reason is the largest communities are meant to be popular, so they may contain people of very different interests. So on Figs. 4 and 5 one can mention a good separability of not so popular communities.

Above we used Lanczos method for computing spectral layouts but failed to compute eigenvectors of LiveJournal and YouTube graphs in a reasonable time because of large sizes of these graphs. In contrast to Lanczos method our stochastic algorithm has linear complexity in number of edges and could be stopped at any moment while sub-optimal representations would be still useful.

4.3 Link Prediction

Since function (5) is associated with estimation of link probability, it is natural to use obtained representations in link prediction task.

In order to perform evaluation on this task we assume that each link prediction method returns a score for each unordered pair of nodes: the higher value means the more likely a link between them. Introducing a separator turns this method into a classifier, which shows if the probability of a link is higher than random. Area under the curve (AUC) is a standard metric to measure the quality of a classifier.

To assess a link prediction method quality the links of the network are randomly divided into two parts: observed links and missed links. Links not from network are called non-existing links. The observed links are a known for algorithm information, while no information about missed links is allowed to be used in prediction. So the task for a link prediction algorithm is to separate missed links and non-existing links.

Unfortunately, there are nearly $|V|^2$ non-existing links, so it is impossible to compute AUC analytically. To solve this problem, we can choose randomly one missed link, one non-existing link, and compare their scores. The AUC value can be estimated unbiased as follows:

$$AUC = \frac{n_{greater} + 0.5n_{equal}}{n_{total}}$$

n_{total} is a number of such comparisons, $n_{greater}$ is number of comparisons where a missed link has a greater score, n_{equal} is a number of comparisons where both links has an equal score.

On the Fig. 6 we can see the AUC measure estimation for citation cit-HepPh network [16] during the optimization process. Train set plot is AUC for separating observed links from the others. We can see it tends to 1, which means that obtained representations efficiently compress the network structure. Test set plot is AUC for separating missed links from non-existing links. "Fair" Z_u means that on each step for testing we use not a NCE estimation Z_u, but analytically computed $Z_u = \ln \sum_{w \in V} In_u^T Out_w$. We can see on the plot, that NCE estimates Z_u accurately.

Since we used constant learning rate, the convergence of the optimization can be assessed by this plot. Also it is clear that 200 epochs are enough to get high quality representations, but reasonable quality representations can be obtained much faster.

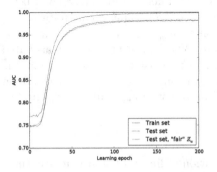 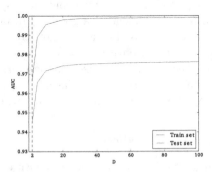

Fig. 6. Dependence of AUC on dimensionality of representations on LiveJournal [16]

The dependence of train set AUC and test set AUC on dimension of representations for LiveJournal social network [16] after BLM learning is shown on

Table 1. Link prediction methods performance. Time was measured for LiveJournal social network [16] on one core of Intel(R) Xeon(R) E5-2670 2.60GHz CPU. Time for BLM does not take into account the training effort.

	BLM	Jaccard	LRW (T steps)	SRW (T steps)
Score function evaluation cost	$O(D)$	$O(\frac{|E|}{|V|})$	$O(|E|T)$	$O(|E|T)$
Parameters for one evaluation	$D = 30$		$T = 3$	$T = 3$
Time for one evaluation, sec	10^{-6}	$4.65 \cdot 10^{-5}$	3.13	3.13

Table 2. Link prediction, AUC

Dataset [16]	BLM(30)	Jaccard	LRW(3)	SRW(3)
soc-LiveJournal	0.975	0.938	0.986	0.985
soc-Pocek	0.978	0.850	0.966	0.967
web-Google	0.961	0.945	0.977	0.978
web-BerkStan	0.979	0.960	0.996	0.996
cit-HepPh	0.983	0.962	0.988	0.989

the Fig. 6. We see that the higher the dimension D, the smaller the increase in AUC measure. We assume $D = 30$ is large enough to capture most interactions in the network.

Baselines. We selected several representative baseline algorithms to compare with according to the survey of link prediction methods [17]. There are two methods for link prediction, namely local and global.

Local similarity indexes are methods that don't use any information about global network structure. They have nearly equal accuracy. We decided to use Jaccard metric for our test, because it is almost the easiest and rapidly computable algorithm to implement. Jaccard similarity metric between two nodes and is a ration of their common neighbours to total number of their neighbours.

Global similarity indexes use information about global network structure. According to the survey [17] random walks are the most accurate technique known for link prediction. The point in these algorithms is computing probability for a random walker, who starts from a fixed node and moves randomly, to appear in another fixed node. We included in the evaluation state of the art random walk methods namely local random walk (LRW) and superposed random walk (SRW) [18].

Results. In this paper we compare AUC measure of Jaccard similarity, LRW with 3 steps, SRW with 3 steps (it is the most popular number of steps according to the survey [17]), and our method with $D = 30$. We selected 5 % of network links randomly as test set and the rest as train set available for algorithms. Results are presented in Table 2. One may see that representations obtained by our method solve link prediction task with a comparable to the state of the art quality. Experiments show that for soc-LiveJournal using $\nu = 10$ in BLM leads to the same results two times faster.

The results on soc-Pocek differs from the results on the other networks. The most methods degrade on this network, especially Jaccard. That can be explained by the sparsity of the network: soc-Poces has average clustering coefficient 0.109, while average clustering coefficients of the other networks are from 0.274 to 0.597. The investigation of the dependence between the quality of link prediction algorithms and the sparsity of the network is a direction for further research, but it is out of scope of this paper.

As one can see in Table 1, LRW and SRW method are not fast enough to make more than 10^5 compares in AUC. So we estimated AUC with $n_{total} = 10^5$ and in the following table we truncated values to the last reliable digit.

5   Discussion and Future Work

In this paper we described a novel scalable method for learning node representations in directed graphs. Initial results showed that our method is useful in graph analysis and allows to learn high-quality dense feature vectors which implicitly compress structure of the graph. Our study is still ongoing and we plan to investigate how representations learned by BLM could be used for various predictions on graphs such as community membership or estimating node attributes e.g. age or interests listed in a social network profile. Another direction of research is to consider other possible formulations of similarity between representations used in softmax equation (1). Finally, it would be interesting to adapt the model to undirected networks.

Acknowledgements. This work was supported by RFBR grant 14-01-31361.

References

1. Koller, D., Friedman, N.: Probabilistic Graphical Models: Principles and Techniques - Adaptive Computation and Machine Learning. The MIT Press, Cambridge (2009)
2. Krizhevsky, A., Hinton, G.E.: Using very deep autoencoders for content-based image retrieval. In: ESANN (2011)
3. Graves, A., Mohamed, A.R., Hinton, G.E.: Speech recognition with deep recurrent neural networks. CoRR abs/1303.5778 (2013)
4. Mikolov, T., Sutskever, I., Chen, K., Corrado, G.S., Dean, J.: Distributed representations of words and phrases and their compositionality. In: Advances in Neural Information Processing Systems 26: 27th Annual Conference on Neural Information Processing Systems 2013, Lake Tahoe, Nevada, USA, 5–8 December 2013, pp. 3111–3119 (2013)
5. Gutmann, M., Hyvärinen, A.: Noise-contrastive estimation of unnormalized statistical models, with applications to natural image statistics. J. Mach. Learn. Res. **13**(1), 307–361 (2012)

6. Chung, F.R.K.: Spectral Graph Theory. CBMS Regional Conference Series in Mathematics, vol. 92. American Mathematical Society (1996)

7. Saerens, M., Fouss, F., Yen, L., Dupont, P.E.: The principal components analysis of a graph, and its relationships to spectral clustering. In: Boulicaut, J.-F., Esposito, F., Giannotti, F., Pedreschi, D. (eds.) ECML 2004. LNCS (LNAI), vol. 3201, pp. 371–383. Springer, Heidelberg (2004)

8. Perrault-Joncas, D.C., Meila, M.: Directed graph embedding: an algorithm based on continuous limits of laplacian-type operators. In: Shawe-Taylor, J., Zemel, R.S., Bartlett, P.L., Pereira, F.C.N., Weinberger, K.Q. (eds.) NIPS, pp. 990–998 (2011)

9. Chung, F.: Laplacians and the cheeger inequality for directed graphs. Ann. Comb. 9, 1–19 (2005)

10. Menon, A.K., Elkan, C.: Link prediction via matrix factorization. In: Gunopulos, D., Hofmann, T., Malerba, D., Vazirgiannis, M. (eds.) ECML PKDD 2011, Part II. LNCS, vol. 6912, pp. 437–452. Springer, Heidelberg (2011)

11. Recht, B., Re, C., Wright, S., Niu, F.: Hogwild: A lock-free approach to parallelizing stochastic gradient descent. In: Shawe-Taylor, J., Zemel, R., Bartlett, P., Pereira, F., Weinberger, K. (eds.) Advances in Neural Information Processing Systems 24, pp. 693–701. Curran Associates, Inc. (2011)

12. Duchi, J., Hazan, E., Singer, Y.: Adaptive subgradient methods for online learning and stochastic optimization. J. Mach. Learn. Res. 12, 2121–2159 (2011)

13. Kunegis, J.: Cora Citation Network Dataset - KONECT, October 2014

14. Mislove, A., Marcon, M., Gummadi, K.P., Druschel, P., Bhattacharjee, B.: Measurement and analysis of online social networks. In: Proceedings of the 5th ACM/Usenix Internet Measurement Conference (IMC 2007), San Diego, CA, October 2007

15. Lancichinetti, A., Fortunato, S., Radicchi, F.: Benchmark graphs for testing community detection algorithms. Phys. Rev. E 78, 046110 (2008)

16. Leskovec, J., Krevl, A.: SNAP datasets: Stanford large network dataset collection, June 2014. http://snap.stanford.edu/data

17. Lü, L., Zhou, T.: Link prediction in complex networks: a survey. Phys. A Stat. Mech. Appl. 390, 1150–1170 (2011)

18. Liu, W., Lü, L.: Link prediction based on local random walk. EPL (Europhys. Lett.) 89(5), 58007 (2010)

Distorted High-Dimensional Binary Patterns Search by Scalar Neural Network Tree

Vladimir Kryzhanovsky and Magomed Malsagov[✉]

Scientific Research Institute for System Analysis, Russian Academy of Sciences, Moscow, Russia
{Vladimir.Krizhanovsky,Magomed.Malsagov}@gmail.com

Abstract. The paper offers an algorithm (SNN-tree) that extends the binary tree search algorithm so that it can deal with distorted input vectors. Perceptrons are the tree nodes. The algorithm features an iterative solution search and stopping criterion. Unlike the SNN-tree algorithm, popular methods (LSH, k-d tree, BBF-tree, spill-tree) stop working as the dimensionality of the space grows ($N > 1000$). In this paper we managed to obtain an estimate of the upper bound on the error probability for SNN-tree algorithm. The proposed algorithm works much faster than exhaustive search (26 times faster at $N = 10000$).

Keywords: Nearest neighbor searching · Perceptron · Search tree · Hierarchical classifier · Multi-class classification

1 Introduction

The paper considers the problem of 1-st nearest neighbor search in a high-dimensional ($N > 1000$) configuration space, where a query point is a distorted version of one of the reference points. The components of the reference vectors take either $+1$ or -1 equiprobably, so on average, vectors are the same distance apart from each other and distributed evenly. We measure the distance between two points with the Hamming distance. In this case popular algorithms become either unreliable or computationally infeasible.

In [1] we investigated the following algorithms: k-dimensional trees (k-d trees) [2], spill-trees [3], LSH (Locality-sensitive Hashing) [4]. We have found that k-d trees for $N > 100$ requires one or two orders of magnitude more computations than exhaustive search (BBF-trees (best bin first) [5] were used). As dimensionality N grows, the error probability of the LSH algorithm approximates one. In the event when the working point coincides with a reference, the spill-tree algorithm works faster than the exhaustive search (by an order of magnitude), but slower than the binary tree by approximately five orders of magnitude. The paper examines the case when the distance between query point and reference one is greater than $0.1\,N$. In these conditions the spill-tree algorithm is slower than the exhaustive search and thus its use makes no sense.

In [1] we offered a tree-like algorithm with perceptrons at tree nodes. Going down the tree is accompanied with the narrowing of the search area. The tree-walk continues until the stop criterion is satisfied. The algorithm works faster than the exhaustive search even when the dimensionality increases (for example, at $N = 2048$ it is 12 times faster).

© Springer International Publishing Switzerland 2015
M.Y. Khachay et al. (Eds): AIST 2015, CCIS 542, pp. 208–217, 2015.
DOI: 10.1007/978-3-319-26123-2_20

In this paper we estimated the upper bound on the error probability of the algorithm. The error probability drops exponentially as the dimensionality of the problem N grows. For example, at $N \geq 500$ the error probability cannot be measured, i.e. the proposed algorithm can be considered exact in this range. Thus, the exact algorithm that excels exhaustive search in speed was obtained.

2 Problem Statement

The algorithm we offer tackles the following problem. Let there be M binary N-dimensional patterns:

$$\mathbf{X}_{\mu} \in R^{N}, x_{\mu i} = \{\pm 1\}, \mu \in [1;M]. \tag{1}$$

In this paper we consider the case when reference vectors are *bipolar* vectors generated in a random fashion. Generated independently of one another, the components of the reference vectors take $+1$ or -1 with equal probability (density coding). We use bipolar vectors rather than binary ones since they are more convenient to use in perceptrons as the mean of their components is zero.

A binary vector $\mathbf{X}$ is an input of the system[1]. It is necessary to find any reference vector $\mathbf{X}_{\mu}$ belonging to a predefined vicinity of input vector $\mathbf{X}$. In mathematical terms the condition looks like:

$$\left| \mathbf{X} \mathbf{X}_{\mu}^{T} \right| \geq \left(1 - 2b_{\max} \right) N, \tag{2}$$

where $b_{\max} \in [0; 0.5)$ is a predefined constant that determines the size of the vicinity. In this problem query point $\mathbf{X}$ is one of the patterns $\mathbf{X}_{\mu}$, bN components of which being flipped randomly, where $0 < b < b_{\max}$.

We will show below that from a statistical point of view the algorithm solves a more complex problem: it can find a pattern closest to an input vector, i.e. the 1-st nearest neighbor. The Hamming distance is used to determine the closeness of vectors. Note, that the Hamming distance plays the same role as dot products for bipolar vectors in contrast with binary vectors.

3 The Point of the Algorithm

The idea of the algorithm is that the search area becomes consecutively smaller. In the beginning the whole set of patterns is divided into two nonoverlapping subsets. A subset that may contain an input vector is picked using the procedure described below. The subset is divided into another two nonoverlapping subsets, and a subset that may contain the input vector is chosen again. The procedure continues until each subset consists of a single pattern. Then the input vector is associated with one of the remaining patterns using the same procedure.

---

[1] Here, $\mathbf{X}$ is a row-vector.

The division of the space into subsets and search for a set containing a particular vector can be quickly done by a simple perceptron with a "winner takes all" decision rule. Each set is controlled by a perceptron trained with the use of patterns of corresponding subset. Each output of the root perceptron points to a tree node of the next level. The perceptron of the descendant node is trained on a subset of patterns corresponding to one output of the root perceptron. The descent down a particular branch of the tree brings us to a pattern that can be regarded as a solution. At each stage of the descent we pick a branch that corresponds to the perceptron output with the highest signal. It is important to note that the same vector $\mathbf{X}$ is passed to each node rather than the result of work of the preceding-node perceptron.

4   The Process of Learning

Each node of the tree is trained independently on its own subset of reference points. A root perceptron of the tree is trained on all M patterns. Each descendant of a root node is trained on $M/2$ patterns. The nodes of the i-th layer are trained on $M/2^{i-1}$ patterns, $i = 1, 2, \ldots k; k = \log_2 M$ is the number of layers in the tree.

All nodes have the same structure – a single-layer perceptron [6] that has N input bipolar neurons and 2 output neurons each of which takes one of the three values $y_i \in \{-1, 0, +1\}, i = 1, 2$.

Let us consider the operation of one node using a root element as an example (all nodes are identical to each other). The Hebb rule is used to train the perceptron:

$$\hat{\mathbf{W}} = \sum_{\mu=1}^{M^*} \mathbf{Y}_\mu^T \mathbf{X}_\mu, \tag{3}$$

where $\hat{\mathbf{W}}$ is a $2 \times N$ -matrix of synaptic coefficients, and $\mathbf{Y}_\mu$ is a two-dimensional vector that defines the required response of the perceptron to the μ -th reference vector $\mathbf{X}_\mu$, M^* - number of patterns for each node ($M^* = M$ for the root node). $\mathbf{Y}_\mu$ may take one of the following combinations: $(-1,0)$, $(+1,0)$, $(0,-1)$, and $(0,+1)$. If the first component of $\mathbf{Y}_\mu$ is nonzero, the reference vector $\mathbf{X}_\mu$ is assigned to the left branch. Otherwise, it is assigned to the right branch. Since the patterns are generated randomly (and therefore distributed evenly), the way they are divided into subsets is not important. So, when training the tree, the set of patterns is always divided into two equal portions corresponding to the left and right branches of the tree, so that the four possible values of $\mathbf{Y}_\mu$ should be distributed evenly among all patterns so that $\sum_{\mu=1}^{M^*} \mathbf{Y}_\mu \to 0$, i.e. the sign of nonzero components of $\mathbf{Y}_\mu$ may be chosen randomly.

The perceptron works in the following way. The signal on output neurons is first calculated:

$$\mathbf{h} = \hat{\mathbf{W}}\mathbf{X}. \tag{4}$$

Then the "winner takes all" criterion is used: a component of vector **h** with the largest absolute value is determined. Since we compare components of **h** between each other in absolute value, we introduced dot product in (2). If it is the first component, the reference vector should be sought for in the left branch, otherwise in the right branch.

The number of operations needed to train the whole tree is

$$\Theta = 2MN \log_2 M. \tag{5}$$

5   The Search Algorithm

Before we start describing the search algorithm, we should introduce a few notions concerning the algorithm.

Pool of Losers. When vector X is presented to a perceptron, it produces certain signals at the outputs. An output that gives the largest signal is regarded as a winner, the others as losers. The pool of losers keeps the amplitude of the output-loser and the location of the corresponding node.

Pool of Responses. After the algorithm comes to a solution (tree leaf), the number of a pattern associated with the leaf and the amplitude of the output signal of a perceptron corresponding to the solution are stored in the pool of responses. So each pattern has its leaf in the tree.

Search Stopping Criterion. If the algorithm comes to a tree leaf and the signal amplitude becomes greater than a threshold value, the search stops. It means that condition (2) holds.

Location of a node is a unique identifier of the node.

Descending the tree is going down from one node to another until the leaf is reached. The branching algorithm is as follows:

1. The input neurons of a perceptron associated with a current tree node are initiated by input vector **X**. Output signals of the perceptron h_L and h_R are calculated.
2. The output with a higher signal and the descendent node related to this output (descendent-winner) are determined. The signal amplitude of the loser output and location of the corresponding descendent-node are stored in the pool of losers.
3. If we reach a tree leaf, we go to step 5, otherwise to step 4.
4. Steps 1 to 4 are repeated for the descendent-winner.
5. The result is put in the pool of responses. At this point the branching algorithm stops.

Now we can formulate our algorithm. Process of descending different tree branches is repeated until the stopping criterion is met. The stages of the algorithm can be described as follows:

1. We descend the tree from the root node to a leaf. The pool of losers and pool of responses are filled in during the process.

2. We check the stopping criterion (2) for the leaf, i.e. we check if the scalar product of vector **X** and the pattern related to the leaf is greater than a predefined threshold. If the criterion is met, we go to step 4, otherwise to step 3.
3. If the criterion fails, we pick a node with the highest signal amplitude from the pool of losers and repeat steps 1 to 3 starting the descend from this node now.
4. We pick a pattern with the highest signal amplitude in the pool of responses, and regard it as a solution.

6   Example of the Algorithm Operation

Let us exemplify the operation of the algorithm. Figure 1 shows a step-by-step illustration of the algorithm for a tree built around eight patterns ($M = 8$). Step 1: the tree root (node 0) receives input vector **X**. The root perceptron generates signals h_L and h_R at its outputs. Let $|h_L| > |h_R|$, then h_R and the location of the descendant-node connected to the right output (node 2) are placed in the pool of losers. Step 2: vector **X** is fed to the node-winner (node 1). A winning node is determined again and the loser is put in the

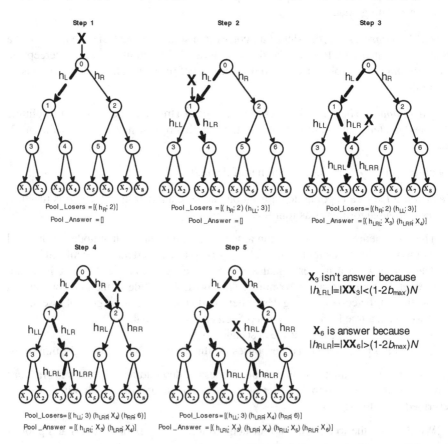

Fig. 1. An example of the algorithm operation.

pool (e.g. h_{LL} and node 3). Step 3: after reaching leaves, we put patterns $(\mathbf{X}_3$ and $\mathbf{X}_4)$ associated with the leaves and signal amplitudes $h_{LRL} = \mathbf{XX}_3$ and $h_{LRR} = \mathbf{XX}_4$ in the pool of responses. Then we find if the patterns meet criterion (2). In our case the criterion is not met, and the algorithm continues. Step 4: if neither pattern gives the solution, we pick the highest-signal node from the pool of losers (e.g. node 2 with signal h_R). Step 5: now the descent starts from this node (node 2) and goes on until we reach the leaves, the pool of losers taking new elements and pair $(h_R; 2)$ leaving the pool of losers. Here $|h_{RLL}| > |h_{RLR}|$ and $(1 - 2b_{max}) N < |h_{RLR}|$, i.e. criterion (2) is true for pattern $\mathbf{X}_6$. The pattern becomes the winner and the algorithm stops. If the criterion never works during the operation of the algorithm, the pattern from the pool of responses that has the highest signal amplitude is regarded as winner.

7   Estimation of the Error Probability

It is hard to obtain a precise estimate of the error probability for the proposed algorithm as for now. However, it is possible to get its upper bound.

SNN-tree algorithm can fail in case when there is more than one pattern that satisfies criterion (2) in the set. Formally, the probability of this event can be written as:

$$P^* = 1 - \Pr\left[\bigcap_{m=1}^{M-1} \left|\mathbf{XX}_m^T\right| < (1 - 2b_{max})N\right]. \tag{6}$$

Presence of such patterns does not always lead to the algorithm failure. Therefore, probability (6) can be used as an estimation of the upper bound on the proposed algorithm failure.

Equation (6) can be calculated exactly by formula:

$$P^* = 1 - \left\{1 - \sum_{k=0}^{b_{max}N} \frac{C_N^k}{2^{N-1}}\right\}^{M-1}. \tag{7}$$

However, it is not possible to use formula (7) at large values of N ($N > 200$). For high dimensions, it is better to use approximation:

$$P^* < \frac{2M}{\sqrt{2\pi\tilde{N}}} \exp\left(-\frac{\tilde{N}}{2}\right), \quad \tilde{N} = N(1 - 2b_{max})^2. \tag{8}$$

Equation (8) shows that the error probability exponentially decreases as the problem dimensionality N grows. For example, at $N = 500$ and $b_{max} = 0.3$ power of the exponent is -40, which explains the fact that experimental error probability for high dimensions could not be measured in work [7]. In fact, SNN-tree can be considered exact for high dimensional problems.

Figure 2 shows dependence of the error probability on dimensionality N at $b_{max} = 0.3$ and $M = N$. As expected, the error probability of the algorithm (markers) is smaller than probabilities calculated using (7) and (8) (solid lines). Therefore, expressions (7) and (8) can be used for algorithm reliability estimation. Moreover, it can be seen that expression (8) is a sufficient approximation of (7).

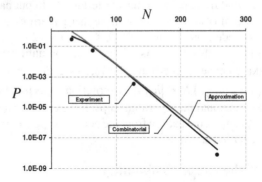

Fig. 2. The algorithm error probability.

8   Estimation of the Computational Complexity

Estimation of the proposed algorithm computational complexity is a quite sophisticated problem that was not solved yet. In this section, results of computational modeling are presented.

It was shown in work [7] that the problem in hand could be solved using only these two algorithms: exhaustive search and SNN-tree. Conducted research shows that the proposed algorithm works faster than exhaustive search, however it errors may occur. According to the results from the previous sections, the error probability at dimensionality $N \geq 500$ is so small that it can be neglected. Therefore, even a small speed advantage of SNN-tree over exhaustive search makes it preferable.

Experiments show (Fig. 3) that as dimensionality N grows the speed advantage of SNN-tree over exhaustive search increases. For example, at $N = M = 2\,000$ and $b = 0.2$ SNN-tree is faster than exhaustive search in 12 times, and at $N = M = 10\,000$ acceleration reaches 26 times. Note b_{max} is a fixed value, but b is a fraction of distorted components

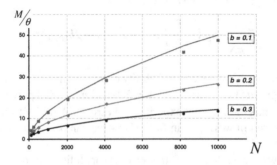

Fig. 3. The speed advantage of SNN-tree over exhaustive search (markers - experiment; solid lines – estimation).

in the input vector. When this fraction b increases the algorithm slows down, since perceptrons at the nodes make errors more often. Using experimental result, we get the empirical estimation of the average number of scalar product operations needed for SNN-tree search:

$$\theta = M \exp\left\{1.3 + b - (0.44 - 0.4b)\log_2 N\right\}. \tag{9}$$

Solid lines in Fig. 3 were built using Eq. (9). This equation allows estimating average speed advantage of SNN-tree over exhaustive search. Using (9), it is possible to predict advantage of SNN-tree at large values of parameters (Table 1).

Table 1. The speed advantage of SNN-tree over exhaustive search for $M = N = 10^5$ and $b_{max} = 0.3$ using Eq. (9).

b	M/θ
0.1	189
0.2	88
0.3	41

9 Conclusion

The paper considers the problem of nearest-neighbor search in a high-dimensional configuration space. The use of most popular methods (k-d tree, spill-tree, BBF-tree, LSH) proved to be inefficient in this case. We offered a tree-like algorithm that solves the given problem (SNN-tree).

In this work, theoretical estimate of the upper bound on the error probability of SNN-tree algorithm was obtained. This estimate shows that the error probability decreases as the dimensionality of the problem grows. Since even at $N > 500$ the error is less than 10^{-15}, it does not seem possible to measure it experimentally. Therefore, it is safe to say that SNN-tree is an exact algorithm. Research investigations of the computational complexity of the algorithm shows that the speed advantage of SNN-tree algorithm over exhaustive search increases as the dimensionality N grows.

So, we can conclude that SNN-tree algorithm represents an efficient alternative to exhaustive search.

The research is supported by the Russian Foundation for Basic Research (grant 12-07-00295 and 13-01-00504).

Appendix A

It is necessary to calculate the following probability:

$$P^* = 1 - \Pr\left[\bigcap_{m=1}^{M-1} \left|\mathbf{X}\mathbf{X}_m^T\right| < (1 - 2b_{max})N\right]. \tag{A1}$$

Let scalar products $\mathbf{XX}_m^T$ and $\mathbf{XX}_\mu^T$ be independent random quantities, $m \neq \mu$.

$$P^* = 1 - \prod_{m=1}^{M-1} \Pr\left[\left|\mathbf{XX}_m^T\right| < (1 - 2b_{\max})N\right]. \tag{A2}$$

Now, it is necessary to calculate the probability that the product of each pattern by input vector is smaller than the threshold.

Scalar product $\mathbf{XX}_m^T$ is a discrete quantity, which values lie in $[-N;N]$. Let k be the number of components with the opposite sign in vectors $\mathbf{X}$ and $\mathbf{X}_m$. Then its probability function is:

$$\Pr\left[\mathbf{XX}_m^T = (N - 2k)\right] = \frac{C_N^k}{2^N}. \tag{A3}$$

Random variable $\mathbf{XX}_m^T$ is symmetrically distributed with zero mean, so

$$\Pr\left[\left|\mathbf{XX}_m^T\right| < (1 - 2b_{\max})N\right] = 1 - 2\sum_{k=0}^{b_{\max}N} \frac{C_N^k}{2^N}. \tag{A4}$$

From A2 and A4 we can conclude that

$$P^* = 1 - \left\{1 - 2\sum_{k=0}^{b_{\max}N} \frac{C_N^k}{2^N}\right\}^{M-1}. \tag{A5}$$

Appendix B

Scalar product

$$\xi = \mathbf{XX}_m^T = \sum_{i=1}^{N} x_i x_{mi} \tag{B1}$$

consists of a large number of random quantities. Therefore, at big dimensions ($N > 100$) its distribution can be approximated by Gaussian law with the following probability moments:

$$\bar{\xi} = 0 \quad \text{b} \quad \sigma^2(\xi) = N. \tag{B2}$$

Therefore, probability (A1) can be described by integral expression:

$$P^* \sim 1 - \left\{1 - \frac{2}{\sqrt{2\pi N}} \int_{-\infty}^{-(1-2b_{\max})N} e^{-\frac{\xi^2}{2N}} d\xi\right\}^{M-1}. \tag{B3}$$

Using the following approximation

$$\int_x^\infty e^{-t^2}\, dt \approx \frac{e^{-x^2}}{2x}, x \gg 1,$$ (B4)

obtain the final estimation of probability (A1.1):

$$P^* < \frac{2M}{\sqrt{2\pi\tilde{N}}} \exp\left(-\frac{\tilde{N}}{2}\right), \quad \tilde{N} = N(1 - 2b_{\max})^2.$$ (B5)

References

1. Kryzhanovsky, V., Malsagov, M., Tomas, J.A.C.: Hierarchical classifier: based on neural networks searching tree with iterative traversal and stop criterion. Opt. Mem. Neural Netw. (Inf. Opt.) **22**(4), 217–223 (2013)
2. Friedman, J.H., Bentley, J.L., Finkel, R.A.: An algorithm for finding best matches in logarithmic expected time. ACM Trans. Math. Softw. **3**, 209–226 (1977)
3. Liu, T., Moore, A.W., Gray, A, Yang, K.: An investigation of practical approximate nearest neighbor algorithms. In: Proceeding of Conference, Neural Information Processing Systems (2004)
4. Indyk, P., Motwani, R.: Approximate nearest neighbors: towards removing the curse of dimensionality. In: Proceedings of 30th STOC, pp. 604–613 (1998)
5. Beis, J.S., Lowe, D.G.: Shape indexing using approximate nearest-neighbour search in high-dimensional spaces. In: Proceedings of IEEE Computer Society Conference on Computer Vision and Pattern Recognition, pp. 1000–1006 (1997)
6. Kryzhanovsky, B., Kryzhanovskiy, V., Litinskii, L.: Machine learning in vector models of neural networks. In: Koronacki, J., Ras, Z.W., Wierzchon, S.T., et al. (eds.) Advances in Machine Learning II. Studies in Computational Intelligence, vol. SCI 263, pp. 427–443. Springer, Berlin (2010). (Dedicated to the memory of Professor Ryszard S. Michalski)
7. Kryzhanovsky, V., Malsagov, M., Zelavskaya, I., Tomas, J.A.C.: High-dimensional binary pattern classification by scalar neural network tree. In: Proceedings of International Conference on Artificial Neural Networks, pp. 169–177 (2014)

Hybrid Classification Approach to Decision Support for Endoscopy in Gastrointestinal Tract

Vyacheslav V. Mizgulin[1], Dmitry M. Stepanov[2(✉)], Stepan A. Kamentsev[3], Radi M. Kadushnikov[2], Evgeny D. Fedorov[4], and Olga A. Buntseva[5]

[1] Ural Federal University Named After the First President of Russia B. N. Yeltsin, Yekaterinburg, Russia
[2] SIAMS Ltd., Yekaterinburg, Russia
distep2@gmail.com
[3] Naumen Company, Moscow, Russia
[4] Pirogov Russian National Research Medical University (RNRMU), Moscow, Russia
[5] Lomonosov Moscow State University, Moscow, Russia

Abstract. This paper provides a new classification approach combining different methods for image and text analysis. In this work the approach is applied endoscopic image of gastrointestinal tract and appropriate text reports. We propose to extract useful information about gastrointestinal tract images from text descriptions using semantic analysis. The text mining algorithm was validated on real text descriptions of endoscopic surveys.

Keywords: Image analysis · Semantic analysis · Hybrid approach · Cancer diagnostic · Decision support · Endoscopy

1 Introduction

First applications of decision support systems in the area of gastrointestinal endoscopy were aimed at increasing identification frequency of neoplasms in course of routine endoscopic research by means of processing the received images, i.e. increasing image contrast and sharpness, and performing structure-based rough assessment of malignant potential for the revealed neoplasms [1, 2]. Modern endoscopic technologies allow obtaining high resolution images where the thin structure of neoplasm is distinguishable. Using such endoscopic images experts can predict a histologic structure of a tumor.

In recent years computer-based methods were applied in researches of stomach neoplasm thin structure. Osawa et al. [3], as well as Dohi et al. [4] presented efficiency of computer analysis to endoscopic images in order to allow determine the boundary of tumor and surrounding mucosa. Miyaki et al. [5] investigated possibilities of applying computer-based analysis to endoscopic images received with zoom-magnification chromoscopy in order to allow distinction of benign tumors and early stomach cancer. Lee et al. [6] presented preliminary results of analyzing stomach suspicious neoplasm images obtained with zoom-magnification narrow band imaging endoscopy. Authors used a neural network for classification of stomach

M.Y. Khachay et al. (Eds.): AIST 2015, CCIS 542, pp. 218–223, 2015.
DOI: 10.1007/978-3-319-26123-2_21

mucous membrane images. Kubota et al. [7] conducted research of cancer invasion depth on extensive material, and defined conditions, under which their analytical system worked with maximum accuracy and demonstrated good results.

The general line of the existing approaches to gastrointestinal endoscopy decision support consists in specialization on certain pathologies, inspection methods, and, above all – application of decision support not in real time, but after inspection. The reason for that is a need of thorough control over the parameters of algorithms used to analyze each image, and a fact that this control can be performed by a trained professional. As a result, from the economic point of view, application of these decision support systems increases cost of endoscopy service.

The purpose of the work was to develop a system for decision in the area of gastro-intestinal endoscopy with the following features:

- self-training capabilities in order to allow diagnosis of various pathologies using images obtained with various endoscopic methods;
- real time operation, in order to allow decision-making in course of inspection, and not after wards;
- completely automated implementation work of analysis algorithms that does not require additional operator training.

The work featured extraction of data required for development of analysis algorithms and image classification from textual reports prepared by medical professionals after endoscopic inspections.

There are multiple approaches for text semantic analysis, described for example, in Vorontsov [8]. In order to solve the problem of supporting medical decisions making, the research was based upon the successful experience of Korepanova et al. [9].

2   Analysis of Endoscopy Images

Analysis of endoscopic images is complicated by the presence of various artifacts that appear in course the image acquisition, such as highlights (appear because of an almost concentric arrangement of illuminator and camera, and wet mucous surface of stomach as well), floating scale, geometric deformations (associated with a wide-angle lens), and uneven brightness. The presence of artifacts requires image preprocessing, image-specific selection of analysis parameters, and use of invariant methods.

Selection of artifacts and image brightness equalization was performed using algorithm described in [10]. It allowed drawing skeletons of gastric mucosa pit-patterns on preprocessed images. Elementary part of a skeleton is referred to as branch. It is hard to determine distinct distribution shapes of topological characteristics for skeleton branches corresponding with different diagnoses. For example, Fig. 1 shows distributions of the following characteristics: tortuosity, elongation factor, runout thickness, and coordination number. Measuring topological characteristics for the set of 800 endoscopic images did not allow clustering the set according to diagnoses that where known for each image. Generally the number of possible diagnoses far exceeds the number and variability of possible analyzed features.

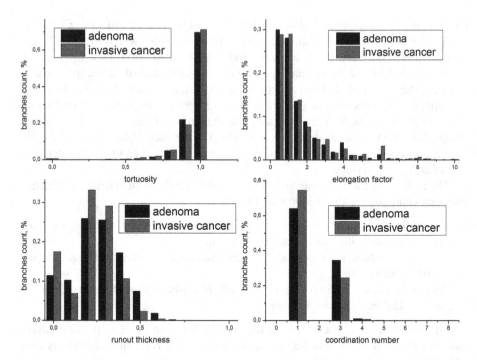

Fig. 1. Distribution of tortuosity, elongation factor, runout thickness and coordination number for skeleton branches

3 Semantic Analysis of Endoscopy Reports

Additional information was available besides the endoscopic images. Medical professionals always write text reports after carrying out endoscopic inspections. In many medical institutions these reports are stored in electronic form. The sequence of text reports semantic analysis is provided below.

Step 1. Preprocessing (Apache Tika is used). At this stage textual and metadata is extracted from documents in different formats.

Step 2. Morphological analysis and lemmatization including extraction of entities, cases, parts of speech, and reduction to an initial form without dictionary on the basis of hypotheses (mystem is used). At this stage the text is transformed into a set of tokens with corresponding lemmas.

Step 3. Latent semantic analysis using randomized singular value decomposition algorithm [11]. For SVD established the following parameters: desired number of SVD triples (dimensions) is 10, and accuracy parameter is 1e-9. The other parameters are set by default. Logarithmic entropy is used as a function. At this step classification space is created.

Step 4. Extraction of keywords for processed documents by comparing word and document vectors.

Step 5. Clustering documents with committee and hierarchical agglomerative clustering algorithms. The algorithms were implemented based on [12].
Step 6. Extraction of cluster keywords.
Step 7. Visualization of n-dimensional classification space using multidimensional scaling algorithm.

The output of semantic analysis included: classification space of documents; keywords for each of the documents; clusters of documents containing different diagnoses; cluster keywords.

In order to verify the algorithm two selections of endoscopy reports with known diagnosis were used. The classification space was built, and diagnosis vectors were found using the first selection. When the endoscopy reports from the second selection were translated into the classification space, the closest diagnosis vector had to be found. If the diagnosis was defined properly, the corresponding endoscopy report was marked as successful. After checking all of the reports from the second selection, accuracy was found as successful cases to the total number of reports ratio.

4   Results and Discussion

Figure 2 show results of clustering two hundred endoscopy reports. Number of clusters was less than the number of diagnoses, confirming the presence of disputes. Results demonstrate the fact that the most dangerous diagnoses, such as early and invasive cancer, do not overlap with other cases. It can be considered that decision support is more useful to diagnose adenomas, intestinal metaplasia and polyps.

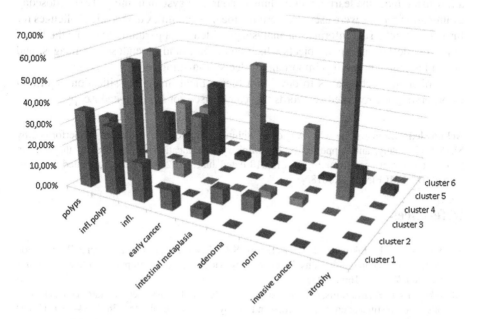

Fig. 2. Distribution of clustered documents by clusters and diagnoses

Results of semantic analysis and classification accuracy estimation are given in Table 1.

Table 1. Semantic analysis and classification accuracy estimation results

Diagnosis	Accuracy
Invasive cancer	89 %
Inflammation	84 %
Inflammation polyps	80 %
Early cancer	78 %
Intestinal metaplasia	69 %
Polyps	68 %
Adenoma	33 %

5   Conclusions

Development of a decision support system for gastrointestinal tract endoscopy possessing the declared features is complicated by a variety of object types and object properties that can be encountered in endoscopy practice. Large number of classifications used in endoscopy diagnostics doesn't allow building a selection with size that will allow using machine learning algorithms in course of system training. Text endoscopy examination reports were used to determine the most complex diagnostic challenges for further research. Text information analysis provided an opportunity to create the list of keywords describing main objects which could be found on endoscopic images, and determine object properties important for decision-making.

Further research goal is to develop a hierarchical image classification algorithm incorporating the revealed keywords that can be used for stomach endoscopy.

Acknowledgements. The work was done within the framework of the project performed by SIAMS Company, and supported by the Ministry of Education and Science of the Russian Federation (Grant agreement 14.576.21.0018 dated June 27, 2014). Project (applied research) unique ID RFMEFI57614X0018.

References

1. Maroulis, D.E., Iakovidis, D.K., Karkanis, S.A., Karras, D.A.: CoLD: a versatile detection system for colorectal lesions in endoscopy video-frames. Comput. Methods Programs Biomed. **70**, 151–166 (2003)
2. Iakovidis, D.K., Maroulis, D.E., Karkanis, S.A.: An intelligent system for automatic detection of gastrointestinal adenomas in video endoscopy. Comput. Biol. Med. **36**, 1084–1103 (2006)

3. Osawa, H., Yamamoto, H., Miura, Y., Ajibe, H., Shinhata, H., Yoshizawa, M., Sunada, K., Toma, S., Satoh, K., Sugano, K.: Diagnosis of depressed-type early gastric cancer using small-caliber endoscopy with flexible spectral imaging color enhancement. Dig. Endosc. **24**(4), 231–236 (2012)
4. Dohi, O., Yagi, N., Wada, T., Yamada, N., Bito, N., Yamada, S., Gen, Y., Yoshida, N., Uchiyama, K., Ishikawa, T., Takagi, T., Handa, O., Konishi, H., Wakabayashi, N., Kokura, S., Naito, Y., Yoshikawa, T.: Recognition of endoscopic diagnosis in differentiated-type early gastric cancer by flexible spectral imaging color enhancement with indigo carmine. Digestion **86**(2), 161–170 (2012)
5. Miyaki, R., Yoshida, S., Tanaka, S., Kominami, Y., Sanomura, Y., Matsuo, T., Oka, S., Raytchev, B., Tamaki, T., Koide, T., Kaneda, K., Yoshihara, M., Chayama, K.: Quantitative identification of mucosal gastric cancer under magnifying endoscopy with flexible spectral imaging color enhancement. J. Gastroenterol. Hepatol. **28**(5), 841–847 (2013)
6. Lee, T.C., Lin, Y.H., Uedo, N., Wang, H.P.: Computer-aided diagnosis in endoscopy: a novel application toward automatic detection of abnormal lesions on magnifying narrow-band imaging endoscopy in the stomach. Conf. Proc. IEEE Eng. Med. Biol. Soc. **2013**, 4430–4433 (2013)
7. Kubota, K., Kuroda, J., Yoshida, M., Ohta, K., Kitajima, M.: Medical image analysis: computer-aided diagnosis of gastric cancer invasion on endoscopic images. Surg. Endosc. **26**(5), 1485–1489 (2012)
8. Vorontsov, K., Potapenko, A.: Tutorial on probabilistic topic modeling: additive regularization for stochastic matrix factorization. In: AIST 2014, CCIS 436, pp. 29–46 (2014)
9. Korepanova, N., Kuznetsov, S.O., Karachunskiy, A.I.: Matchings and decision trees for determining optimal therapy. In: AIST 2014, CCIS 436, pp. 101–110
10. Dunaeva, O.A., Malkova, D.B., Machin, M.L., Edelsbrunner, H.: Segmentation of clinical endoscopic images based on the classification of topological vector features
11. Library for Computing Singular Value Decompositions. http://tedlab.mit.edu/~dr/SVDLIBC/
12. Pantel, P.A.: Clustering by Committee. Thesis, Department of Computing Science. Edmonton, Alberta (2003)

User Similarity Computation for Collaborative Filtering Using Dynamic Implicit Trust

Falguni Roy[1], Sheikh Muhammad Sarwar[1]([✉]), and Mahamudul Hasan[2]

[1] Institute of Information Technology, University of Dhaka, Dhaka, Bangladesh
bit0230@iit.du.ac.bd, smsarwar@du.ac.bd
[2] Department of Computer Science and Engineering,
University of Dhaka, Dhaka, Bangladesh
munna09bd@gmail.com

Abstract. Collaborative filtering is one of the most prominent techniques in Recommender System (RS) to retrieve useful information by using most similar items or users. However, traditional collaborative filtering approaches face many limitations like data sparsity, semantic similarity assumption, fake user profiles and they often do not care about user's evolving interests; such flaws lead to user's dissatisfaction and low performance of the system. To cope with these limitations, we propose a new dynamic trust-based similarity approach. We compute trust score of the users by means of implicit trust information between them. The experimental results demonstrate that the proposed approach performs better than the existing trust-based recommendation algorithms in terms of accuracy by dealing with the aforementioned limitations.

Keywords: Recommender Systems (RS) · Collaborative Filtering (CF)

1 Introduction

Recommender systems (RS) are among the most popular forms of web information customization systems, which uses some of the classical information retrieval techniques. RS are mainly used in E-commerce and entertainment-based websites; the aim of a recommender system is to predict interested items that seem to be useful or interesting for a user taking service of the system. Collaborative filtering and content-based filtering are among the most popular recommender system techniques. Content-based filtering requires rich content descriptions of the items and a profile of the user's preferences; prediction for an unknown item is made by matching item description with user's preferences [1]. However, since item descriptions are analyzed by the machine, it becomes difficult to retrieve multimedia information like machine perception of the content such as color, texture etc., which usually differs greatly with human perception and, thus, to evaluate the quality of the items [2,3].

On the other hand, collaborative filtering (CF) needs regular users' participation with an easy way to represent users' choices to the system and an algorithm

© Springer International Publishing Switzerland 2015
M.Y. Khachay et al. (Eds.): AIST 2015, CCIS 542, pp. 224–235, 2015.
DOI: 10.1007/978-3-319-26123-2_22

that is able to compare people with those who made similar choices. Based on the methodology, collaborative filtering can be categorized as either memory-based or model-based method. Model based CF provides item recommendation by developing a model of user ratings using different machine learning algorithms such as Bayesian networks, neural networks, genetic algorithms and clustering techniques. Memory-based CF uses a rating matrix and provides recommendation for a specific item based on the relationship between the target user with other users by applying special statistical techniques to the rating matrix [4]. Based on the definition of similarity between two users, CF can be further divided into user-based and item-based approaches.

A relatively new system, named trust-based recommender system, uses the concept of trust among users along with the traditional mechanism of similarity detection techniques to improve system's accuracy and reliability. Although CF provides better prediction for the target user with higher user and item coverage, it still suffers from many limitations such as data sparsity, cold start (user and/or item) and malicious insiders attacks [5]. Cold start problems caused by the presence of large amount of new users and/or items in the database [6] implying the data sparsity of user-item rating matrix and similarity user-user matrix as a consequence. Traditional collaborative filtering treats users' similarity as symmetric and does not pay attention to user's changing interests over time, which causes unreliable recommendations [2,7]. A trust-based recommender system can help to cope with many limitations of existing systems, which use traditional CF [8]. In recommender system, trust can be classified as either explicit or implicit. Explicit trust can be obtained explicitly from users and it is generally referred to as *"web of trust"* or *"trust statement"* by the users within the recommender system [9]. In explicit trust, the users have to explicitly state their faith on others and it causes extra burden on the users. Existing literature describes how explicit trust may cause user dissatisfaction and noisy trust [8]. On the contrary, implicit trust resolves all of these limitations. Implicit trust is usually deduced from the user's rating behavior within the system. By analyzing each user's rating pattern, implicit trust is established between the users.

The purpose of this paper is to propose a new similarity calculation method based on dynamic implicit trust among users, which will improve prediction accuracy and reliability by treating similarity as asymmetric with the involvement of user's changing interests. The proposed method consists of several steps where similarity and trust between users are determined; then both are combined to form a more effective and accurate similarity computation algorithm for finding out most similar neighbors of target users for further execution by the system.

The paper is organized as follows: In Sect. 2, we provide necessary background and discuss related work. In Sect. 3, we present our trust-based approach. In Sect. 4, we report the results of experimental evaluation. Section 5 concludes the paper.

2   Backgound and Related Work

Similarity computation is the basic step of any CF algorithm. Accuracy of prediction depends highly on how precisely similarity is being calculated. In [2], the pros and cons of a variety of similarity measures are mentioned. Pearson correlation coefficient (PCC) is reported as one of the most used methods to calculate similarity between users. However, PCC only takes into account the ratings of common items rated by both users and their average ratings. Mean squared difference estimates the similarity by calculating the mean difference of the item ratings and Jaccard computes the similarity by determining the proportion of common and uncommon items rated by both users. All of these methods only consider the items rated by both users, while measuring similarity between two users without taking into account their changing interests and treat them as equally similar [2]. To resolve these problems, trust is introduced in the RS. According to the sociological definition, trust requires a belief and an oral commitment. As a consequence, it is very difficult to define and model trust between users formally. In recommender system, trust is defined on the basis of personal background, history of interaction, context, similarity, reputation, trust statement etc. in the system [10]. Guo *et al.* defines trust in recommender system as "Trust is defined as one's belief towards the ability of others in providing valuable ratings" [11]. According to trust theory, trust has four distinct properties [12]:

- Trust is asymmetric.
- Trust is transitive and this property of trust is very important because it helps to find new trustable neighbors of a target user who are not directly connected with the target user.
- Trust is context dependent.
- Trust is dynamic; it means trust between two people can increase or decrease with the experience over time.

Trust-based RS takes advantage of trust network to achieve recommendations for a user on the basis of those people whom the user trusts. Trust network is a directed graph whose nodes define the system's users and the edges between two nodes define trust relation between them. The weight of the edges determines the degree of trust assigned by themselves to others. Usually, the degree of trust is measured from the trust information. Many researchers have suggested their explicit trust-based RS with its effectiveness to improve systems performance. For example, Massa *et al.* proposed an explicit trust-based RS which took users' feedback directly; users could express their *"web of trust"* for others by identifying valuable and consistent users on the basis of their reviews and ratings [9]. Avesain *et al.* presented a domain-specific explicit trust-based RS and the targeted domain was the ski mountaineering [13]. In their system, users could share their opinion about different ski routes and express their trust in others' opinions. Users build their trust in others relying on *"trust statement"* and the trust score of each user depends on the trust statements of others on him/her and their trust scores. However, explicit trust model suffers from few limitations on defining trust score. First, explicit trust model requires additional manual labor and user effort to receive RS service. Second, for simplicity and privacy issue, explicit trust scores are defined as binary format, which also bound the users to express their degree of trust to a user. Third, the amount of trust information is comparatively small and easily could be noisy in terms of users' choice. On contrast, implicit trust is more reliable to use and the trust information of implicit trust can be inferred from previous rating information [6,14]. By analyzing rating patterns, including historical behavior of users in rating terms, it is possible to measure trust score between the users and identify reliable users whose ratings are useful for recommendation. For example, Shambour et al. computed predicted ratings using simple version of Resnick's prediction formula based on a single user only, computed trust based on mean squared distance (MSD) and propagated trust based on the MoleTrust matric [6]. Hwang et al. almost did it in similar way except that they computed the trust score by averaging the prediction error on co-rated items [15]. Lathia et al. proposed a trust method to define degree of trust by tracking the value of ratings provided by other users [16]. In this method, a user who produces even inverse (polar) ratings is more trustworthy than the one who is not agreeing to share opinions. Papagelis et al. define trust through user similarity computed by Pearson correlation coefficient. O'Donovan et al. proposed another trust computation method in [1]: a rating provided by users is correct if the absolute difference between the predicted rating and the actual rating is smaller than a threshold. After that they define two kinds of trust methods based on available rating; profile-level trust and item-level trust. However, all these methodologies use users rating information for trust computation and apart from being transitive and in some cases they are also asymmetric. However, these trust computation methodologies are not dynamic and/or context dependent [8,12]. A user's trust in others could increase through good experiences and decrease by negative experiences over time. Moreover, these methodologies do not pay any attention to changes of trust and this may cause low performance of trust-based recommender systems.

Several researchers tried to enchange traditional CF on the basis of users changing interests. They proposed different models to capture patterns of users' preference changes according to rating time. Yi Ding et al. proposed a method to predict user's future interests precisely by applying time weights and parametrization for every item cluster [17]. Qian et al. used a nonlinear forgetting function to define users' interest by assigning different time weights to the users' ratings [18]. Some researchers also tried to model dynamic trust using hybrid techniques for RS. For example, Zhuo et al. proposed a dynamic trust method based on Ant Colony algorithm [19]. Punam et al. also introduced a new approach for dynamic trust by adopting the same algorithm [20]. In this approach, implicit trust was measured from the rating matrix and ant colony algorithm was used for updating trust based on time. Abhishek et al. measured trust using confidence and similarity method and updated trust using special dance strategy of honey bees of bee colony algorithm [21]. Zhimin et al. proposed a methodology based on a combination of trust evolution, user similarity and interests [7]. The author used time weight based on non-linear forgetting function to assess user interest. In this paper, our approach is to define the time weight at the beginning of the implicit trust construction and ensure that the constructed trust is *asymmetric*, potentially *transitive* and *dynamic*.

3   Proposed Approach

This section discusses the proposed implicit trust-based similarity model. Here, we describe our proposed methodology of similarity determination of a target user with others.

3.1   Trust-Based Similarity Computation

The main goal of this paper is to design an effective, reliable recommender system by integrating dynamic implicit trust metric with similarity metric into the traditional CF process. The proposed system consists of the following modules: Similarity Computation module (SC), Trust Computation module (TC) and Combined Trust and Similarity Computation module ($CTSC$). The system takes rating matrix as input and at the end returns trust-based similarity matrix. Detailed process of these modules is included in the following subsections.

3.2   Similarity Computation Module

In this stage, we extract a neighborhood of similar minded users for the target user (U_a). For this purpose, we first determine similarity between the targeted user and all other members (U_b) of the system, where $U_b \in U = \{U_1, U_2, \ldots, U_n\}$ and $U_b \neq U_a$. Similarity is calculated by integrating Pearson Correlation Coefficient (PCC) and Jaccard similarity [6]. PCC is used to measure numerical similarity between two users; it determines how similarly they rate an item. The major drawback of PCC is that it only considers the rating correlation between users but doesn't consider the amount of co-rated items by both users. To resolve this drawback, Jaccard is introduced to take co-rated items by users into account. The basic idea here is that, two people are more similar if they have rated more the same items.

$$PCC(a,b) = \frac{\sum\limits_{i \in I_a \cap I_b} (r_{a,i} - \bar{r_a}) \cdot (r_{b,i} - \bar{r_b})}{\sqrt{\sum\limits_{i \in I_a \cap I_b} (r_{a,i} - \bar{r_a})^2} \cdot \sqrt{\sum\limits_{i \in I_a \cap I_b} (r_{b,i} - \bar{r_b})^2}} \tag{1}$$

$$Jaccard(a,b) = \frac{|I_a \cap I_b|}{|I_a \cup I_b|} \tag{2}$$

where $r_{a,i}$, $r_{b,i} \in [1,5]$ represent the ratings of target user U_a and recommender U_b for item i respectively, while $\bar{r_a}$ and $\bar{r_b} \in [1,5]$ represent the average rating of all items in the system that target user U_a and recommender U_b rated separately. $|I_a|$ and $|I_b|$ represent the number of items that both user U_a and U_b rated individually and $|I_a \cap I_b|$ determines the number of items that were commonly rated by them. By combining PCC and Jaccard method [6], we get another similarity method that resolves each methods drawback. The combined method is denoted as $JPCC$ [6].

$$JPCC(a,b) = PCC(a,b) \cdot Jaccard(a,b). \tag{3}$$

3.3   Trust Computation Module

In this section, we propose a new method for determining the implicit trust between users as an integration of Mean Square Difference (MSD) and Confidence. Implicit trust is populated by defining the similarity or degree of similarity between users [8]. MSD is used to define the degree of similarity between users [6] and Confidence determines how much a target user should rely on other's rating [21]. Any RS database contains ratings from different types of users with various kinds of ratings on different time stamps. Some users may seem to contribute more than other by rating more items but that does not justify their rating process as all ratings cannot be trusted. Before defining the MSD for each pair of users, each time the recommendation process is performed separately by using recommender as target's sole recommendation partner. Because if a recommender has delivered high accurate recommendations to the target user in the past, then the recommender should acquire a high trust score from the target user [15]. To perform the recommendation process, we use Resnick's prediction equation [22]. And most of the time both users rate the same item at different time. Some of the cases, their rating time difference for the same item is too high. However, traditional methods do not pay any attention on user's rating time and treat the users being similar even if their co-rated items by both users have high rating time difference and in the meantime the recommender's preferences has changed. For this reason, at the time of performing prediction for recommendation, we try to penalize recommender's prediction effects on a target user according to the predicted item rating's time inspired by forgetting curve from Psychology. We use exponential decay function to penalize prediction effects of a recommender according to his/her rating time. If the predicted item is rated too early by the recommender, then its effects should be very low at the prediction time. For this reason, we modify Resnick's prediction equation with the decay function to compute the predicted rating by Eq. 4. In Eq. 4 and the rest ones in the following sections, we consider user U_a as the user for whom we will be predicting ratings, i.e. *target user*; we consider user U_b as the user with whom user U_a has similarity or trust-based relationship, i.e. *recommender user*.

$$p_{a,i} = \bar{r_a} + (r_{b,i} - \bar{r_b})e^{-T\lambda} \tag{4}$$

In Eq. 4, $r_{b,i} \in [1,5]$ denotes the rating of the item i by the recommender U_b, and $\bar{r_a}$ and $\bar{r_b} \in [1, 5]$ denotes the mean rating of both users U_a and U_b, respectively. $p_{a,i}$ denotes the predicted rating of the item i for the target user U_a and predicted by the recommender U_b. T denotes the time interval, $T = T_r - T_i$, where T_r is the most recent rating time of the recommender user U_b, which means the last time that the user U_b assigned rating to an item and T_i defined the exact time when user U_b rates the item i. λ is a constant parameter which defines the decay rate. As the prediction accuracy defines the trustworthiness between the users, we used MSD [6,23] to define the degree of similarity between them from the prediction error of co-rated items.

$$MSD = 1 - \frac{\sum\limits_{i \in I_a \cap I_b} (p_{a,i} - r_{a,i})^2}{|I_a \cap I_b|} \tag{5}$$

Here $p_{a,i}$ denotes the predicted rating of item i, which is measured by Eq. 4 and $r_{a,i}$ refers to the actual rating of the item i of the target user U_a. $I_a \cap I_b$ denotes the co-rated items by both user.

However, the similarity-based trust has several major drawbacks, which are described in previous research works [1,8,14,15]. There are such cases when both users rate a small amount of common items; then according to MSD, they can even appear as highly trustworthy to each other. To resolve all of those drawbacks, we incorporate the confidence value in our method. According to [24], *"Confidence expresses the reliability of the affiliation between the users based on the number of co-rated items and influenced when the amount of co-rated items are changed"*. Higher confidence of user U_a to user U_b specifies that user U_b is highly reliable to user U_a in a sense that their co-rated items compose a significant fraction of the number of items rated by user U_b. The confidence of a target user U_a for whom we will be generating recommendation with respect to a recommender user U_b is calculated by Eq. 6. As one can see, confidence of U_a to U_b, $Confidence(a,b)$, may be a completely different value than $Confidence(b,a)$.

$$Confidence(a, b) = \frac{|I_a \cap I_b|}{|I_b|} \tag{6}$$

Here $I_a \cap I_b$ represent commonly rated items by both users and I_b denotes the whole set of items that recommender U_b rated in the system.

3.4 New Implicit Trust Measurement Method

After determining MSD and $Confidence$ between users, we combine these in Eq. 7, measure implicit trust between them, and create a new trust matrix.

$$\begin{aligned} Trust(a, b) &= MSD \cdot Confidence \\ &= \frac{|I_a \cap I_b| - \sum\limits_{i \in I_a \cap I_b} (p_{a,i} - r_{a,i})^2}{|I_a \cap I_b|} \cdot \frac{|I_a \cup I_b|}{|I_b|} \\ &= \frac{|I_a \cap I_b| - \sum\limits_{i \in I_a \cap I_b} (p_{a,i} - r_{a,i})^2}{|I_b|} \end{aligned} \tag{7}$$

The computed implicit trust supports the following properties:

- **Transitivity**: Transitive property says that if target user U_a trusts recommender U_b and U_b trusts another user U_c in the system, then it can be inferred that the target user U_a also could trust in user U_c to some extent. The trust value, computed by this method, is (potentially) transitive because we could build indirect trust connection between users with this trust value.
- **Asymmetry**: In reality, trust is a personal and subjective issue. So, two users who are involved in a trust-based relationship, might not trust each other to the same extent and it is a common phenomenon. In the proposed methodology, we used *Confidence*, which is calculated by the commonly rated items by users divided by the amount of recommender's rated items and we also penalize recommender's rating effects in Eq. 4 for MSD calculation by the recommender's item rating time. In addition, the calculated trust score of the proposed method is also combination of MSD and *Confidence*. Thus, our trust score should not be same for both users because of difference in the amount of individually rated items and the rating times for every user. For this reason, the calculated trust score is asymmetric.
- **Dynamic**: Usually trust is built in a gradual way and changed by the time going on with good or bad experience with the trusted user. Trust can be increased with good experience and decreased with bad experience. In our proposed method, we use recommender's item rating time in Eq. 4, which does not make the calculated trust constant for both users. The trust score changes over time depending on the activity of users in the recommender system.

3.5   Method for Tuning Constant (λ)

Interest varies from user to user and it's specially depends on time. Even the same user's duration of interests could vary according to the type of things he or she likes. The duration for which a user's preference for a specific item lasts is generally determined by his/her present choices; so the value of old ratings is questionable. On the other side, the same old ratings have different impact on different users. So, the decay rate, which is a constant rate to lessen the effect of the old data, depends on the duration of user preferences for particular items. This determines that the decay rate varies with different users. If a target user's preference lasts longer for a specific item, then the decay rate will be lower.

   In this paper, λ is a personalized parameter denoting the decay rate. It is a constant value for a user but it would vary from user to user. In this paper, we included λ as one of the parameters of our model and our contribution lies in obtaining λ. In order to tune λ, we take all the previous rated items of a user into account. Furthermore, we compute time interval T for each rated item as $T = T_r - T_i$, where T_r is the time of the most recent rating made by the user and T_i is the rating time for item i. Then we determine T_{median} from all the time intervals of the rated items for a specific user. A regular user of the system rates a lot of items from the beginning to the present and the time interval of each item differs with other items. As a result, time intervals for all the items result in a skew distribution and the median value is better suited for skewed distributions to derive the central tendency. After determining T_{median} for each user, we define λ for every user as:

$$\lambda = \frac{1}{T_{median}} \qquad (8)$$

For the lower the value of T_{median}, the value of λ will be higher and for the higher the value of λ, the old rated data's effect at the prediction time decays faster. For this reason, old rated items are treated as low important information at the prediction time compared to recently rated items.

3.6 Combined Trust and Similarity Computation

At this module, we integrate TC and SC values between a recommender U_b and a target user U_a to get actual similarity of them. To integrate these values, we use arithmetic mean and denote the integrated method by $TJPCC(a,b)$.

$$TJPCC(a,b) = \frac{JPCC(a,b) + Trust(a,b)}{2}. \qquad (9)$$

4 Experimental Results

4.1 Dataset

The experimental evaluation has been carried out using Movielens dataset, which is developed by GroupLens and Internet Movie Database (IMDB) (Inc). This dataset contains information about users movie ratings and available online[1]. The dataset used has following features:

- It contains over 1,000,209 ratings from 6040 user to 3952 movies.
- Every user ratings lie on the scale of 1–5.
- Each user has rated minimum 50 movies.

4.2 Evaluation Metrics

To measure the accuracy of the recommendations by using our proposed method, we use the most popular evaluation metric: Mean Absolute Error (MAE). It determines the accuracy of recommendations [22] and is defined as the average of absolute deviations between the system's predicted rating against the actual rating assigned by the user for a set of items. A lower MAE value represents a higher recommendation accuracy. Given the set of actual/predicted pairs $(r_{a,i}, p_{a,i})$ for all the movies (n_a) rated by user U_a, the MAE for user U_a is computed as:

$$MAE = \frac{\sum_{i=1}^{n_a} |r_{a,i} - p_{a,i}|}{n_a}. \qquad (10)$$

---

[1] http://grouplens.org/datasets/movielens.

4.3   Performance Results

Table 1 shows the comparison of prediction accuracy of the proposed method with different traditional trust methods. Through a series of experiments, we compare the recommendation quality of the proposed method with different traditional trust-based methods like TFS [6], JMSD [25], O'Donovan-Trust (denoted as O'D-Trust) [1] and Resnick-UCF (denoted as R-UCF) [22] by using MAE with respect to different neighborhood sizes denoted as K-NB.

Table 1. MAE of proposed and existing trust-based similarity methods

K-NB	TJPCC	TFS [6]	JMSD [25]	O'D-Trust [1]	R-UCF [22]
10	**0.680**	0.722	0.780	0.780	0.815
20	**0.675**	0.710	0.770	0.760	0.785
30	**0.669**	0.707	0.765	0.758	0.775
40	**0.661**	0.702	0.765	0.757	0.772
50	**0.658**	0.698	0.765	0.755	0.770
70	**0.655**	0.692	0.765	0.755	0.760
90	**0.654**	0.690	0.765	0.755	0.760

Table 1 has shown that our proposed method performs better than other trust-based collaborative filtering algorithms with different neighborhood size and leads to fewer errors with the increasing neighborhood size.

4.4   Discussion

The proposed method deals with the following weakness of existing collaborative filtering algorithms:

(i) The data sparsity in user-item rating matrix also causes sparseness in the user similarity matrix which is reduced in our proposed method by combining JPCC(a,b) and Trust(a,b).

(ii) The trust score, computed by our proposed method, supports asymmetric property of trust, which means that the degree of trust between two users is not necessary the same. As a consequence, the proposed method provides two different similarity values for a user-pair and it is based on the trust value of a user in another user.

(iii) The proposed method uses the rating time of recommender's items at the TC module to pay attention to the recommender's changing interests and the recommender's recent preferences compared to his/her old preferences at the time of trust computation, which will effects on similar users definition.

5   Conclusion

In this paper, we have presented a similarity computation method for collaborative filtering in which dynamic trust acts as a catalyst. Unlike the non-dynamic trust-based algorithms, we utilize a time-based trust computation function, which considers a users change of taste over a certain time period. Thus our main contribution lies in defining a framework, which considers trust, time and similarity in a single function. Moreover, our model unifies three properties of trust in a single framework, and we provide a detailed discussion of how our method fulfills those properties. The experimental results depicts the achieved improvement of our method over the existing trust-based CF algorithms.

References

1. O'Donovan, J., Smyth, B.: Trust in recommender systems. In: Proceedings of the 10th International Conference on Intelligent User Interfaces, pp. 167–174. ACM (2005)
2. Cacheda, F., Carneiro, V., Fernández, D., Formoso, V.: Comparison of collaborative filtering algorithms: limitations of current techniques and proposals for scalable, high-performance recommender systems. ACM Trans. Web (TWEB) 5(1), 1–33 (2011). Article No. 2
3. Pazzani, M.J.: A framework for collaborative, content-based and demographic filtering. Artif. Intell. Rev. 13(5–6), 393–408 (1999)
4. Sarwar, B., Karypis, G., Konstan, J., Riedl, J.: Item-based collaborative filtering recommendation algorithms. In: Proceedings of the 10th International Conference on World Wide Web, pp. 285–295. ACM (2001)
5. Bellogín, A., Cantador, I., Díez, F., Castells, P., Chavarriaga, E.: An empirical comparison of social, collaborative filtering, and hybrid recommenders. ACM Trans. Intell. Syst. Technol. (TIST) 4(1), 1–29 (2013). Article No. 14
6. Shambour, Q., Lu, J.: A trust-semantic fusion-based recommendation approach for e-business applications. Decis. Support Syst. 54(1), 768–780 (2012)
7. Chen, Z., Jiang, Y., Zhao, Y.: A collaborative filtering recommendation algorithm based on user interest change and trust evaluation. JDCTA 4(9), 106–113 (2010)
8. Guo, G., Zhang, J., Thalmann, D., Basu, A., Yorke-Smith, N.: From ratings to trust: an empirical study of implicit trust in recommender systems. In: Proceedings of the 29th Annual ACM Symposium on Applied Computing. SAC 2014, pp. 248–253. ACM, New York (2014)
9. Massa, P., Bhattacharjee, B.: Using trust in recommender systems: an experimental analysis. In: Jensen, C., Poslad, S., Dimitrakos, T. (eds.) iTrust 2004. LNCS, vol. 2995, pp. 221–235. Springer, Heidelberg (2004)
10. Golbeck, J.: Tutorial on using social trust for recommender systems. In: Proceedings of the Third ACM Conference on Recommender Systems, pp. 425–426. ACM (2009)
11. Guo, G.: Integrating trust and similarity to ameliorate the data sparsity and cold start for recommender systems. In: Proceedings of the 7th ACM Conference on Recommender Systems, pp. 451–454. ACM (2013)
12. Fazeli, S., Loni, B., Bellogin, A., Drachsler, H., Sloep, P.: Implicit vs. explicit trust in social matrix factorization. In: Proceedings of the 8th ACM Conference on Recommender systems, pp. 317–320. ACM (2014)

13. Avesani, P., Massa, P., Tiella, R.: Moleskiing. it: a trust-aware recommender system for ski mountaineering. Int. J. Infonomics **20**, 1–10 (2005)
14. Yuan, W., Shu, L., Chao, H.C., Guan, D., Lee, Y.K., Lee, S.: Itars: trust-aware recommender system using implicit trust networks. IET Commun. **4**(14), 1709–1721 (2010)
15. Hwang, C.-S., Chen, Y.-P.: Using trust in collaborative filtering recommendation. In: Okuno, H.G., Ali, M. (eds.) IEA/AIE 2007. LNCS (LNAI), vol. 4570, pp. 1052–1060. Springer, Heidelberg (2007)
16. Lathia, N., Hailes, S., Capra, L.: Trust-based collaborative filtering. In: Karabulut, Y., Mitchell, J., Herrmann, P., Jensen, C.D. (eds.) Trust Management II, vol. 263, pp. 119–134. Springer, US (2008)
17. Ding, Y., Li, X.: Time weight collaborative filtering. In: Proceedings of the 14th ACM International Conference on Information and Knowledge Management, pp. 485–492. ACM (2005)
18. Wang, Q., Sun, M., Xu, C.: An improved user-model-based collaborative filtering algorithm? J. Inf. Comput. Sci. **8**(10), 1837–1846 (2011)
19. Zhuo, T., Zhengding, L., Kai, L.: Time-based dynamic trust model using ant colony algorithm. Wuhan Univ. J. Nat. Sci. **11**(6), 1462–1466 (2006)
20. Bedi, P., Sharma, R.: Trust based recommender system using ant colony for trust computation. Expert Syst. Appl. **39**(1), 1183–1190 (2012)
21. Kaleroun, A.: Hybrid Bee Colony Trust Mechanism in Recommender System. Ph.D thesis, Thapar University (2014)
22. Resnick, P., Iacovou, N., Suchak, M., Bergstrom, P., Riedl, J.: Grouplens: an open architecture for collaborative filtering of netnews. In: Proceedings of the 1994 ACM Conference on Computer Supported Cooperative Work, pp. 175–186. ACM (1994)
23. Shardanand, U., Maes, P.: Social information filtering: algorithms for automating. In: Proceedings of the SIGCHI Conference on Human Factors in Computing Systems, pp. 210–217. ACM Press/Addison-Wesley Publishing Co. (1995)
24. Papagelis, M., Plexousakis, D., Kutsuras, T.: Alleviating the sparsity problem of collaborative filtering using trust inferences. In: Herrmann, P., Issarny, V., Shiu, S.C.K. (eds.) iTrust 2005. LNCS, vol. 3477, pp. 224–239. Springer, Heidelberg (2005)
25. Bobadilla, J., Serradilla, F., Bernal, J.: A new collaborative filtering metric that improves the behavior of recommender systems. Knowl.-Based Syst. **23**(6), 520–528 (2010)

Similarity Aggregation for Collaborative Filtering

Sheikh Muhammad Sarwar[1], Mahamudul Hasan[2], Masum Billal[2],
and Dmitry I. Ignatov[3(✉)]

[1] Institute of Information Technology, University of Dhaka, Dhaka, Bangladesh
smsarwar@du.ac.bd
[2] Department of Computer Science and Engineering,
University of Dhaka, Dhaka, Bangladesh
{munna,billalmasum93}@gmail.com
[3] National Research University Higher School of Economics, Moscow, Russia
dignatov@hse.ru

Abstract. In this paper we show how several similarity measures can be combined for finding similarity between a pair of users for performing Collaborative Filtering in Recommender Systems. Through aggregation of several measures we find super similar and super dissimilar user pairs and assign a different similarity value for these types of pairs. We also introduce another type of similarity relationship which we call medium similar user pairs and use traditional JMSD for assigning similarity values for them. By experimentation with real data we show that our method for finding similarity by aggregation performs better than each of the similarity metrics. Moreover, as we apply all the traditional metrics in the same setting, we can assess their relative performance.

Keywords: Recommender Systems · Collaborative Filtering · Similarity measures · Similarity fusion

1 Introduction

Recommendation is a social process through which people close to a target user suggest her movies, songs, food etc. However, this social process has become a prevalent component in the virtual world as well because of the tremendous growth of information in the World Wide Web. Unlike social recommendation process, recommendation in the virtual world is rather implicit. It means that people do not directly get suggestion from their peers, rather a computational process helps to generate recommendations for them by automatically identifying a cluster of people who behave similarly. Naturally, a person takes recommendations or suggestions from another person if they both have similar choices or preferences. But, in the virtual world we have access to the preference of millions of users and hence it is possible to get recommendation as a service by assessing similarity of a specific user and a group of users computationally. In the literature this process is referred to as collaborative filtering.

© Springer International Publishing Switzerland 2015
M.Y. Khachay et al. (Eds.): AIST 2015, CCIS 542, pp. 236–242, 2015.
DOI: 10.1007/978-3-319-26123-2_23

One of the major tasks of a Recommender System (RS) is a prediction, i.e. a process through which a RS predicts the rating of a specific item for a user. Rating scale can vary in different ways for different systems. Usually, a rating scale takes integer values from 1 to 5 or from 1 to 10. So, two entities are associated with a rating; one is the user and the other is an item. When a system is being used by several users and consists of several items, a user-item matrix holding the rating data for all the items can be formed. This matrix is the major source for finding similarity between different users in the system. So, the basic philosophy is to analyze the previous ratings of two users and based on these values try to asses similarity of the users' preferences and use that to predict ratings for items which have not yet been rated. The most notable part of CF algorithms refers to the group of metrics used to determine the similarity between each pair of users, among which the Pearson Correlation Coefficient (PCC) is one of the most popular similarity measures [1].

Apart from PCC, there are several similarity measures having inherent advantages and drawbacks. Popular methods include cosine similarity, constrained Pearson correlation coefficient (CPCC), sigmoid function based Pearson correlation coefficient (SPCC), adjusted cosine measure (ACOS), Jaccard similarity and mean squared differences (MSD) [2]. Furthermore, Jaccard and MSD can be combined by multiplication to form a new measure, which is referred to as JMSD [2]. In this paper we hypothesize that to get the most out of the measures we need to combine them in some way as all do not perform well in different situations. Specifically, we state the importance of using different measures for computing similarity of different user pairs. Practically it is rather hard to develop a working heuristic to select a proper similarity measure for a specific user pair. In order to achieve this goal to some extent, we introduce the notion of support; it is defined as the number of measures endorsing the similarity relation between two users. We specifically handle the cases where the relation between a couple of users have high support, low support or average support. As a result, we do not specifically develop a new measure, rather we show how to reap the benefits of existing measures to design an approach which performs better than each of them.

2 Proposed Method

In our experimentation we have used 8 different similarity measures; PCC, SPCC, CPCC, ACOS, COS, JMSD, MSD and Jaccard. All of them are described in section. There are many papers on these measures reporting their individual performances in various tasks [3], but in this paper we implement all the metrics individually under the same experimental setup and report their MAE (Mean Absolute Error). MAE determines the accuracy of recommendations by defining the average absolute deviation between the system's predicted rating against the actual rating assigned by the user [4]. A lower MAE value corresponds to a higher recommendation accuracy. Given the set of actual/predicted pairs $(r_{u,i}, p_{u,i})$ for all the movies (M_u) rated by user u, the MAE for user u is computed as:

$$MAE = \frac{\sum\limits_{i \in M_u} |r_{u,i} - p_{u,i}|}{|M_u|}. \tag{1}$$

2.1 Computing Support Matrix

Using 8 similarity measures in total, we calculate a support value performing the following steps:

1. For a single measure, we calculate the similarity between every pair of users.
2. Then we calculate the median from this similarity measure among all user pairs.
3. Using the median as a threshold, we classify the whole similarity space into two binary classes 0 and 1. Values higher than the median fall into class 1, while the rest fall to class 0.
4. Now, we introduce the notion of support. We assert that if the similarity class of two users is 1, then their similarity relation is supported by the measure we have used to compute similarity. Hence, we increment the support count for that pair of users by 1.
5. We continue this process for the all eight matrices and increment the support value of those two users who satisfy the rule above.
6. Finally, as an outcome of this process, we retrieve a user by user support matrix $S \in \mathbb{R}^{n \times n}$, where n is the number of users in the system, and $S_{uv} \in \{0, 1, \ldots, 8\}$.

2.2 Finding Super Similar, Average Similar and Super Dissimilar User Pairs

Now, we introduce the notion of super similar, medium similar, and super dissimilar users using our support matrix S.

Definition 1. *Super Similar Users: If $S_{uv} \geq 5$, for a pair of users u and v, then we denote them as super similar users.*

Definition 2. *Super Dissimilar Users: If $S_{uv} \leq 2$, for a pair of users u and v, then we denote them as super dissimilar users.*

Definition 3. *Medium Similar Users: Relationship classes that neither belong to super similar or super similar falls into the classes for medium similar, that is $S_{uv} \in \{3, 4\}$.*

The threshold for choosing super similar user pairs comes from an empirical analysis, which is shown in Table 1. In the table, we show MAE values for different support settings. In order to find out a proper value of support for super similar users, we enumerate the values of support from greater than or equal to 0 to greater than or equal to 8 and check the MAE. We set user-user similarity value as 1 (*i.e.* we make them super similar) for a specific set of support values (for example, greater than or equal to 5), and we set 0 for all the support values below that specific set of support values. As a result, only super similar users are having a full influence on each other, while other users who are not super similar do not have any effect. It indicates that a user-user pair for which we have support value less than a specific threshold value are totally unrelated. We can see that for support value being greater than or equal to 5 the respective MAE value is comparatively lower. The plot, which is based on the table and shown in Fig. 1, makes more sense since it explains the reasoning for our definition of super similar users in Sect. 2.1. However, when super similar users have support value greater than or equal to 5, super dissimilar users should have support value equal or less than 4. But, here we make a finer distinction and define medium similar and super dissimilar users for better performance of the system.

Table 1. MAE values for different support thresholds of super similar users

Minimal support	0	1	2	3	4	5	6	7	8
MAE		0.6883	0.6875	0.6870	0.6822	0.6758	**0.6731**	0.6744	0.6937	0.7679

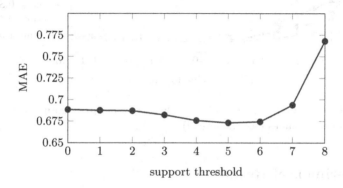

Fig. 1. A graph for finding the optimal similarity values for super similar and super dissimilar users

2.3 Prediction Function

Our prediction function is typical for collaborative filtering; however, it is based on similarity defined in our own way. To calculate the predicted rating p_u^i for user u of an item i, the following Deviation From Mean (DFM) as aggregation approach is used [4]:

$$p_{u,i} = \bar{r}_u + \frac{\sum_{v \in N_u} sim(u,v) \cdot (r_{v,i} - \bar{r}_v)}{\sum_{v \in N_u} sim(u,v)} \tag{2}$$

In Eq. 2, N_u is a set of k most similar users to a given user u, $\bar{r}_u$ represents the average of ratings made by the given user u and $\bar{r}_v$, $r_{v,i}$ are the average of ratings and rating of item i made by the neighbor v, respectively. In Eq. 2 we set $sim(u,v) = 0.9$ for $S_{uv} \geq 5$ and $sim(u,v) = -0.3$ for $S_{uv} \leq 2$. Finally, we set

$sim(u,v) = JMSD(u,v)$ if $S_{uv} = 3$ or $S_{uv} = 4$. Now, we describe the reasoning behind the usage of the aforementioned values.

In Fig. 2 we show the MAE values we obtain for setting different values for super dissimilar users keeping the similarity value for super similar users constant. If we observe the graphs closely, we can see that MAE comes down to the lowest value and then rises. Moreover, we can see that if we take –0.3 as the similarity value for all the super dissimilar users and 0.9 as similarity value for all super similar users, it results in a good MAE.

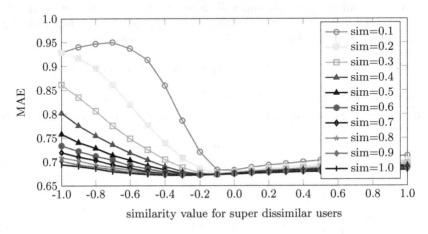

Fig. 2. MAE curves for different similarity values of super dissimilar users parametrized by super similarity values (see the legend)

3   Experimental Result

We have tested our hypothesis using MovieLens dataset. We used the training data with 80 % of the available ratings and 20 % of the rating data was set as the test set. Details of the dataset and testing procedure can be found in [5].

Table 2. MAE values for 8 different similarity measures

PCC	SPCC	CPCC	ACOS	COS	JMSD	MSD	JACCARD
0.688	0.687	0.685	0.687	0.687	**0.680**	0.688	0.682

In Table 2, we show the MAE values for all the measures implemented by us and we can see that JMSD performs better than all the other metrics. However, in Table 3 we show that the proposed approach – super similar (with similarity 0.9) combined with average user (with the same similarity as JMSD value) and super dissimilar (with similarity –0.3) performs better than JMSD. We also show the performance JMSD combined with super similar and super dissimilar users respectively. Note that for all the metrics, including ours, we multiply a confidence value with similarity value as multiplying confidence produces better

Table 3. MAE values for different combinations of similarity ranges

Super similar ($\geq$5) + super dissimilar ($<$5) (no medium similarity)	Super similar + JMSD	Super dissimilar + JMSD	JMSD	Super similar + medium similar + super dissimilar
0.673	0.675	0.735	0.680	**0.668**

result for all the metrics. More details on confidence value can be found in [6], but we provide its Formula 3 below:

$$conf(u, v) = \frac{|I_u \cap I_v|}{|I_v|}. \tag{3}$$

Here, $|I_u \cap I_v|$ is the number of common ratings between user u and user v, and $|I_v|$ is the number of assigned ratings by user v.

4 Conclusion

This paper is our initial footstep of proving the fact that a specific metric or similarity value might be suitable for a specific set of users. Here we performed our experimentation using three groups of user-user pairs: super similar, medium similar and super dissimilar. We show through the experimentation that among the existing metrics JMSD outperforms others in terms of MAE. However, our hybrid approach by aggregation outperforms JMSD using the the same measure.

Since we had a look only at user-based measures, the important venue of our future work could be similarity fusion with the item-based measures. In fact, our heuristic approach is performed better in terms of MAE than similarity fusion based approach of that type reported in [7]. We hope that to this end we can use similarity measures from Formal Concept Analysis to exploit interplay between objects (users) and items (attributes) of the proposed support matrix [8].

Acknowledgment. The first three authors were partially supported by their university. The last author was partially supported by the Russian Foundation for Basic Research grants no. 13-07-00504 and 14-01-93960 and made a contribution within the project "Data mining based on applied ontologies and lattices of closed descriptions" supported by the Basic Research Program of the National Research University Higher School of Economics. We also deeply thank the reviewers and Konstantin Vorontsov for their comments and remarks that helped.

References

1. Ortega, F., Sánchez, J.L., Bobadilla, J., Gutiérrez, A.: Improving Collaborative Filtering-based recommender systems results using Pareto dominance. Inf. Sci. **239**, 50–61 (2013)

2. Liu, H., Hu, Z., Mian, A., Tian, H., Zhu, X.: A new user similarity model to improve the accuracy of Collaborative Filtering. Knowl. Based Syst. **56**, 156–166 (2014)
3. Ahn, H.J.: A new similarity measure for Collaborative Filtering to alleviate the new user cold-starting problem. Inf. Sci. **178**(1), 37–51 (2008)
4. Bobadilla, J., Ortega, F., Hernando, A., Alcalá, J.: Improving Collaborative Filtering recommender system results and performance using genetic algorithms. Knowl. Based Syst. **24**(8), 1310–1316 (2011)
5. Sarwar, B., Karypis, G., Konstan, J., Riedl, J.: Analysis of recommendation algorithms for e-commerce. In: Proceedings of the 2nd ACM Conference on Electronic Commerce, EC 2000, New York, USA, pp. 158–167, ACM (2000)
6. Kaleroun, A.: Hybrid bee colony trust mechanism in recommender system. Ph.D. thesis, Thapar University (2014)
7. Wang, J., de Vries, A.P., Reinders, M.J.T.: Unifying user-based and item-based collaborative filtering approaches by similarity fusion. In: Proceedings of the 29th Annual International ACM SIGIR Conference on Research and Development in Information Retrieval, pp. 501–508 (2006)
8. Eklund, P.W., Ducrou, J., Dau, F.: Concept similarity and related categories in information retrieval using formal concept analysis. Int. J. Gen. Syst. **41**(8), 826–846 (2012)

Distributed Coordinate Descent
for L1-regularized Logistic Regression

Ilya Trofimov[1(✉)] and Alexander Genkin[2]

[1] Yandex, Moscow, Russia
trofim@yandex-team.ru
[2] AVG Consulting, Brooklyn, USA
alexander.genkin@gmail.com

Abstract. Logistic regression is a widely used technique for solving classification and class probability estimation problems in text mining, biometrics and clickstream data analysis. Solving logistic regression with L1-regularization in distributed settings is an important problem. This problem arises when training dataset is very large and cannot fit the memory of a single machine. We present d-GLMNET, a new algorithm solving logistic regression with L1-regularization in the distributed settings. We empirically show that it is superior over distributed online learning via truncated gradient.

Keywords: Large-scale learning · Logistic regression · L1-regularization · Sparsity

1 Introduction

Logistic regression with L1-regularization is the method of choice for solving classification and class probability estimation problems in text mining, biometrics and clickstream data analysis. Despite the fact that logistic regression can build only linear separating surfaces, the testing accuracy of it, with proper regularization is often good for high dimensional input spaces. For several problems the testing accuracy has shown to be close to that of nonlinear classifiers such as kernel methods [1]. At the same time training and testing of linear classifiers is much faster. It makes the logistic regression a good choice for large-scale problems. A desirable trait of model is sparsity, which is conveniently achieved with L1 or elastic net regularizer.

A broad survey [2] suggests that coordinate descent methods are the best choice for L1-regularized logistic regression on the large scale. Widely used algorithms that fall into this family are: BBR [3], GLMNET [4], newGLMNET [5]. Software implementations of these methods start with loading the full training dataset into RAM.

Completely different approach is online learning [6–8]. This kind of algorithms do not require to load training dataset into RAM and can access it sequentially (i.e. reading from disk). Balakrishnan and Madighan [6], Langford et al. [7] report

© Springer International Publishing Switzerland 2015
M.Y. Khachay et al. (Eds.): AIST 2015, CCIS 542, pp. 243–254, 2015.
DOI: 10.1007/978-3-319-26123-2_24

that online learning performs well when compared to batch counterparts (BBR and LASSO).

Nowadays we see the growing number of problems where both the number of examples and the number of features are very large. Many problems grow beyond the capabilities of a single computer and need to be handled by distributed systems. Approaches to distributed training of classifiers naturally fall into two groups by the way they split data across computing nodes: by examples [9] or by features [10]. We believe that algorithms that split data by features can achieve better sparsity while retaining similar or better performance and competitive training speed with those that split by examples. Our experiments so far confirm that belief.

Parallel block-coordinate descent is a natural algorithmic framework if we choose to split by features. The challenge here is how to combine steps from coordinate blocks, or computing nodes, and how to organize communication. When features are independent, parallel updates can be combined straightforwardly, otherwise they may come into conflict and not yield enough improvement to objective; this has been clearly illustrated by Bradley et al. [11]. Bradley et al. [11] proposed Shotgun algorithm based on randomized coordinate descent. They studied how many variables can be updated in parallel to guarantee convergence. Ho et al. [12] presented distributed implementation of this algorithm compatible with Stale Synchronous Parallel Parameter Server.

Richtárik and Takáč [13] use randomized block-coordinate descent and also exploit partial separability of the objective. The latter relies on sparsity in data, which is indeed characteristic to many large scale problems. They present theoretical estimates of speed-up factor of parallelization. Peng et al. [10] proposed a greedy block-coordinate descent method, which selects the next coordinate to update based on the estimate of the expected improvement in the objective. They found their GRock algorithm to be superior over parallel FISTA and ADMM.

In contrast, our approach is to make parallel steps on all blocks, then use combined update as a direction and perform a line search. We show that sufficient data for line search have the size $O(n + p)$, where n is the number of examples, p is the number of features, so it can be performed on one machine. Consequently, that's the amount of data sufficient for communication between machines. Overall, our algorithm fits into the framework of CGD method proposed by Tseng and Yun [14], which allows us to prove convergence. Block-coordinate descent on a single machine is performed as a step of GLMNET [4].

When splitting data by examples, online learning comes in handy. A classifier is trained in online fashion on each subset, then parameters of classifiers are averaged and used as a warmstart for the next iteration, and so on [9,15]. We performed an experimental comparison of our algorithm with distributed online learning.

Our main contributions are the following:

- We propose a new parallel coordinate descent algorithm for L1-regularized logistic regression and guarantee its convergence (Sect. 2)

- We demonstrate how our algorithm can be efficiently implemented on the distributed cluster architecture (Sect. 3)
- We empirically show effectiveness of our implementation in comparison with distributed online learning via truncated gradient (Sect. 4)

The C++ implementation of our algorithm, which we call d-GLMNET, is publicly available at https://github.com/IlyaTrofimov/dlr.

2 Parallel Coordinate Descent Algorithm

In case of binary classification the logistic regression estimates the class probability given the feature vector $\mathbf{x}$

$$P(y = +1|\mathbf{x}) = \frac{1}{1 + \exp(-\boldsymbol{\beta}^T\mathbf{x})}$$

This statistical model is fitted by maximizing the log-likelihood (or minimizing the negated log-likelihood) at the training set. Some penalty is often added to avoid overfitting and numerical ill-conditioning. In our work we consider L1-regularization penalty, which provides sparsity in the model. Thus fitting the logistic regression with L1-regularization leads to the optimization problem

$$\boldsymbol{\beta}^* = \underset{\boldsymbol{\beta} \in R^n}{\operatorname{argmin}} f(\boldsymbol{\beta}) \tag{1}$$

$$f(\boldsymbol{\beta}) = L(\boldsymbol{\beta}) + \lambda\|\boldsymbol{\beta}\|_1 \tag{2}$$

where $L(\boldsymbol{\beta})$ is the negated log-likelihood

$$L(\boldsymbol{\beta}) = \sum_{i=1}^{n} \log(1 + \exp(-y_i\boldsymbol{\beta}^T\mathbf{x}_i)) \tag{3}$$

$y_i \in \{-1, +1\}$ are labels, $\mathbf{x}_i \in \mathbb{R}^p$ are input features, $\boldsymbol{\beta} \in \mathbb{R}^p$ is the unknown vector of weights for input features. We will denote by nnz the number of non-zero entries in all x_i.

The first part of the objective - $L(\boldsymbol{\beta})$ is convex and smooth. The second part is L1-regularization term - $\lambda\|\boldsymbol{\beta}\|_1$ is convex and separable, but non-smooth. Hence one cannot use directly efficient optimization techniques like conjugate gradient method or L-BFGS which are often used for logistic regression with L2-regularization.

Our algorithm is based on building local approximations to the objective (2). A smooth part (3) of the objective has quadratic approximation [4]

$$L_q(\boldsymbol{\beta}, \Delta\boldsymbol{\beta}) \stackrel{\text{def}}{=} L(\boldsymbol{\beta}) + \nabla L(\boldsymbol{\beta})^T \Delta\boldsymbol{\beta} + \frac{1}{2}\Delta\boldsymbol{\beta}^T \nabla^2 L(\boldsymbol{\beta})\Delta\boldsymbol{\beta}$$

$$= \frac{1}{2}\sum_{i=1}^{N} w_i(z_i - \Delta\boldsymbol{\beta}^T\mathbf{x}_i)^2 + C(\boldsymbol{\beta}) \tag{4}$$

where

$$z_i = \frac{(y_i + 1)/2 - p(\mathbf{x}_i)}{p(\mathbf{x}_i)(1 - p(\mathbf{x}_i))}$$
$$w_i = p(\mathbf{x}_i)(1 - p(\mathbf{x}_i))$$
$$p(\mathbf{x}_i) = \frac{1}{1 + e^{-\boldsymbol{\beta}^T \mathbf{x}_i}}$$

The core idea of GLMNET and newGLMNET is iterative minimization of the penalized quadratic approximation to the objective

$$\underset{\Delta\boldsymbol{\beta}}{\operatorname{argmin}} \left\{ L_q(\boldsymbol{\beta}, \Delta\boldsymbol{\beta}) + \lambda \|\boldsymbol{\beta} + \Delta\boldsymbol{\beta}\|_1 \right\} \tag{5}$$

via cyclic coordinate descent. This form (4) of approximation allows to make Newton updates of the vector $\boldsymbol{\beta}$ without storing the Hessian explicitly. Also the approximation (5) has a simple closed-form solution with respect to a single variable $\Delta\beta_j$

$$\Delta\beta_j^* = \frac{T\left(\sum_{i=1}^n w_i x_{ij} q_i, \lambda\right)}{\sum_{i=1}^n w_i x_{ij}^2} - \beta_j \tag{6}$$

$$T(x, a) = \operatorname{sgn}(x) \max(|x| - a, 0)$$

$$q_i = z_i - \Delta\boldsymbol{\beta}^T \mathbf{x}_i + (\beta_j + \Delta\beta_j) x_{ij}$$

In order to adapt the algorithm to the distributed settings we replace the full Hessian with its block-diagonal approximation $\tilde{H}$. More formally: let us split p input features into M disjoint sets S_k

$$\bigcup_{k=1}^M S_k = \{1, \ldots, p\}$$

$$S_m \cap S_k = \emptyset, k \neq m$$

Denote by $\tilde{H}$ a block-diagonal matrix

$$(\tilde{H})_{jl} = \begin{cases} (\nabla^2 L(\boldsymbol{\beta}))_{jl}, & \text{if } \exists m : j, l \in S_m \\ 0, & \text{otherwise} \end{cases} \tag{7}$$

Let $\Delta\boldsymbol{\beta} = \sum_{m=1}^M \Delta\boldsymbol{\beta}^m$, where $\Delta\beta_j^m = 0$ if $j \notin S_m$. Then

$$L_q(\boldsymbol{\beta}, \Delta\boldsymbol{\beta}^m) = L(\boldsymbol{\beta}) + \nabla L(\boldsymbol{\beta})^T \Delta\boldsymbol{\beta}^m + \frac{1}{2}\Delta(\boldsymbol{\beta}^m)^T \nabla^2 L(\boldsymbol{\beta}) \Delta\boldsymbol{\beta}^m$$

$$= L(\boldsymbol{\beta}) + \nabla L(\boldsymbol{\beta})^T \Delta\boldsymbol{\beta}^m + \frac{1}{2} \sum_{j,k \in S_m} (\nabla^2 L(\boldsymbol{\beta}))_{jk} \Delta\beta_j \Delta\beta_k$$

$$= \frac{1}{2} \sum_{i=1}^N w_i (z_i - (\Delta\boldsymbol{\beta}^m)^T \mathbf{x}_i)^2 + C(\boldsymbol{\beta})$$

by summing this equation over m

$$\sum_{m=1}^{M} L_q(\boldsymbol{\beta}, \Delta\boldsymbol{\beta}^m) = \sum_{m=1}^{M} \left(L(\boldsymbol{\beta}) + \nabla L(\boldsymbol{\beta})^T \Delta\boldsymbol{\beta}^m + \frac{1}{2} \sum_{j,k \in S_m} (\nabla^2 L(\boldsymbol{\beta}))_{jk} \Delta\beta_j \Delta\beta_k \right)$$
$$= ML(\boldsymbol{\beta}) + \nabla L(\boldsymbol{\beta})^T \Delta\boldsymbol{\beta} + \frac{1}{2} \Delta\boldsymbol{\beta}^T \tilde{H} \Delta\boldsymbol{\beta} \qquad (8)$$

From the Eq. (8) and separability of L1 penalty follows that solving the approximation to the objective

$$\operatorname*{argmin}_{\Delta\boldsymbol{\beta}} \left\{ L(\boldsymbol{\beta}) + \nabla L(\boldsymbol{\beta})^T \Delta\boldsymbol{\beta} + \frac{1}{2} \Delta\boldsymbol{\beta}^T \tilde{H} \Delta\boldsymbol{\beta} + \lambda ||\boldsymbol{\beta} + \Delta\boldsymbol{\beta}||_1 \right\}$$

is equivalent to solving M independent sub-problems

$$\operatorname*{argmin}_{\Delta\boldsymbol{\beta}^m} \left\{ L_q(\boldsymbol{\beta}, \Delta\boldsymbol{\beta}^m) + \sum_{j \in S_m} |\beta_j + \Delta\beta_j^m| \, \middle| \, \Delta\beta_j^m = 0 \text{ if } j \notin S_m \right\} \qquad (9)$$

and can be done in parallel over M machines. This is the main idea of the proposed algorithm d-GLMNET. We describe a high-level structure of d-GLMNET in the Algorithm 1.

Algorithm 1. *Overall procedure of d-GLMNET*
$\boldsymbol{\beta} \leftarrow 0$
Split $\{1, \ldots, p\}$ *into* M *disjoint sets* $S_1, \ldots, S_M$.
Repeat until convergence:

1. *Do in parallel over M machines*
2. *Minimize* $L_q(\boldsymbol{\beta}, \Delta\boldsymbol{\beta}^m) + ||\boldsymbol{\beta} + \Delta\boldsymbol{\beta}^m||_1$ *with respect to* $\Delta\boldsymbol{\beta}^m$
3. $\Delta\boldsymbol{\beta} \leftarrow \sum_{m=1}^{M} \Delta\boldsymbol{\beta}^m$
4. *Find* $\alpha \in (0, 1]$ *by the line search procedure (Algorithm 3)*
5. $\boldsymbol{\beta} \leftarrow \boldsymbol{\beta} + \alpha\Delta\boldsymbol{\beta}$

return $\boldsymbol{\beta}$

The downside of using line search is that it can hurt sparsity. We compute the regularization path (Sect. 4.2) by running Algorithm 1 with decreasing L1 penalty, and the algorithm starts with $\boldsymbol{\beta} = 0$, so absolute values of $\boldsymbol{\beta}$ tend to increase. However there may be cases when $\Delta\beta_j = -\beta_j$ for some j on step 2 of Algorithm 1, so β_j can to go back to 0. In that case, if line search on step 3 selects $\alpha < 1$, then the opportunity for sparsity is lost.

To retain the sparsity our algorithm takes two precautions. First, line search is prevented if $\alpha = 1$ guarantees sufficient decrease in the objective value (step 1 of Algorithm 3). Second, there is a complication in the convergence criterion. It starts by checking if relative decrease in the objective is sufficiently small or maximum number of iteration has been reached. If that turns out true, the algorithm checks if setting α back to 1 would not be too much of an increase in the objective. If that is also true, the algorithms updates β with $\alpha = 1$ and then stops.

Algorithm 2 presents our approach for solving sub-problem (9). d-GLMNET makes one cycle of coordinate descent over input features for approximate solving (9). Despite the fact that GLMNET and newGLMNET use multiple passes we found that our approach works well in practice. We also use $\tilde{H} + \nu I$ with small $\nu = 10^{-6}$ instead of $\tilde{H}$ in (8). The fact that matrix $\tilde{H} + \nu I$ is positive definite is essential for the proof of convergence (see Sect. 2.1).

Algorithm 2. *Solving quadratic sub-problem at machine m*
$\Delta\beta^m \leftarrow 0$
Cycle over j in S_m:

1.   *Minimize $L_q(\beta, \Delta\beta^m) + ||\beta + \Delta\beta^m||_1$ with respect to $\Delta\beta_j^m$ using (6)*

return $\Delta\beta^m$

Like in other Newton-like algorithms a line search should be done to guarantee convergence. The Algorithm 3 describes our line search procedure. We found that selecting α_{init} by minimizing the objective (2) (step 2, Algorithm 3) speeds up the convergence of the Algorithm 1. We used $b = 0.5, \sigma = 0.01, \gamma = 0$ in line search procedure for numerical experiments.

Algorithm 3. *Line search procedure*

1. *If $\alpha = 1$ yields sufficient relative decrease in the objective, return $\alpha = 1$.*
2. *Find $\alpha_{init} = \mathrm{argmin}_{\delta < \alpha \leq 1} f(\beta + \alpha\Delta\beta), \delta > 0$.*
3. *Armijo rule: let α be the largest element of the sequence $\{\alpha_{init} b^j\}_{j=0,1,\dots}$ satisfying*

$$f(\beta + \alpha\Delta\beta) \leq f(\beta) + \alpha\sigma D$$

where $0 < b < 1, 0 < \sigma < 1, 0 \leq \gamma < 1$, and

$$D = \nabla L(\beta)^T \Delta\beta + \gamma\Delta\beta^T \tilde{H}\Delta\beta + \lambda(||\beta + \Delta\beta||_1 - ||\beta||_1)$$

return α

2.1   Convergence

Algorithm d-GLMNET falls into the general framework of block-coordinate gradient descent (CGD) proposed by Tseng and Yun [14]. CGD is about minimization of a sum of a smooth function and separable convex function: in our case, negated log-likelihood and L1 penalty. At each iteration CGD solves penalized quadratic approximation problem

$$\underset{\Delta\beta}{\mathrm{argmin}} \left\{ L(\beta) + \nabla L(\beta)^T \Delta\beta + \frac{1}{2}\Delta\beta^T H \Delta\beta + \lambda||\beta + \Delta\beta||_1 \right\} \qquad (10)$$

where H is positive definite, iteration specific. For convergence it also requires that for some $\lambda_{max}, \lambda_{min} > 0$ for all iterations

$$\lambda_{min} I \preceq H \preceq \lambda_{max} I \tag{11}$$

At each iteration updates are done over some subset of features. (That would always be all features in our case, so the rules of subset selection are irrelevant). After that a line search by the Armijo rule should be conducted. Then Tseng and Yun [14] prove that $f(\boldsymbol{\beta})$ converges as least Q-linearly and $\boldsymbol{\beta}$ converges at least R-linearly.

d-GLMNET inherits the properties of newGLMNET, for which Yuan et al. [5] already proved that it belongs to the CGD framework and inferred the convergence results. That's why we only give the sketch of the proof, outlining the difference. newGLMNET algorithm in (10) for H uses full Hessian $H = \nabla^2 L(\boldsymbol{\beta}) + \nu I$, and Yuan et al. [5] proves (11) for that. Instead, d-GLMNET uses block-diagonal approximation $H = \tilde{H} + \nu I$, where $\tilde{H}$ is defined in (7). That's why CGD iteration (10) for the full set of features is block separable and can be parallelized. To prove (11) for block-diagonal H denote its diagonal blocks by $H^1, \ldots, H^M$ and represent an arbitrary vector $\mathbf{x}$ as a concatenation of subvectors of corresponding size: $\mathbf{x}^T = (\mathbf{x}_1^T, \ldots, \mathbf{x}_M^T)$. Then we have

$$\mathbf{x}^T H \mathbf{x} = \sum_{m=1}^{M} \mathbf{x}_m^T H^m \mathbf{x}_m$$

Notice that $H^m = \nabla^2 L(\boldsymbol{\beta}^m) + \nu I$, where $\nabla^2 L(\boldsymbol{\beta}^m)$ is a Hessian over the subset of features S_m. So for each H^m property (11) is already proved in [5]. That means $\lambda_{min} ||\mathbf{x}_m||^2 \leq \mathbf{x}_m^T H^m \mathbf{x}_m \leq \lambda_{max} ||\mathbf{x}_m||^2$ for $m = 1, \ldots, M$, and we obtain the required

$$\lambda_{min} ||\mathbf{x}||^2 \leq \mathbf{x}^T H \mathbf{x} \leq \lambda_{max} ||\mathbf{x}||^2.$$

3   Scalable Software Implementation

Typically most of datasets are stored in "by example" form, so a transformation to "by feature" form is required for d-GLMNET. For large datasets this operation is hard to do on a single machine. We use a Map/Reduce cluster [16] for this purpose. This transformation typically takes 1–5 % of time relative to the regularization path calculating (Sect. 4.2). Training dataset partitioning over machines is done by means of a Reduce operation. We did not implemented d-GLMNET completely in the Map/Reduce programming model since it is ill-suited for iterative machine learning algorithms [9,17].

In d-GLMNET machine m solves at each iteration the sub-problem (9). The machine m stores the part X_m of training dataset corresponding to a subset S_m of input features. $X_m = \{L_j | j \in S_m\}$ where $L_j = \{(i, x_{ij}) | x_{ij} \neq 0\}$. Our program expects that input file is already in "by feature" representation, see Table 1. This format of input file allows to read training dataset sequentially from the disk and make coordinate updates (6) while solving sub-problem (9). Our program stores into the RAM only vectors: $\mathbf{y}$, $(\exp(\boldsymbol{\beta}^T x_i))$, $(\Delta\boldsymbol{\beta}^T x_i)$, $\boldsymbol{\beta}$, $\Delta\boldsymbol{\beta}$. Thus the total memory footprint of our implementation is $O(n + p)$.

Table 1. Input file format

feature_id	(example_id, value)	(example_id, value)	...	feature_id	(example_id, value)	...

Algorithm 4 presents a high-level structure of our software implementation[1]. We consider this as a general framework for distributed block-coordinate descent, which can be used with various types of updates during step 2.

Algorithm 4. *Distributed coordinate descent*
Repeat until convergence:

1. *Do in parallel over M machines*
2.   *Read part of training dataset X_m sequentially; make updates of $\Delta\boldsymbol{\beta}^m$,* $(\Delta(\boldsymbol{\beta}^m)^T x_i))$
3. *Sum up vectors $\Delta\boldsymbol{\beta}^m$, $(\Delta(\boldsymbol{\beta}^m)^T x_i)$ using MPI_AllReduce:*
4. $\Delta\boldsymbol{\beta} \leftarrow \sum_{m=1}^{M} \Delta\boldsymbol{\beta}^m$
5. $(\Delta\boldsymbol{\beta}^T \mathbf{x}_i) \leftarrow \sum_{m=1}^{M} (\Delta(\boldsymbol{\beta}^m)^T \mathbf{x}_i)$
6. *Find step size α using line search (Algorithm 3)*
7. $\boldsymbol{\beta} \leftarrow \boldsymbol{\beta} + \alpha\Delta\boldsymbol{\beta}$,
8. $(\exp(\boldsymbol{\beta}^T x_i)) \leftarrow (\exp(\boldsymbol{\beta}^T x_i + \alpha\Delta\boldsymbol{\beta}^T x_i))$

Sequential data reading from disk instead of RAM may slow down the program in case of smaller datasets, but it makes the program more scalable. Also it conforms to the typical pattern of a multi-user cluster system: large disks, many jobs started by different users are running simultaneously. Each job might process large data but it is allowed to use only a small part of RAM at each machine.

Solving sub-problem (9) during step 2 in Algorithm 4 requires $O(nnz)$ operations and it is well suited for large and sparse datasets. The communication cost during step 3 in Algorithm 4 is $O(n + p)$.

4 Numerical Experiments

4.1 Datasets and Experimental Settings

We used three datasets for numerical experiments. These datasets are from the Pascal Large Scale Learning Challenge 2008.[2]

---

[1] We used an implementation of MPI_AllReduce from the Vowpal Wabbit project https://github.com/JohnLangford/vowpal_wabbit.
[2] http://largescale.ml.tu-berlin.de/.

- **epsilon** - A synthetic dataset, we used preprocessing and train/test splitting from http://www.csie.ntu.edu.tw/~cjlin/libsvmtools/datasets/binary.html.
- **webspam** - Webspam classification problem, we used preprocessing and train/ test splitting from http://www.csie.ntu.edu.tw/~cjlin/libsvmtools/datasets/binary.html.
- **dna** - Splice cite recognition problem. We did the same preprocessing as in challenge (see ftp://largescale.ml.tu-berlin.de/largescale/dna/) and did train/ test splitting.

Table 2. Datasets summary

Dataset	Size	# examples (train/test)	# features	nnz	Avg nonzeros
epsilon	12 Gb	$0.4 \times 10^6/0.1 \times 10^6$	2000	8.0×10^8	2000
webspam	21 Gb	$0.315 \times 10^6/0.035 \times 10^6$	16.6×10^6	1.2×10^9	3727
dna	71 Gb	$45 \times 10^6/5 \times 10^6$	800	9.0×10^9	200

The datasets are summarized in Table 2. Numerical experiments were carried out at 16 multicore blade servers having Intel(R) Xeon(R) CPU E5-2660 2.20 GHz, 32 GB RAM, connected by Gigabit Ethernet. Each server ran one instance of d-GLMNET or Vowpal Wabbit at once.

4.2 Experimental Protocol for d-GLMNET

We tested d-GLMNET by solving the problem (1) for a set of regularization parameters, see Algorithm 5.

Algorithm 5. *Computing the regularization path*
Find λ_{max} for which entire vector $\beta = 0$.
For $i = 1$ to 20
*   Solve (1) with $\lambda = \lambda_{max} * 2^{-i}$ using previous β as a warmstart*

For each λ we calculated for a corresponding final β the testing quality and the number of non-zero entries. For the "dna" dataset we tested 4 additional regularization parameters $\lambda \in [2730.7, 5461.3]$ because of low density of points in the region with $100 - 300$ non-zero features (Fig. 1c).

4.3 Experimental Protocol for Distributed Online Learning via Truncated Gradient

We compared d-GLMNET with the distributed variant of online learning via truncated gradient. The online learning via truncated gradient was presented in [7]. An idea for adapting it to the distributed settings was presented in [9]. We used the first part of [9, Algorithm 2] which proposes to compute a weighted average of classifiers trained at M machines independently. The second part of this algorithm takes the result of the first part as a warmstart for L-BFGS. As we pointed out earlier L-BFGS it not applicable for solving logistic regression with L1-regularization. This algorithm requires training dataset partitioning by examples over M machines.

The Algorithm 2 from [9] is implemented in the Vowpal Wabbit[3] project. We didn't use feature hashing since it may decrease the quality of a classifier. We tested the same set of regularization parameters as for d-GLMNET, i.e. $\lambda \in \{\lambda_{max}2^{-1}, \lambda_{max}2^{-2}, \ldots, \lambda_{max}2^{-20}\}$[4]. Since online learning has many free parameters we made a full search for "epsilon" and "webspam" datasets. We tested jointly learning rates (raging from 0.1 to 0.5), decays of the learning rate (raging from 0.5 to 0.9) for each λ and allowed Vowpal Wabbit to make 50 passes of online learning. After each pass we saved a vector β. After training we evaluated a quality of all classifiers at the test set and counted the number of non-zero entries in β.

For the biggest dataset "dna" we did 25 passes and used default learning rate (0.1) and decay (0.5). We also tested additional range of regularization parameter $\lambda \in \{10.7, 10.7 \times 2^{-1}, \ldots, 10.7 \times 2^{-9}\}$ since Vowpal Wabbit produced only very sparse classifiers with low testing quality.

Table 3. Execution times

Dataset	d-GLMNET				Vowpal Wabbit
	# iter	Time, sec	Linear search	Avg time per iter, sec	Avg time per iter, sec
epsilon	182	1667	5 %	9	30
webspam	269	6318	6 %	23	50
dna	123	17626	25 %	143	59

4.4 Results

Figure 1 demonstrates results of the experiments: area under Precision-Recall curve on the test set against the number of non-zero components in the β. We compare results for the whole regularization path of d-GLMNET and each parameter combination and pass number for Vowpal Wabbit. The d-GLMNET algorithm is a clear winner: for each data set, each degree of sparsity, it yields the same or better testing quality. We notice that for online learning different combinations of parameters yield very different results. Online learning is often advertised as a very fast method, but the need to perform a search of good parameters lessens this advantage. At the same time the d-GLMNET algorithm has no free parameters except a regularization coefficient.

---

[3] https://github.com/JohnLangford/vowpal_wabbit, we used version 7.5.

[4] The parameter λ in (2) is related to the option - -l1 arg in Vowpal Wabbit by equation $arg = \lambda/n$ where n is the number of training examples.

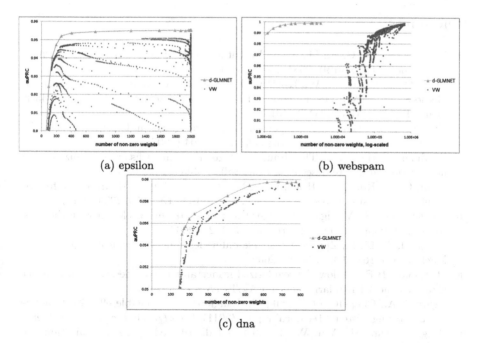

(a) epsilon   (b) webspam

(c) dna

Fig. 1. Testing quality (area under Precision-Recall curve) versus number of non-zero elements in β

Table 3 presents execution times for the whole regularization pass for each dataset, total number of iterations, and average time per iteration. We found that linear search does not hurt much the performance - it takes 5–25 % time at different datasets. There is no direct time comparison between d-GLMNET and Vowpal Wabbit because of the parameter search for the latter. The last column in the table gives average time per iteration for Vowpal Wabbit: this can be compared to the same number for d-GLMNET, because one iteration for both algorithms corresponds to one full pass over the training data set, and has the same computational complexity $O(nnz)$.

Conclusions and Future Work. In this paper we presented d-GLMNET - a new algorithm for solving logistic regression with L1 regularization via distributed coordinate descent and its efficient software implementation. We proved by a set of numerical experiments that it is superior over distributed online learning via truncated gradient. Future work may include extending d-GLMNET to L2 regularization and doing computations on multicore systems.

Acknowledgments. We would like to thank John Langford for the advices on Vowpal Wabbit and Ilya Muchnik for his continuous support.

References

1. Yuan, G.-X., Ho, C.-H., Lin, C.-J.: Recent advances of large-scale linear classification. Proc. IEEE **100**(9), 2584–2603 (2012)
2. Yuan, G.-X., Chang, K.-W., Hsieh, C.-J., Lin, C.-J.: A comparison of optimization methods and software for large-scale L1-regularized linear classification. J. Mach. Learn. Res. **11**, 3183–3234 (2010)
3. Genkin, A., Lewis, D.D., Madigan, D.: Large-scale Bayesian logistic regression for text categorization. Technometrics **49**(3), 291–304 (2007)
4. Friedman, J., Hastie, T., Tibshirani, R.: Regularization paths for generalized linear models via coordinate descent. J. Stat. Softw. **33**(1), 1–22 (2010)
5. Yuan, G.-X., Ho, C.-H., Hsieh, C.-J., Lin, C.-J.: An improved GLMNET for L1-regularized logistic regression. J. Mach. Learn. Res. **13**, 1999–2030 (2012)
6. Balakrishnan, S., Madigan, D.: Algorithms for sparse linear classifiers in the massive data setting. J. Mach. Learn. Res. **1**, 1–26 (2007)
7. Langford, J., Li, L., Zhang, T.: Sparse online learning via truncated gradient. J. Mach. Learn. Res. **10**, 777–801 (2009)
8. McMahan, H.B.: Follow-the-regularized-leader and mirror descent : equivalence theorems and L1 regularization. In: AISTATS 2011 (2011)
9. Agarwal, A., Chapelle, O., Dudík, M., Langford, J.: A reliable effective terascale linear learning system. Technical report (2011). http://arxiv.org/abs/1110.4198
10. Peng, Z., Yan, M., Yin, W.: Parallel and distributed sparse optimization. In: STATOS 2013 (2013)
11. Bradley, J.K., Kyrola, A., Bickson, D., Guestrin, C.: Parallel coordinate descent for L1-regularized loss minimization. In: ICML 2011, Bellevue, WA, USA (2011)
12. Ho, Q., Cipar, J., Cui, H., Kim, J.K., Lee, S., Gibbons, P.B., Gibson, G.A., Ganger, G.R., Xing, E.P.: More effective distributed ML via a stale synchronous parallel parameter server. In: NIPS 2013 (2013)
13. Richtárik, P., Takáč, M.: Parallel coordinate descent methods for big data optimization. Technical report (2012). http://arxiv.org/abs/1212.0873
14. Tseng, P., Yun, S.: A coordinate gradient descent method for nonsmooth separable minimization. Math. Program. **117**, 387–423 (2009)
15. Zinkevich, M., Weimer, M., Smola, A., Li, L.: Parallelized stochastic gradient descent. In: NIPS 2010 (2010)
16. Dean, J., Ghemawat, S.: MapReduce: simplified data processing on large clusters. In: OSDI 2004, San Francisco (2004)
17. Low, Y., Gonzalez, J., Kyrola, A., Bickson, D., Guestrin, C., Hellerstein, J.M.: Graphlab: a new framework for parallel machine learning. In: UAI 2010, Cataline Island, California (2010)

Social Network Analysis

Building Profiles of Blog Users Based on Comment Graph Analysis: The Habrahabr.ru Case

Alexandra Barysheva, Mikhail Petrov, and Rostislav Yavorskiy[✉]

Higher School of Economics National Research University, Moscow, Russia
{asbarysheva,mvpetrov_1}@edu.hse.ru, ryavorsky@hse.ru

Abstract. Our study is aimed at developing a language-independent tool for building user profiles of online community users. To that end the definition of a comment graph, a convenient representation of users interaction, is studied. The set of comment graph characteristics for users that form the basis of the profiling techniques is suggested. Finally, the user profiling method based on cluster analysis is presented. The described method was applied to Habrahabr data set.

Keywords: Blog · Comment graph · User profiling

1 Introduction

This paper focuses on analysis of online communities organised in a form of a blog. A blog (or a weblog) is a type of a website where a person (a blogger) publishes information, news, opinions or thoughts, on a regular basis. A blog consists of entries, known as posts, written by a blogger. In addition, blog users can interact with each other by commenting posts or other comments.

The goal of this work is to provide a method of profiling blog users.

All blogs have the structure of a comment graph. A comment graph is a kind of a social graph (social network) where vertices correspond to users and edges that represent the relation "user A comments a post of user B". The comment graph shows how discussion participants interact with each other.

In order to develop language-independent method for modeling profiles of users, six users attributes on comment graphs – the basis of profiling methods – are described below, as well as the method for retrieving behavioral patterns of users based on clustering technique. Finally we describe the results of application of the presented method to the data set collected from the Russian Information Technology blog service HabraHabr[1]. Also we want to present a tool that visualises online conversations in terms of user roles in online community.

---

[1] http://habrahabr.ru/.

© Springer International Publishing Switzerland 2015
M.Y. Khachay et al. (Eds.): AIST 2015, CCIS 542, pp. 257–262, 2015.
DOI: 10.1007/978-3-319-26123-2_25

2   Related Work

A user profile can include information about user interests, user behavior or personal characteristics like gender, age, etc.

The task of modeling user profile consists of many subtasks and different tools can use various approaches: content-based and tag-based methods, tracing user activity or discovery of user friend-connections etc. Typically, the majority of known profiling methods combine different techniques.

An example of Twitter user classification is presented in [1]. This paper explores the methods of automatic inferring the values of user attributes such as political orientation or ethnicity. Authors apply a machine learning approach which relies on an exhaustive set of features obtained from such observable information as network structure, the user behavior and the linguistic content of the users Twitter feed.

In [2] authors also study algorithms for profiling Twitter users, based on their tweets and community relationships. They propose a hybrid community-based and text-based method for the demographic estimation of Twitter users. These demographic characteristics were evaluated by clustering of followers and tracking the tweet history.

It is important to note that work [3] is more relevant to our study in case of discovering bloggers who take an important role in conversation. The article provided several "definitions of bloggers roles including (1) agitators who stimulate discussion, and (2) summarisers who provide summaries of the discussion." The method proposed by the authors uses analysis of links between blogs.

Actually, quite often papers focus on user activity patterns, ignoring the content of messages users exchange. One of the articles that combines these two approaches is [4]. In fact, the authors analyse both tweeting patterns and social interactions.

Furthermore, some researchers use only linguistic content of posts. For instance, in article [5] Natural Language Processing was used in order to predict age and gender of the blog users. The approach investigates three types of features: content, style of the posts and topics.

The task of user profiling also naturally arises in the field of expertise retrieval, see [6].

3   Definition of a Comment Graph

All user interactions with other users can be represented as a comment graph. This term can be interpreted in two ways. First of all, it is a graph of entire community, let us denote it as *community graph*. It should be emphasised that by community we mean all posts on a particular topic. Secondly, each particular post has its own comment graph (further we refer to this kind of graph as *blog post graph*).

3.1   Community Graph

Community graph is a graph $G_c = (V_c, E_c)$, where V_c include all community members that published at least one comment or post, E_c represents the relation "user A comments a post of user B". The graph is directed, i.e. the edge is directed from a user that leaves a comment to a person who is commented. The graph edges are weighted by the number of comments a user gave to the other user. Loops are allowed.

3.2   Blog Post Graph

Blog post graph is a graph $G_p = (V_p, E_p)$, where V_p is a set of vertices and G_p is a set of edges. V_p consists of a post and all comments to the post. Every vertex is labeled by the author of the comment. E_p represent the relation "comment A is a comment to comment B". The graph is directed and has is a tree structure that reflects the commenting process. The root of such a tree is the initial blog post, therefore it is clear from the graph structure where the conversation starts and ends.

4   User Clustering Algorithm

In this study we want to analyse online discussion participants of blogs in terms of behavioral patterns in commenting process. The main idea is to group people with similar behavior. The task of grouping a set of objects in such a way that objects in the same group are more similar to each other than to those in other groups is the aim of cluster analysis. In this way, we refer the term profile to the description of a cluster a user belongs to.

4.1   User Attributes

First of all user attributes to perform clustering should be chosen. In this work the following characteristics were calculated from the comment graphs.

Community Graph Characteristics

1. In-degree, the number of people that leave comment to users entries.
2. Out-degree, the number of people that were commented by a user.

Blog Post Graph Characteristics

3. The number of posts a user left (how often a user was the author of a post).
4. The number of posts a user commented. It helps to separate users that are interested in only one topic from those who take active part in all discussions of the community.

There are two more indices designed specially for this study:

5. Average distance between a node and a comment (in terms of comments number in between). It shows the level of a users involvement in long chains of comments.
6. Share of terminals. It counts the fraction of user's comments that was not commented among all user's comments.

4.2 Clustering

In this work well-established k-means algorithm [7] with Euclidean metric as distance between users was chosen for user clustering. It allows one to receive disjoint groups of people with similar behavior.

5 Habrahabr Data Set

Habrahabr is an online blogging service that attracts people who are interested in topis from information technologies and related fields. For our study a set of personal blogs was collected. The data set contains 200 blog post graphs from different hubs. It describes the situation in community of popular hubs during two subsequent months. It was figured out that more than 2300 users commented these posts and the most active person posted about 200 comments on average.

6 Results

According to the results of clustering all users in the dataset are split into the following eight groups with clearly defined patterns of behavior. It should be mentioned that the group *Observers* is preselected before clustering; this group consists of users that left less than three comments. To identify the optimal number of clusters we have used the silhouette measure [8]. Before clustering all the attributes has been normalised.

1. Stars.
   (a) *Silent stars* are authors of popular posts who do not participate in the discussions.
   (b) *Communicative stars* are authors of popular posts who are actively involved in the discussions.
2. Chatters are people who keep the discussion going on. Contrary to concluders (who usually terminate the thread, see below) their comments typically induce one or more replying comments.
   (a) *Active chatters* are participants who leave many comments, and reply to almost every comment on their posts.
   (b) *Idle chatters* are people who write few comments, but usually their comments support the subsequent debate.
3. Socialisers are users who do not produce many comments, although the number of people they talk to is notably high.

(a) *Investigators* are participants who communicate with many people within very narrow discussion (few blog posts).

(b) *Socialisers* are those who communicate with many people in various discussions.

4. *Concluders* are participants, who produce little comments and quite often one of their comment is the last one in the discussion branch.

5. *Observers* are community members with very low activity.

Table 1 represents average values for each cluster.

Table 1. Parameters of the clusters.

Cluster name	Number of users	In degree	Out degree	Avg. distance	Share of terminals	Posts left	Posts commented
Silent stars	2	**143,5**	2,0	1,0	13 %	12,0	1,5
Communicative stars	2	85,0	65,0	3,8	44 %	2,5	**9,5**
Active chatters	38	4,0	4,3	**8,9**	23 %	0,0	1,7
Investigators	50	14,0	12,5	4,6	43 %	0,5	**2,5**
Socialisers	122	11,0	13,1	3,5	40 %	0,1	**7,1**
Concluders	361	4,2	4,1	2,7	**61 %**	0,2	2,2
Idle chatters	498	3,8	2,6	2,8	12 %	0,1	1,7
Observers	1277	0,6	1,2	2,1	57 %	0,0	1,1

7   Visualisation Tool

This work is a part of an ongoing project on developing of a tool for visualised monitoring and analysis of professional online communities. The product is developed under open-source license. One can assess the current state of the project by the following link http://github.com/ryavorsky/HabraGraph. For testing purposes the tool is available at http://habragraph.1gb.ru. The project is open for participants.

8   Conclusion

In this paper we have proposed the classification of online community members of the web portal habrahabr.ru. The profiling method is based on their activity in the process of commenting and discussion of blog posts. To make our approach suitable for the analysis of online communities, regardless of the language of communication, we excluded from consideration the content of blogs and comments.

The novelty of the proposed approach is that in addition to the standard features of the graph (such as in- and out-degree) we also take into account the parameters describing the location of comments in the discussion thread: distance from a comment to the initial post, and how often a user's comment was a terminal.

The results can be used to extend the functionality of the blogs in the detailed description of the profile of the participants, which in turn is a tool to motivate members of online communities and monitor their level of involvement. Moreover, it can be a reasonable method of professional community analysis. One can find examples of our analyses of professional communities in [9, 10].

References

1. Pennachiotti, M.A.: Machine learning approach to twitter user classification. In: Fifth International AAAI Conference on Weblogs and Social Media, p. 45 (2011)
2. Kazushi, I.: Twitter user profiling based on text and community mining for market analysis. Knowl.-Based Syst. **51**, 35–47 (2013)
3. Shinsuke, N.: Discovering important bloggers based on analyzing blog threads. In: 2nd Annual Workshop on the Webblogging Ecosystem (2005)
4. Rocha, E.: User profiling on Twitter. Semant. Web: Interoperability, Usability, Applicability **1**(1), 105–110 (2011)
5. Santosh, R.: Author profiling: predicting age and gender from blogs. In: PAN at CLEF (2013)
6. Balog, K., Fang, Y., de Rijke, M., Serdyukov, P., Si, L.: Expertise retrieval. Found. Trends Inf. Retrieval **6**(2–3), 127–256 (2012)
7. MacQueen, J.B.: Some methods for classification and analysis of multivariate observations. In: Proceedings of 5th Berkeley Symposium on Mathematical Statistics and Probability. University of California Press, pp. 281–297 (1967)
8. Rousseeuw, P.: Silhouettes: a graphical aid to the interpretation and validation of cluster analysis. J. Comput. Appl. Math. **20**(1), 53–65 (1987)
9. Krasnov, F., Yavorskiy, R.: Measurement of maturity level of a professional community. Bus. Inf. **1**(23), 64–67 (2013). (in Russian)
10. Yavorskiy, R., Vlasova, E., Krasnov, F.: Connectivity analysis of computer science centers based on scientific publications data for major Russian cities. In: Procedia Computer Science, vol. 31, pp. 892–899 (2014)

Formation and Evolution Mechanisms in Online Network of Students: The Vkontakte Case

Sofia Dokuka$^{(\boxtimes)}$, Diliara Valeeva, and Maria Yudkevich

Center for Institutional Studies, NRU HSE, Moscow, Russia
{sdokuka,dvaleeva,yudkevich}@hse.ru

Abstract. The mechanisms of real-world social network formation and evolution are one of the most important topics in the field of network science. In this study we collect data about the development of the Vkontakte (a popular Russian social networking site) network of first-year students at a Russian university. We analyze the network formation process from the moment of network establishing until its stabilization. Using Conditional Uniform Graph Test, we compare the graph-level indices of the observed network with random same-size networks that were generated according to random, preferential attachment, and small-world algorithms. We propose two explanatory mechanisms of online network growth: the connected component attachment mechanism and the brokerage mechanism.

Keywords: Network growth · Network evolution · Online networks · Student networks · Higher education

1 Introduction and Related Works

Since their invention, online social networking sites (SNS) such as Facebook, Twitter, LinkedIn, and many others have attracted millions of users. Young people actively use SNS to communicate with their friends, look for people with similar interests, and obtain various types of information [1]. In [2–4], the authors show that students positions in social networks might influence their present and future achievements such as academic performance or future success on the job market. To answer the question of how online social networks might influence student achievements, we need a detailed study of network formation and evolution processes.

In this study we address the question how student online social networks are formed from the moment of inception till network stabilization. We trace the student online network from a time when most students do not personally know each other to the moment when their online social network growth is stabilized. We are interested in the formation and evolution of the network as well as in the network properties that are involved in these processes.

Our study brings together two groups of studies established in the literature. In the first group there is a number of works where the analysis of

© Springer International Publishing Switzerland 2015
M.Y. Khachay et al. (Eds.): AIST 2015, CCIS 542, pp. 263–274, 2015.
DOI: 10.1007/978-3-319-26123-2_26

dynamic network evolution is performed without in-depth analysis of node-level characteristics and their changes. Some of these studies focus specifically on online communities and the predictors of their growth and success. For example, Backstrom et al. [5] predict network structures participating in group growth and success. While [5] try to answers how development and sustainability of online communities might be explained, Kairam et al. [6] go further. They estimate the role of network structural characteristics such as group clustering, transitivity, cliques and, analyzing these network properties, distinguish two different types of community growth: diffusion and non-diffusion.

There are also papers that focus on explanation of network evolution using a comparison with random graphs models. Leskovec et al. [7] use different random graph models for the analysis of dynamic online networks and show that the linear preferential attachment mechanism explains well the Flickr, Delicious, and Yahoo!Answers networks. Capocci et al. [8] evaluate the role of preferential attachment in the dynamics of the Wikipedia networks and Mislove et al. [9] analyze the role of preferential attachment and triadic closure in the Flickr networks. Garg et al. [10] study the FriendFeed dynamic network and pose the question of how preferential attachment, triadic closure, average path length, and homophily predict network formation with a specific focus on the joint effect of preferential attachment and triadic closure. They show that the age of nodes, proximity between them, and subscription to common services affect network formation and evolution. Hu et al. [11] estimate the structural evolution of Wealink online networks and reveal that dynamics of network characteristics such as density, clustering, heterogeneity, and modularity develop non-monotonically in comparison with random networks. They observe shrinking of the path length and diameter in the observed online networks.

In the second group of studies dynamic online networks are analyzed together with the information dissemination or behavior contagion of actors. In other words, changes in users characteristics and informational flows between nodes depend on network structure. Borge-Holthoefer et al. [12] study how communications on Twitter depend on the composition and structure of protest networks. They show that changes in network structure are related to changes in information diffusion: an increase in network hierarchy is associated with a decrease in the size of message flows between actors. Bakshy et al. [13] observe information spread on Facebook and reveal that weak ties are more important for the spread of information than strong ones. Lewis et al. [14] studied the coevolution of Facebook friendship ties and student tastes. They find that students who share certain tastes in music and in movies, but not in books, are significantly likely to befriend one another. However, they find little evidence for the diffusion of tastes among Facebook friends. Overall, in this group of papers, specific attention is paid to the role of network microstructures in behavior and information transmission, whereas the network structure in general and its evolution is analyzed in a less detailed way.

In summary, in related works we can find studies on the coevolution of dynamic networks and behavior, on the structure of online communities, and the applicability of different random graph models to the explanation of real-world networks. We contribute to previous studies by analyzing the evolution of the observed network by applying random network growth [15], preferential attachment [16], and small-world [17] network dynamics models in the case of Vkontakte social network (a Russian online networking site similar to Facebook). Comparing graph-level measurements of the observed network with random network properties, we can understand how network models might be applicable to real-world online networks and where they fail to describe the dynamics of the observed network. Additionally, we contribute to related works by analyzing network dynamics in-depth from the very formation to final stabilization moments.

2   Network Growth Models

There are at least three theoretical constructions that allow analysis of the emergence and evolution of social networks: the random network growth model, the preferential attachment model, and the small-world model.

The random network growth model was proposed by Erdos and Renyi [15]. The Erdos-Renyi random graph is obtained by starting with a set of n isolated nodes and adding successive edges between them at random. The Erdos-Renyi networks do not adequately describe degree distribution and clustering coefficient of real networks.

To capture the network degree distribution properties, Barabasi and Albert proposed the preferential attachment network model [16]. The preferential attachment network begins with an initial connected network and then new nodes are added to it. The probability of connecting to the particular node is proportional to the number of edges that the existing nodes already have. The scale-free networks capture well the degree distribution of real-world networks but they cannot describe the high clustering.

The average number of steps along the shortest paths for all possible pairs of network nodes (average path length) tends to be relatively small for real-world networks. The algorithm of small-world network formation was developed by Watts and Strogatz [17]. The small-world network begins from a regular ring lattice with n nodes which is connected to k neighbors. The number of closed triads in the network will be relatively high, resulting in a high clustering coefficient. To provide the short average path length property to the model, the option of edge rewiring is added to the model. This means that the edge disconnects from one of its nodes and randomly connects to another node. However, the small-world network model does not capture the scale-free property and produces unrealistic degree distribution.

Altogether, these models of network formation are well developed and their new transformations are still appearing today. At the same time, these models are not widely applied to the analysis of dynamic online networks. Data about online networks are widely used in network science today because the information about both actors and their connections might be gathered temporally and without large costs. Still, online networks reflect the main network mechanisms that we can observe in other networks (preferential attachment, triadic closure, short average path length, network diameter etc.). In this study, we compare online network evolution mechanisms to the aforementioned random network mechanisms.

3 Data

3.1 Description of the Case

We study online friendship networks of first-year students of Economics Department at one Russian university in the 2014–2015 academic year. In Russia students might enroll in universities by several different trajectories. One of them is participation in Olympiads. Olympiads are all-Russian competitions for talented school students in a wide range of subjects. Students that become winners and awardees of these Olympiads can be accepted to any university of their choice without additional exams free of tuition. The winners and awardees of selected Olympiads have the right of prior enrolment in universities. While on average about 5 % of school graduates are admitted as Olympiad winners, in the university under study (which is one of the top universities in the country) the share of Olympiad winners is about 24 %.

Another way to apply to Russian universities is to submit the results of the Unified State Examination (USE) to different schools. The USE is a standardized test in various subject areas and has a unified grading scale throughout all of Russia. High school graduates are obliged to pass the USE in the Russian language and Mathematics, whereas other subject tests are not obligatory to graduate from school. Universities accept students with the highest USE scores free of tuition. Other students that are not Olympiad winners and have low USE scores, might study with full tuition.

Overall, in 2014, 97 students who studied both at tuition-free and full tuition places were matriculated. In this university, students study in groups of up to 30 students. The division into study groups is random. In first year of study most of the courses are obligatory and students cannot change their classes. Lectures are usually delivered to several groups simultaneously, while seminar classes are delivered to each group separately.

A very small number of students in our sample knew each other before the university because they might have studied at the same school or participated in same Olympiads. However, most of the students see each other for the first time on September 1 that is the official start of classes. We observe a natural experiment with the September 1 as a starting event that promoted student communication. During the first couple of weeks after September 1 students are actively engaged in network formation, whereas before and after this period their networks remain relatively stable.

3.2 Data Collection Procedure

The list of students matriculated to the university is public. Therefore, we found students Vkontakte (Russian Facebook) profiles according to their first and second names in a public list of students. Sometimes there were several profiles of users with the same names. In such cases, we selected users in the age cohort of 17–20 years and additionally checked for their city and university they study in. In cases when we could not find student profiles, we searched for these students in each following wave of data collection. If we could find these students, we added them to our database. After the identification of students by their names, we could obtain their unique ID on Vkontakte (in the URL bar).

We gathered data about student ego-networks using Vkminer application[1]. Vkontakte has an open API system that allows downloading data about friends, pictures, audio, videos, and other open information about Vkontakte user pages. Using the open API, Vkminer application downloads relevant data based on the unique ID of each user. The Vkminer application forms data in Microsoft Excel spreadsheets. To the purpose of this research, we gather data on student friendship ties but do not analyze their full ego-networks. Network analysis and modelling were done in *igraph* package [18] in R statistical environment [19].

Overall, we gathered data on 13 waves with an average time period between waves in one week and fixing for important events that could change the online network structure. Data collection stopped October 14 because there were no significant changes in the network from the middle of September. For the additional control of network growth, we collected one wave on November 13. This wave showed that the network remained stable until November. A description of the main stages of admission process in the observed university, the number of students that participate in our analysis, and the number of friendship links are presented in Table 1.

4   Results

4.1   Descriptive Statistics

The information on the dynamics of the social network size is presented in Table 1. We see that the number of nodes within social network systematically grows but there are several fluctuations. As far as the number of edges is concerned, we see that at the first moments of observation their number is relatively low. It might tell us that enrolled students do not seek to establish links within the online environment before the offline acquaintance. After the beginning of the study (September, 1), the number of friendship links within the online network increases dramatically. In Fig. 1 we present the visualization of the student friendship network at several selected time points.

In the first time period the observed network is rather sparse: there are few relations between nodes. When the study begins, during the first days of September, network dramatically becomes dense and clusterized. From September 1st to September 5th the number of ties between students increases from 33 to 240. This supports the idea about the importance of offline interaction for the establishment of online ties. Closer to November, the network becomes stable (almost no new nodes or edges appear) and it consists of several clusters.

---

[1] http://linis.hse.ru/en/soft-linis/.

Table 1. Description of time periods, number of nodes and edges

Date of data collection	Event	Number of matriculated students	Number of students that participated in the analysis	Number of friendship links between students in Vkontakte network
July 31	Public list of students that were matriculated on tuition-free places according to their Olympiad results	9	7	2
August 5	Public list of students that were matriculated on tuition-free places according to their USE results	73	59	15
August 11	Public list of students that were matriculated on full tuition places	97	64	18
August 27		97	63	17
August 30		97	64	19
September 1	Official start of the academic year	97	58	33
September 5		97	67	240
September 10		97	61	208
September 21		97	71	385
September 29		97	71	404
October 6		97	71	423
October 14		97	71	431

4.2 Modeling Results

The study of network evolution can be implemented in different ways. To investigate the structure of relatively stable evolving networks we can use stochastic actor-oriented models [20] or separate temporal exponential random graph models [21]. However, these models have several limitations and can be applied only in the case of relatively stable network evolution. Therefore, they cannot be applied in our case, in which we trace the network emergence from the very beginning to the moment of saturation.

To investigate the dynamics of the observed social network we use Conditional Uniform Graph test (CUG-test) [22], which allows us to analyze compared network graph-level indices with the same measures of the random networks. We compare our observed network at each timepoint with three random networks of the same size (based on Erdos-Renyi, Barabasi-Albert, and Watts-Strogatz

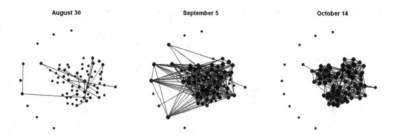

Fig. 1. The network growth. Nodes are students Vkontakte accounts, edges are friendship links between accounts. The size of the node reflects the degree centrality. Nodes are fixed at the same place.

models). The graph-level indices of the random networks were calculated as a mean value from the distribution of 10000 random networks.

We compare the following network indices: clustering coefficient, size of giant component, average path length, network diameter, and degree assortativity. The clustering coefficient reflects the tendency of nodes to group together and form clusters. The size of giant component shows how many nodes can be reached by paths running along the whole network. The measures of average path length and network diameter also reflect the nodes reachability within the given network. Degree assortativity shows the preferential attachment tendency in a network. The results of observed and random comparisons are presented in Figs. 2, 3, 4, 5 and 6.

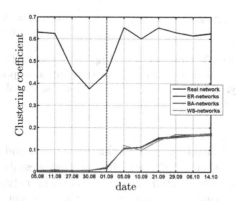

Fig. 2. Clustering coefficient dynamics. Vertical line corresponds to September 1, start of the school year.

We reveal that the observed network behaves in a different way than the simulated random networks. The clustering coefficient of the observed network was much higher ($p < 0.01$) than for any random network (Fig. 2). This means that there are many more close triads within the observed network than could

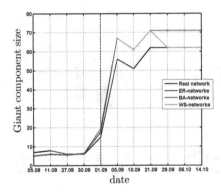

Fig. 3. Giant component size dynamics. Vertical line corresponds to September 1, start of the school year.

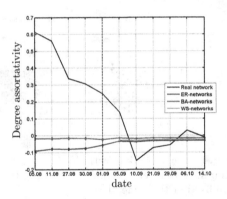

Fig. 4. Degree assortativity dynamics. Vertical line corresponds to September 1, start of the school year.

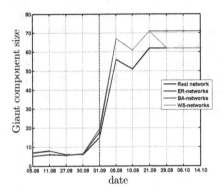

Fig. 5. Average path length dynamics. Vertical line corresponds to September 1, start of the school year.

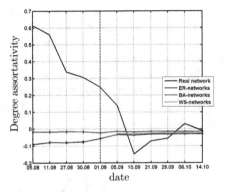

Fig. 6. Network diameter dynamics. Vertical line corresponds to September 1, start of the school year.

be expected by chance. Students tend to cluster in triads and form dense groups with each other. In other words, we observe a well-established tendency of "friends of friends to be connected" in real world social networks [23].

The size of giant connected component is the only network measure that evolves in a similar way for both observed and simulated networks (Fig. 3). Still, the number of nodes in the random networks tend to be higher than in an observed one ($p < 0.01$). This can be explained by taking into consideration that some students did not create links with their peers and remained isolates within the Vkontakte network (see Fig. 1).

The degree assortativity index for the observed network during the first several observation periods was much higher ($p < 0.01$) than can be expected in the case of network randomness (Fig. 4). However, we should take into consideration that there were only few edges within the social network before September 1. After the offline acquaintance, degree assortativity decreased to approximately 0, which means that degree centrality does not play any role in network formation in the late observational steps. We can explain these indices by the absence of hierarchy based on degree after September 1 because students actively befriend each other without any decision based on individual or network characteristics of their peers. Their potential aim during this process is to add the maximum amount of classmates to improve personal position in the network and as a result to gather information from the new environment. At the same time, students before September 1 might be more selective in their decision about forming ties with their future classmates.

The dynamics of average path length and network diameter for the observed network described by random network models are quite acceptable. At the beginning, the average path length and diameter are small because there are few edges in the network. Also, the size of giant connected component is also small, which means that it is easy to reach one node from another. After the September 1 both average path length and network diameter increased in the observed networks (Figs. 4 and 5) because many nodes joined the giant connected component. At the same time, not all of them created many links within this component (Fig. 6). In the first days of September it is difficult for students to reach their peers because the network is not stable and students tend to befriend each other and increase path length. Eventually, as students create many links with their friends within the online environment, the average path length and network diameter decrease. At the late moments of observation the values of average path length and diameter were very similar to the simulated values because the observed network stopped evolving dramatically. After network stabilization, information spread in online network (for example, through reposts on some study information) became more possible due to the decrease in path length and diameter.

The descriptive statistics and CUG-test results give us evidence about two possible mechanisms that are involved in online network formation. The first mechanism is a *connected component attachment mechanism* (Fig. 7). This implies that over time nodes look for an option to join the connected component and create links with other nodes. In other words, the degree centrality x_{i+} of the actor i is becoming higher than 0 ($x_{i+} \geq 1$). During this change from t_1 to t_2, such network characteristics as the size of giant component, average path length, and diameter increase whereas the degree assortativity decrease. This process might be explained by the tendency of students to belong to network core and not to stay on network periphery [24]. Core position and connectedness with peers that belong to the network core might be useful for receiving new information, for discussing study matters with classmates, and for increasing academic performance and social status. The second mechanism is a *brokerage mechanism* (Fig. 8) which assumes that nodes make connections with the friends of their friends and as a result create close triads. During this change from t_1 to t_2, the clustering coefficient increases whereas average path length and network

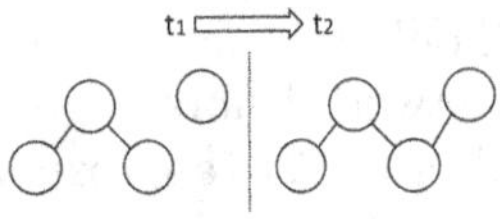

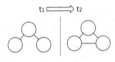

Fig. 7. Connected component attachment mechanism

Fig. 8. Brokerage mechanism

diameter decrease. In other words, the ratio of closed triads to all possible triads becomes higher and closer to 1. The brokerage mechanism refers to the triadic closure network mechanism which is based on the intuition that two people with a common friend tend to become friends with each other. Online social networks provide information about users friends and students tend to arrange new ties with their peers. The resulting dense and highly clustering social network is more efficient in terms of information propagation [25] which is crucial for students who actively use networking sites for information exchange.

4.3   Conclusion

In this study we analyze friendship network evolution on the Vkontakte social networking site. The observed network dynamics are compared with dynamics of random network models: random network growth [15], preferential attachment [16], and small-world [17]. We focus on differences between observed and random networks in graph-level indices such as clustering coefficient, size of giant component, average path length, network diameter, and degree assortativity.

We show that random networks describe well the dynamics of giant connected component size and at certain observational points describe well the dynamics of average path length and network diameter. At the same time, the network clustering coefficient and degree assortativity are not described by random models properly, these measures are significantly higher in the observed networks.

We attribute these results to the presence of attachment to the connected component and brokerage mechanisms in the observed networks. To estimate the extent to which both attachment to connected component and brokerage mechanisms are involved in online network formation, we need more sophisticated network models that can account for these network effects as well. Our future research concerns estimation of such models as well as comparative analysis of network structures in various online networks.

Acknowledgements. The authors thank Olessia Koltsova, Sergey Koltsov, and Vladimir Filippov for the opportunity to use VKminer application. We would like to thank Benjamin Lind for the discussion and feedback on this work. The article was prepared within the framework of the Basic Research Program at the National Research University Higher School of Economics (HSE) and supported within the framework of a subsidy granted to the HSE by the Government of the Russian Federation for the implementation of the Global Competitiveness Program. The financial support from the Government of the Russian Federation within the framework of the implementation of the 5-100 Programme Roadmap of the National Research University Higher School of Economics is acknowledged.

References

1. Subrahmanyam, K., Reich, S.M., Waechter, N., Espinoza, G.: Online and offline social networks: use of social networking sites by emerging adults. J. Appl. Dev. Psychol. **29**, 420–433 (2008)
2. Calvó-Armengol, A., Patacchini, E., Zenou, Y.: Peer effects and social networks in education. Rev. Econ. Stud. **76**, 1239–1267 (2009)
3. Conti, G., Galeotti, A., Mueller, G., Pudney, S.: Popularity. J. Hum. Resour. **48**, 1072–1094 (2013)
4. Fletcher, J.: Friends or family? revisiting the effects of high school popularity on adult earnings. Appl. Econ. **46**, 2408–2417 (2014)
5. Backstrom, L., Huttenlocher, D., Kleinberg, J., Lan, X.: Group formation in large social networks: membership, growth, and evolution. In: Proceedings of the 12th ACM SIGKDD International Conference on Knowledge Discovery and Data Mining, pp. 44–54. ACM (2006)
6. Kairam, S.R., Wang, D.J., Leskovec, J.: The life and death of online groups: predicting group growth and longevity. In: Proceedings of the Fifth ACM International Conference on Web Search and Data Mining, pp. 673–682. ACM (2012)
7. Leskovec, J., Backstrom, L., Kumar, R., Tomkins, A.: Microscopic evolution of social networks. In: Proceedings of the 14th ACM SIGKDD International Conference on Knowledge Discovery and Data Mining, pp. 462–470. ACM (2008)
8. Capocci, A., Servedio, V.D., Colaiori, F., Buriol, L.S., Donato, D., Leonardi, S., Caldarelli, G.: Preferential attachment in the growth of social networks: the internet encyclopedia wikipedia. Phys. Rev. E **74**, 036116 (2006)
9. Mislove, A., Koppula, H.S., Gummadi, K.P., Druschel, P., Bhattacharjee, B.: Growth of the flickr social network. In: Proceedings of the First Workshop on Online Social Networks, pp. 25–30. ACM (2008)
10. Garg, S., Gupta, T., Carlsson, N., Mahanti, A.: Evolution of an online social aggregation network: an empirical study. In: Proceedings of the 9th ACM SIGCOMM Conference on Internet Measurement Conference, pp. 315–321. ACM (2009)
11. Hu, H., Wang, X.: Evolution of a large online social network. Phys. Lett. A **373**, 1105–1110 (2009)
12. Borge-Holthoefer, J., Baños, R.A., González-Bailón, S., Moreno, Y.: Cascading behaviour in complex socio-technical networks. J. Complex Netw. **1**, 3–24 (2013)
13. Bakshy, E., Rosenn, I., Marlow, C., Adamic, L.: The role of social networks in information diffusion. In: Proceedings of the 21st International Conference on World Wide Web, pp. 519–528. ACM (2012)
14. Lewis, K., Gonzalez, M., Kaufman, J.: Social selection and peer influence in an online social network. Proc. Natl. Acad. Sci. **109**, 68–72 (2012)
15. Erdos, P., Rényi, A.: On the evolution of random graphs. Publ. Math. Inst. Hung. Acadamy Sci. **38**, 343–347 (1961)
16. Barabási, A.L., Albert, R.: Emergence of scaling in random networks. Science **286**, 509–512 (1999)
17. Watts, D.J., Strogatz, S.H.: Collective dynamics of "small-world" networks. Nature **393**, 440–442 (1998)

18. Csardi, G., Nepusz, T.: The igraph software package for complex network research. Inter. J. Complex Syst. **1695**, 1–9 (2006)
19. Statistical Package, R.: R: a language and environment for statistical computing. R Foundation for Statistical Computing, Vienna, Austria (2009)
20. Snijders, T.A., Van de Bunt, G.G., Steglich, C.E.: Introduction to stochastic actor-based models for network dynamics. Soc. Netw. **32**, 44–60 (2010)
21. Krivitsky, P.N., Handcock, M.S.: A separable model for dynamic networks. J. R. Stat. Soc. Ser. B (Statistical Methodology) **76**, 29–46 (2014)
22. Anderson, B.S., Butts, C., Carley, K.: The interaction of size and density with graph-level indices. Soc. Netw. **21**, 239–267 (1999)
23. Goodreau, S.M., Kitts, J.A., Morris, M.: Birds of a feather, or friend of a friend? using exponential random graph models to investigate adolescent social networks. Demography **46**, 103–125 (2009)
24. Vaquero, L.M., Cebrian, M.: The rich club phenomenon in the classroom. Scientific reports 3 (2013)
25. Burt, R.S.: Structural holes and good ideas. Am. J. Sociol. **110**, 349–399 (2004)

Large-Scale Parallel Matching
of Social Network Profiles

Alexander Panchenko[1]([⊠]), Dmitry Babaev[2], and Sergei Obiedkov[3]

[1] FG Language Technology, TU Darmstadt, Darmstadt, Germany
panchenko@lt.informatik.tu-darmstadt.de
[2] Tinkoff Credit Systems Inc., Moscow, Russia
dmitri.babaev@gmail.com
[3] National Research University Higher School of Economics, Moscow, Russia
sergei.obj@gmail.com

Abstract. A profile matching algorithm takes as input a user profile of one social network and returns, if existing, the profile of the same person in another social network. Such methods have immediate applications in Internet marketing, search, security, and a number of other domains, which is why this topic saw a recent surge in popularity.

In this paper, we present a *user identity resolution* approach that uses minimal supervision and achieves a precision of 0.98 at a recall of 0.54. Furthermore, the method is computationally efficient and easily parallelizable. We show that the method can be used to match *Facebook*, the most popular social network globally, with *VKontakte*, the most popular social network among Russian-speaking users.

Keywords: User identify resolution · Entity resolution · Profile matching · Record linkage · Social networks · Social network analysis · Facebook · Vkontakte

1 Introduction

Online social networks enjoy a tremendous success with general public. They have even become a synonym of the Internet for some users. While there are clear global leaders in terms of the number of users, such as Facebook[1], Twitter[2] and LinkedIn[3], these big platforms are constantly challenged by a plethora of niche and/or local social services trying to find their place on the market. For instance, VKontakte[4] is an online social network, similar to Facebook in many respects, that enjoys a huge popularity among Russian-speaking users.

Current situation leads to the fact that many users are registered in several social networks. People use different services in parallel as they provide

---

[1] http://www.facebook.com.
[2] http://www.twitter.com.
[3] http://www.linkedin.com.
[4] http://www.vk.com.

© Springer International Publishing Switzerland 2015
M.Y. Khachay et al. (Eds.): AIST 2015, CCIS 542, pp. 275–285, 2015.
DOI: 10.1007/978-3-319-26123-2_27

complimentary features and user bases. For instance, one common pattern for Russian-speaking users is to communicate with Russian-speaking peers with help of Vkontakte and with foreign friends with help of Facebook. Another common pattern is to use LinkedIn for professional and Facebook for private contacts.

Publicly available user information can help in building the next generation of personalised web services, such as search, recommendation systems, targeted marketing, and messaging, to name a few. For instance, Bartunov et al. [1] suggest to use profile matching to perform automatic contact merging on mobile phones. Actually, a similar technology is already integrated in the Android mobile operative system[5]. On the other hand, profile information may be subject to de-anonymization attacks, undesirable for a user [2–4]. No wonder several researchers from information retrieval and security communities recently tried to study methods of user profile correlation across online social networks [4–10].

As information about a single user can be scattered across different networks, integration of data from various platforms can lead to a more complete user representation. Therefore, in many applications it makes sense to build an *integral profile*, featuring information from several sources. In order to do so, it is necessary to perform *user identity resolution*, i.e., to find the same person across various networks. In this paper, we propose a simple, yet efficient method for matching profiles of online social networks.

The contribution of our work is two-fold:

1. We present a new method for matching profiles of social networks. The method has only four meta-parameters. Unlike most existing approaches (see Sect. 2), the method is easily parallelisable and can be used to process the profiles from an entire social network in a matter of hours. We provide an open-source implementation of the method.[6]
2. We present results of the largest matching experiment to date known to us. While most prior experiments operated on datasets ranging from thousands to hundreds of thousands of profiles, we performed a match of 3 million profiles of Facebook (FB) to 90 million profiles of VKontakte (VK), demonstrating that third parties can perform matching on the scale of entire social network. To the best of our knowledge, we are the first to present a matching of FB to VK.

2   Related Work

2.1   Profile Matching

Bartunov et al. [1] developed a probabilistic model that relies on profile attributes and friendship links. The algorithm was tested on roughly 2 thousand Twitter users and 9 thousand Facebook users. The method achieves F-measure up to 0.89 (precision of 1.0 and recall of 0.8). However, the this is a *local* identity resolution

---

[5] https://www.android.com.

[6] https://github.com/dmitrib/sn-profile-matching.

method, that requires profiles to be ego-networks of the seed user. From the other hand, in this paper we present a *global* identity resolution method that can potentially match any user of one network with any user of another network.

Veldman [6] conducted a set of extensive experiments with profile matching algorithms. She used profile similarity metrics based on both attributes (name, email and birth date) and friendship relations. The author performed experiments on 2 thousand profiles of LinkedIn and Hyves social networks.

Malhotra et al. [9] used 30 thousand of paired Twitter and LinkedIn profiles to train several supervised models based on attributes, such as name, user id and location. The authors report an F-measure up to 0.98 with precision up to 0.99.

Sironi [5] also used supervised models based on features stemming from similarity of profile attributes. This experiment was done on 34 thousand of Facebook, Twitter and LinkedIn profiles where 2 thousand were paired. Their approach yields precision and recall around 0.90.

Narayanan and Shmatikov [7] proposed an approach that establishes connections between users based on their friendship relations. This incremental method requires a small initial number of matched profiles and access to a graph of friendship links. The authors used the method to match 224 thousand Twitter users with 3.3 million Flickr users and observed an error rate of 12 %.

Balduzzi et al. [2] showed that matching can be done effectively based on email addresses.

Jain et al. [10] developed a system that takes as input a Twitter account and finds a corresponding Facebook account. The system relies on profile, content, self-mention and network-based similarity metrics.

Goga et al. [4] present a comprehensive study on profile matching technology. The authors try to correlate accounts of Facebook, Twitter, Google+, Flickr, and MySpace to check a feasibility of a de-anonymization attack. They show that up to 80 % of Twitter, Facebook and Google+ profiles from their ground truth can be matched with a nearly zero false positive rate. Their matching method is based on features extracted from user names, locations and pictures unified with help of a binary classifier taking as input two profiles. Two key differences of this method from ours are the following. First, Goga et al. [4] perform no candidate selection. Therefore, in this approach all pairwise comparisons should be done, which is not efficient if one deals with the entire social network. Second, this approach uses no features based on friends similarity, which are core of our approach.

2.2   Name Similarity Matching

Our method heavily relies on name similarity matching. In its simplest form a name can be considered as a string. There is a large body of literature on how to define string similarity [11] and use it to extract similar names from a data set with some works focusing specifically on personal names; see a survey and experimental comparison in [12]. According to this survey, one of the best algorithms for approximate name matching is the algorithm from [13], which uses a prefix tree to efficiently compute the Levenshtein distance. In [14], three generations of name matching methods are identified, with only third-generation methods showing good results in terms of both precision and recall.

3   Dataset

Two social networks were used in our profile matching experiment. One is the biggest Russian social network VKontakte; the other is Facebook, which is also very popular among Russian-speaking users.

In our experiments, we used publicly available data from VK and FB. The matching algorithm is based on name similarity and the friendship relation: each profile is represented by the first and/or last name of the user and by a list of names of his or her friends in the social network. No other characterising features of profiles were used.

3.1   VKontakte

We collected about 90 million VK profiles that set Russia as their current location. We gathered first and second name of each user along with list of her friends using the "users.get" method of the social network API[7]. Therefore, we can assume that in our experiment VK friend lists are *complete*.

3.2   Facebook

We deal with 3 million public Facebook profiles from Russia. User's name can be obtained via the official API[8], but not list of her friends. That is why friend lists were generated from events displayed in user's feed. Users A and B were considered as friends if a message "A and B are now friends" appeared in feeds of A and B. Profile feeds were collected via the "user/feed" method of the FB API. The problems with this approach is that (1) access to users's feed can be restricted by privacy settings; (2) one need to download all wall posts to gather list of friends, which is not always possible due to API restrictions and requires multiple API calls. Therefore, we should assume that in our experiment FB friend lists are *incomplete*.

3.3   Test Data

VKontakte provides a field where a used can specify a link to her FB page. We gathered about 850 thousand known VK-FB profile pairs. However, only 92,488 Facebook users were found in our Facebook dataset out of these 850 thousand profiles. These pairs were used as a ground truth to check correctness of the matching algorithm. A subset of the test data used in our experiments is publicly available[9].

---

[7] https://vk.com/dev/users.get.
[8] https://developers.facebook.com/docs/graph-api.
[9] https://github.com/dmitrib/sn-profile-matching.

3.4   Name Romanisation

Names of Russian FB and VK users can be spelled in both Latin and Cyrillic alphabets i.e. "Alexander Ivanov" or "Александр Иванов". To enable correct name matching, all user names in both networks were converted to Latin script using the Russian-Latin BGN transliteration rules[10].

4   Profile Matching Algorithm

The algorithm consists of three phases:

1. *Candidate generation.* For each VK profile we retrieve a set of FB profiles with similar first and second names.
2. *Candidate ranking.* The candidates are ranked according to similarity of their friends.
3. *Selection of the best candidate.* The goal of the final step is to select the best match from the list of candidates.

Each profile from VK network is processed independently and hence this operation can be easily parallelised (we rely on MapReduce framework[11]). It is possible to perform matching in both directions (VK→FB and FB→VK). However, all profiles from the target network must be stored at each computational node. Therefore, direction of matching VK → FB minimises the memory footprint of such nodes. Below we describe each step of the method in detail.

4.1   Candidate Generation

It is computationally inefficient to calculate similarity of each VK profile with each FB profile. This operation would require about $1.3 \cdot 10^{20}$ pairwise comparisons. This first step reduces the search space retrieving FB users with names similar to the input VK profile. Two names are considered similar if the first letter is the same and the edit distance [14] between names is less than two. This should be true for both first and last names.

We use an index based on Levenshtein Automata [15] to perform fuzzy match between a VK user name and all FB user names. In particular, we relied on the Lucene implementation of this approach[12].

However, the edit distance does not provide a complete solution for name matching, since many first names have several rather different variants, e.g., "Robert" and "Bob", or "Mikhail" and "Misha". One way to address this problem is to use a dictionary of proper names prepared by linguists, e.g., [16], to decide whether two names are synonyms. However, such dictionaries often skip some name variants. For example, the entry for the Russian name "Alexander"

---

[10] http://earth-info.nga.mil/gns/html/romanization.html.
[11] http://hadoop.apache.org.
[12] org.apache.lucene.util.automaton.LevenshteinAutomata.

in [16] includes "Sanya", but not "Sanek". In addition, they do not include variants based on similarity with names from other languages: e.g., "Alejandro" is not in the entry for "Aleksandr" in [16].

Therefore, we decided to build our own dictionary using pairs of profiles known to belong to the same person. We do this by taking the transitive closure of the symmetric binary relation over names given by these pairs. Every two names from the same equivalence class are considered to be synonyms. The fundamental deficiency of this approach is that, being a variant is not an equivalence relation, since transitivity does not always hold. For example, two different Russian names, "Alexander" and "Alexey", are often abbreviated as "Alex". With our approach, this results in declaring "Alexander" and "Alexey" variants of each other, which they are not. Nevertheless, we let this happen and use shared friends to disambiguate between persons erroneously declared to have similar names.

Another problem is that some people use totally unrelated first names (such as "Andrey" and "Vladimir" or even "Max" and "Irina") in different networks. We solve this problem by removing "strange" pairs based on the number of times such a pair occurs in the list (unique or infrequent pairs can safely be removed). The final list of synonym clusters was quickly checked manually.

While candidate generation step greatly reduces search space, a person that indicated different name or a pseudonym in two social networks will not be recognised with our approach. From the other hand, in this situation a person is probably prefers to hide his or her identity and therefore it is more appropriate to perform no matching for this user at all.

4.2   Candidate Ranking

The higher the number of friends with similar names in VK and FB profiles, the larger the similarity of these profiles. Two friends are considered to be similar if:

- First two letters of their last names match, and
- The similarity between their first names and the similarity between their last names are both greater than thresholds α and β, correspondingly. We empirically set α to 0.6 and β to 0.8. String similarity sim_s is calculated as follows:

$$sim_s(s_i, s_j) = 1 - \frac{lev(s_i, s_j)}{\max(|s_i|, |s_j|)},$$

where lev is edit distance of string s_i and s_j. At this step we use the standard algorithm for calculation of Levenstein distance[13], not Levenstein Automata.

Matching friends with rare names should be weighted higher than a match of friend with matching friends with common names. Indeed, two unrelated profiles can easily have several friends with similar common names.

---

[13] org.apache.lucene.search.spell.LevensteinDistance.

Probability of a user with first name s^f and second name s^s, provided than these events are independent is $P(s^f, s^s) = P(s^f)P(s^s) = \frac{|s^f|}{N} \frac{|s^s|}{N}$. Here $|s^f|$ and $|s^s|$ are frequencies of respectively first and second names and N is the total number of profiles. Thus, expectation of name frequency equals to $\frac{|s^f| \cdot |s^s|}{N}$. In our approach, contribution of each friend to similarity sim_p of two profiles p_{vk} and p_{fb} is inverse of name expectation frequency, but not greater than one:

$$sim_p(p_{vk}, p_{fb}) = \sum_{j:sim_s(s_i^f, s_j^f) > \alpha \wedge sim_s(s_i^s, s_j^s) > \beta} \min(1, \frac{N}{|s_j^f| \cdot |s_j^s|}).$$

Here s_i^f and s_i^s are first and second names of a VK profile, correspondingly, while s_j^f and s_j^s refer to a FB profile.

4.3 Best Candidate Selection

FB candidates are ranked according to their similarity sim_p to an input profile p_{vk}. There are two thresholds the best candidate p_{fb} should pass to match:

- its score should be higher than the *similarity threshold* γ:

$$sim_p(p_{vk}, p_{fb}) > \gamma.$$

- it should be either the only candidate or score ratio between it and the next best candidate p'_{fb} should be higher than the *ratio threshold* δ:

$$\frac{sim_p(p_{vk}, p_{fb})}{sim_p(p_{vk}, p'_{fb})} > \delta.$$

The δ threshold enforces the fact that a VK user has only one account in FB. On the other hand, one FB profile can be linked with several VK profiles. Still, in this case only the match with the highest score is kept.

5   Results and Discussion

We performed matching of VKontakte and Facebook profiles (c.f. Table 1) with the approach described above. Results of the candidate ranking step were saved. At this point, we conducted a series of experiments varying the similarity threshold γ and the ratio threshold δ. Results of these experiments in terms of precision and recall with respect to the test collection (see Sect. 3.3 are presented in Fig. 1). The bold line denotes the best precision at certain level of recall.

As one may observe, our method yields very good results achieving precision of 0.97 at recall of 0.58. Furthermore, a configuration of the approach yielding 99 % precision recalls roughly 50 % of relevant profiles.

In order to perform the final matching of VK and FB we chose a version of the algorithm that provides precision of 0.98 and recall of 0.54 (see Table 2). Results

Table 1. Statistics of VKontakte and Facebook.

	VKontakte	Facebook
Number of users in our dataset	89,561,085	2,903,144
Number of Russian-speaking users[a]	100,000,000	13,000,000
User overlap	29 %	88 %

[a]http://www.comscore.com/Insights/Data-Mine/Which-Sites-Capture-The-Most-Screen-Time-in-Russia and http://vk.com/about provide statistics on number of Russian-speaking users.

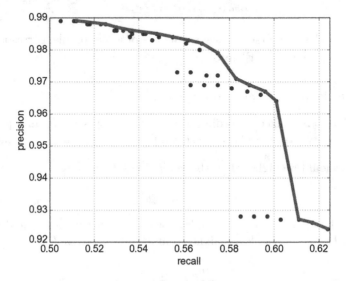

Fig. 1. Precision-recall plot of our matching method. Here we perform a grid search of two method parameters: profile *similarity threshold* $\gamma \in [1; 4]$ and profile similarity *ratio threshold* $\delta \in [3; 6]$. The bold line denotes the best precision at a given recall.

were obtained in 4 h on a Hadoop cluster with 100 nodes of type m2.xlarge (2 vCPU, 17 GB RAM) on the AWS EC2 cloud[14]. The mentioned above configuration of the method mentioned above retrieved 644,334 VK profiles of FB users. Thus, we found corresponding VK pages of 22 % Facebook users present in our collection.

While our approach makes only few errors, reaching precision of 0.98, it is not able to match a significant fraction of 40–50 % of user profiles. The key factors hampering correct retrieval are the following:

– In our method, we perform fuzzy search with name synonyms that can lead to semantic drift. For instance, "Maria" is expanded with its alias "Masha". According to fuzzy search "Masha" and "Misha" are related. But the latter is a shortcut for "Michael" in Russian.

---

[14] http://aws.amazon.com.

Table 2. Matching of user profiles of Facebook and VKontakte social networks. The upper table presents four main parameters of our profile matching method. The lower part of the table presents results of the final matching of the two networks.

Parameter	Value
First name similarity threshold, α	0.8
Second name similarity threshold, β	0.6
Profile similarity threshold, γ	3
Profile ratio threshold, δ	5
Number of matched profiles	644,334 (22 % of 2,903,144 FB users)
Expected precision	0.98
Expected recall	0.54

- Implementation of the Levenstein Automata used in our experiments retrieves candidates with distance lower or equal than two. Thus, people with long names and surnames can be missed during candidate generation.
- People often intentionally indicate different names in two social networks or use different aliases. Our approach is not designed to identify and match such profiles.
- First letter mismatch. Different variants of the same name/surname in Russian can start from different letters. Furthermore, transliteration can lead to such mismatches as well, e.g. surname "Ефимов" can be spelled in Latin as "Efimov" or "Yefimov".
- People often use transliterated versions of their names in one network, but stick to the original Cyrillic versions in the other. The method always works with transliterated names, but our transliteration can be quite different from the one done by a user.
- Due to nature of the friend collection method used, some FB friends can be absent in our dataset.

In order to improve performance of the method, one would need to tackle the problems mentioned above.

6 Conclusion

In this paper, we presented a new user identity matching method. Unlike most previous approaches, our method is able to work on the scale of real online social networks, such as Facebook, matching tens of millions of users in several hours on a medium-sized computational cluster. The method yields excellent precision (up to 98 %). At the same time it is able to recall up to 54 % of correct matches.

The method was used to perform the most largest-scale matching experiment up to date. We matched 90 millions of VKontakte users with 3 million of Facebook users.

A prominent direction for the future work, is to use supervised learning in order to improve *candidate ranking*. One way pioneered by [4] is to use a binary classifier predicting if two profiles match; profile similarity in this case would be the confidence of positive class. Learning to rank methods [17] is another way to cast profile matching as a supervised problem. The supervised models provide a convenient framework where name similarity features, used in our method, can be mixed with attribute-, network-, and image-based features.

Acknowledgements. This research was conducted as part of a project funded by Digital Society Laboratory LLC. We thank Prof. Chris Biemann and three anonymous reviewers for their thorough comments that significantly improved quality of this paper.

References

1. Bartunov, S., Korshunov, A., Park, S.T., Ryu, W., Lee, H.: Joint link-attribute user identity resolution in online social networks. In: Proceedings of the Sixth SNA-KDD Workshop at KDD (2012)
2. Balduzzi, M., Platzer, C., Holz, T., Kirda, E., Balzarotti, D., Kruegel, C.: Abusing social networks for automated user profiling. In: Jha, S., Sommer, R., Kreibich, C. (eds.) RAID 2010. LNCS, vol. 6307, pp. 422–441. Springer, Heidelberg (2010)
3. Wondracek, G., Holz, T., Kirda, E., Kruegel, C.: A practical attack to de-anonymize social network users. In: 2010 IEEE Symposium on Security and Privacy (SP), pp. 223–238. IEEE (2010)
4. Goga, O., Perito, D., Lei, H., Teixeira, R., Sommer, R.: Large-scale correlation of accounts across social networks. Technical report, International Computer Science Institute (2013)
5. Sironi, G.: Automatic alignment of user identities in heterogeneous social networks. Master's thesis, Politechnico di Milano, Italy (2012)
6. Veldman, I.: Matching profiles from social network sites: Similarity calculations with social network support. Master's thesis, University of Twente, Italy (2009)
7. Narayanan, A., Shmatikov, V.: De-anonymizing social networks. In: 2009 30th IEEE Symposium on Security and Privacy, pp. 173–187. IEEE (2009)
8. Raad, E., Chbeir, R., Dipanda, A.: User profile matching in social networks. In: 13th International Conference on Network-Based Information Systems (NBiS), pp. 297–304. IEEE (2010)
9. Malhotra, A., Totti, L., Meira Jr., W., Kumaraguru, P., Almeida, V.: Studying user footprints in different online social networks. In: Proceedings of the 2012 International Conference on Advances in Social Networks Analysis and Mining (ASONAM 2012), pp. 1065–1070. IEEE Computer Society (2012)
10. Jain, P., Kumaraguru, P., Joshi, A.: @I seek 'fb.me': identifying users across multiple online social networks. In: Proceedings of the 22nd International Conference on World Wide Web Companion, International World Wide Web Conferences Steering Committee, pp. 1259–1268 (2013)
11. Boytsov, L.: Indexing methods for approximate dictionary searching: comparative analysis. J. Exp. Algorithmics (JEA) **16**, 1–1 (2011)
12. Du, M.: Approximate name matching. NADA, Numerisk Analys och Datalogi, KTH, Kungliga Tekniska Högskolan. Stockholm: un (2005)

13. Navarro, G., Baeza-Yates, R., Marcelo Azevedo Arcoverde, J.: Matchsimile: a flexible approximate matching tool for searching proper names. J. Am. Soc. Inf. Sci. Technol. **54**(1), 3–15 (2003)
14. Lisbach, B., Meyer, V.: Linguistic Identity Matching. Springer, Heidelberg (2013)
15. Schulz, K., Mihov, S.: Fast string correction with Levenshtein-automata. Int. J. Doc. Anal. Recogn. **5**, 67–85 (2002)
16. Petrovsky, N.: Dictionary of Russian personal names. http://www.gramota.ru/slovari/info/petr M.: In Russian Dictionaries (2000)
17. Trotman, A.: Learning to rank. Inf. Retrieval **8**(3), 359–381 (2005)

Identification of Autopoietic Communication Patterns in Social and Economic Networks

Dmitry B. Berg[1,2,3] and Olga M. Zvereva[3]([✉])

[1] Institute of Industrial Ecology, Ekaterinburg, Russia
bergd@mail.ru
[2] International Alexander Bogdanov Institute, Ekaterinburg, Russia
[3] Ural Federal University, Pr. Mira, 19, Yekaterinburg 620002, Russian Federation
OM-Zvereva2008@yandex.ru

Abstract. Communications develop the basis for social and economic system functioning. In every communication act system agents exchange information, senses, money, services, industrial goods, energy, etc. Economic agent communications form the network. One of the most important system characteristic is its ability to reproduce itself (autopoiesis), which is performed by circular communications in the closed contours. The main goal of this work is to consider the technology of autopoietic patterns identification in social and economic networks. A new approach to initial data collection for evaluation of social communications is proposed. In this study the data collected while "KOMPAS TQM" system implementation was analyzed. The SNA methods and instruments were used for revealing autopoietic patterns and their subsequent analysis.

Keywords: Autopoietic pattern · Economic network · Social Network Analysis · Leontief's intersectoral equilibrium

1 Introduction

Communications develop the basis for social and economic system functioning [1]. In every communication act system agents can exchange information, senses, money, services, industrial goods, energy, etc., with each other [2]. A set of economic agent communications forms a network [3]. If agents are connected by means of permanent relations (productive, social, etc.), then this network will be stable in the time. The network topological structure can be very complicated. The well-known network taxonomies are based on the structural geometric theory and anyone can reveal a star, line or ring type of structures. As a rule, real networks appear to be the aggregations of these elementary structures.

One of the most important system characteristic is its ability to reproduce itself (autopoiesis) which is performed according to Maturana and Varella's [4] theory by cycled communications in closed contours. An autopoietic contour is a ring in the topological sense but not every ring seems to be an autopoietic contour. Revealing the conditions under which the autopoiesis appears in real

© Springer International Publishing Switzerland 2015
M.Y. Khachay et al. (Eds.): AIST 2015, CCIS 542, pp. 286–294, 2015.
DOI: 10.1007/978-3-319-26123-2_28

economic and social network contours is a real challenge. Other important issues are autopoietic contour interactions in the network and existing autopoietic patterns identification.

Presented study was carried out by using the framework UCINET 6 for Windows, which supports the SNA (Social Network Analysis) methodology. Comparable in size (agent/node number) networks of economic and social types were investigated.

2 Economic Agent Communications

Economic agent communication network analysis was performed on the basis of data collected for the set of the 12 real business entities, which have the partner relations in manufacturing and consuming their goods and services. The set was discussed in details in [5]. Manufacture and consumption characteristics correspond to the economics of municipality with population of 10000 people. As the initial data, the authors of [5] had used the annual reports of similar type enterprises. These reports include the average manufacture volumes, average annual wages, and average numbers of employees.

Table 1. Consumption (payment) matrix of municipal economic agents

Product volumes												
A	1	2	3	4	5	6	7	8	9	10	11	12
1	0	0	0	0	0	0	496000	1340680	0	0	0	0
2	405	0	0	53993	590322	0	0	0	0	0	0	0
3	0	0	0	79100	0	0	0	0	0	0	0	8227
4	0	0	0	0	0	0	0	0	0	0	0	109460
5	0	0	0	0	0	0	0	0	0	0	0	79389
6	0	0	0	0	0	0	0	0	0	0	0	28775
7	0	0	0	0	0	4081	0	0	0	0	0	0
8	0	8726	419318	0	0	0	0	0	0	0	0	0
9	0	0	0	0	0	0	0	0	0	0	0	2352
10	0	0	0	0	0	0	0	0	0	0	33	9000
11	0	0	0	0	0	0	0	0	0	0	0	16280
12	14000	17000	40450	59000	170300	48000	16000	30000	19200	7000	7730	0

Mutual agent needs in products aggregated for a year period are presented in Table 1, where "A = Agent". The table column and row numbers are in accordance with the agent numbers. Each table row reflects product deliveries of the agent, whose number is equal to the row number to all network agents: each matrix element value is equal to product volume delivered to the agent, whose number is equal to the column number. Each table column contains product consumptions of a single agent. All table values are expressed in terms of money, so Table 1 can be considered as a table of agents mutual payments.

Enterprise list of the network under discussion looks as follows:

1. agricultural farm (cereal and industrial crops cultivating);
2. meat and dairy farm (meat and dairy production);
3. poultry farm (poultry breeding and egg production);
4. meat processing plant (production of semi-finished goods from meat and poultry);
5. dairy plant (production of dairy goods);
6. bakery (bakery product manufacturing);
7. flour mill (flour production);
8. feed mill (feed production);
9. furniture factory (home and office furniture production);
10. autoservice workshop (car maintenance and repair);
11. trucking company (shipping operations).

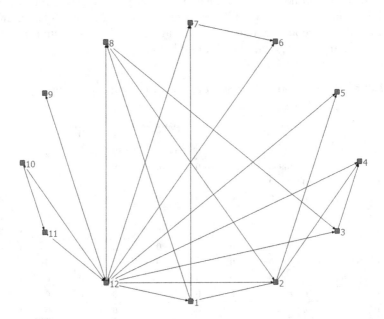

Fig. 1. Municipal enterprise network based on the payment matrix from Table 1 (made in UCINET 6 for Windows)

Population is considered to be the separate twelfth sector of the model. The 12th agent consumes the products of other agents and provides them with the necessary labor resource. The result network is visualized in Fig. 1.

It is possible to detect a number of the closed contours in this economic communication network. Some of them are described in Table 2, where "N = Number", "Circ = Circularity" and "AP = Autopoiesis". The link weights are also presented in this table. The network analysis reveals that only some closed

Table 2. Closed contours in the economic network

N	Contour	Circ	Link weights	AP
1	$1 \to 8 \to 2 \to 1$	yes	$1340680 + 419318 + 405$	yes
2	$1 \to 7 \to 6 \to 12 \to 1$	yes	$496000 + 4081 + 27514 + 14000$	yes
3	$1 \to 8 \to 3 \to 12 \to 1$	yes	$1340680 + 419318 + 8227 + 14000$	yes
4	$2 \to 4 \to 12 \to 2$	yes	$53993 + 109460 + 17000$	yes
5	$2 \to 5 \to 12 \to 2$	yes	$590322 + 79389 + 17000$	yes
6	$3 \to 12 \to 8 \to 3$	yes	$8227 + 30000 + 419318$	yes
7	$4 \to 12 \to 3 \to 4$	yes	$109460 + 40450 + 79100$	yes
8	$6 \to 12 \to 7 \to 6$	yes	$27514 + 16000 + 4081$	yes
9	$11 \to 12 \to 10 \to 11$	yes	$16280 + 7000 + 33$	yes
10	$12 \to 1 \to 8 \leftarrow 12$	no	$14000 + 1340680 + 30000$	no
11	$12 \to 2 \to 1 \leftarrow 12$	no	$17000 + 405 + 14000$	no
12	$8 \to 2 \to 4 \to 12 \to 3 \leftarrow 8$	no	$8726 + 53993 + 109460 + 40450 + 419318$	no

contours are circular (the corresponding graph is a digraph). One of the closed but not circular contours is the 10th contour $12 \to 1 \to 8 \leftarrow 12$. Despite the fact that the population (12) works at the agricultural farm (1) and at the feed mill (8) too and these enterprises are connected through the cereal shipping, this contour is not circular because people buy nothing both in the farm and in the mill (according to Table 1).

All the circular network contours are of the autopoietic type since the agents consume products of each other in a cycle. As an example, the agricultural farm (1) delivers cereals to the feed mill (8), which in its turn provides the meat and dairy farm (2) with feed. The farm (2) delivers part of its product to the agricultural cooperative (1). Most of the other autopoietic contours include population (12), and it seems to be the central node in the network, which integrates separate autopoietic contours in one autopoietic pattern. This hypothesis is proved by the SNA since all centrality measures [6–8] of this node have the maximum values in the network. Centrality in-degree is 0.636 and out-degree is 1.0 (connected with all other agents). The value of Freeman's betweenness centrality is equal to 88.833 this measure shows that more than 88 % of the shortest paths from one agent to another in this network includes the 12th agent.

It is important to underline that in the consumption (payment) matrix (Table 1) there is no balance between delivery and consumption values, i.e. an agent can deliver its product in a larger volume than it consumes other agent products, and vice versa, an agent can consume more than can give itself (the row element sum is not equal to the element sum of the column with the same number). The source of this imbalance is the "autopoiesis lack" in this network. Some agents consume products, which are not manufactured in this network. Moreover, they can deliver their products not only to the other network agents but to the agents, which are not included in this municipal network (the links with external agents are not presented in Table 1). That is why more product volume can be moved through the link between 2 agents in one circular contour than it is necessary for the autopoiesis support. This product excess is used by succeeding agents in order to manufacture their product for the agent external to this network.

3   Balanced Matrix Development

Based on the set of revealed closed circular contours and the Leontief's intersectoral equilibrium [9] the balanced matrix was derived from the matrix presented in Table 1. It is shown as Table 3. It is a real balanced matrix because every agent delivers to the other network agents the same volume of its production as it receives from them. Comparison of Tables 1 and 3 demonstrates that each link can reflect a specific autopoietic communication and also an ordinary product shipping. For example, the communication link formed while the 2nd agent product consumption (meat and dairy farm) by the 4th agent (meat processing plant) has the weight equal to 53993 thousand rubles a year as soon as the autopoietic component of this communication link is only 2154 thousand rubles a year, i.e. 25 times less.

Table 3. Consumption matrix of municipal economic network autopoietic pattern

Product volumes												
Agent	1	2	3	4	5	6	7	8	9	10	11	12
1	0	0	0	0	0	0	1841	10644	0	0	0	0
2	16	0	0	2154	23555	0	0	0	0	0	0	0
3	0	0	0	61452	0	0	0	0	0	0	0	8227
4	0	0	0	0	0	0	0	0	0	0	0	109460
5	0	0	0	0	0	0	0	0	0	0	0	79389
6	0	0	0	0	0	0	0	0	0	0	0	28775
7	0	0	0	0	0	2255	0	0	0	0	0	0
8	0	8725	63558	0	0	0	0	0	0	0	0	0
9	0	0	0	0	0	0	0	0	0	0	0	2352
10	0	0	0	0	0	0	0	0	0	0	26	6974
11	0	0	0	0	0	0	0	0	0	0	0	7756
12	12469	17000	6131	45844	55834	26520	414	61639	2352	7000	7730	0

All the theoretical results being discussed were verified by the experiments made with the help of agent-based [10, 11] software model discussed in details in [12].

4   Social Agent Communications

Social communication network structure appears as a result of pair interactions of actors who proved to be social agents. In a contrast to economic system in social agent communications there is an exchange of some information or actions, which are of special sense for agents [2]. Under certain conditions a chain of pair communications in a social group can become closed and this results in an autopoietic pattern formation. An autopoietic pattern has to support the reproduction of relations and senses arising while these communications, *i.e.*, this group obtains the property of self-identification.

Questionnaire is considered to be the most common source of social communication data [13]. Nevertheless, its usage has also well-known difficulties arising from its design, delivery, and completing. In our research project data collected while "KOMPAS TQM" [14] system implementation was processed. "KOMPAS TQM" system is a quality management system [15], and it was introduced into the educational process in one of the departments of Ural Federal University. This system supports the process of regular communication result evaluation. System users enter positive or negative marks from a certain range. Every mark must be followed by a comment. Thus, the marks entered into the system reflect the real communications between system users and confirmed by comments. The mark sign (" $+''$ *or* "$-''$) characterizes the "information receiver" attitude to the communication result in the whole (positive or negative) while numerical value (from 1 to 5 points) reflects the strength (weight) of this communication (*i.e.*, the usefulness degree for a receiver).

For the further research of the social communication network three-week evaluation sample for the 13–student academic group was constructed. The matrix with student total marks as elements is shown as Table 4. Those marks that were put and received for external communications are not shown in it. This matrix became the initial data for sociogram construction (see Fig. 2) where each agent[1] gender attribute is reflected: the male agents are shown as blue squares and female ones as red circles.

It appears that in the case of just the same network sizes (13 and 12 accordingly) the social communication network is more cohesive. Its density [7] is 0.66 while in the economic communications network the same index is 0.21. It can be explained by the fact that in the social communications there is a smaller number of barriers than in communications of the economic type. In a social group, its member communicates almost with all group members as soon as to work and to learn together and do not communicate is almost impossible. In an economic system, communications are determined by its technological process.

In a social communication network, we can find closed contours ("closed paths" [16] in the SNA terminology), which have a property of circularity, thus, can be considered as autopoietic ones. In order to be sure that these communications will be reproduced later the corresponding links must have positive weights. We do not insist on the version of strong ties (links with significant weights) due to the theory proposed by M.S. Grannovetter [17] considering "the strength of weak ties".

If there is a link with a negative weight, it means that this communication has failed, and will not be repeated for the next time, as a result, this contour

---

[1] In SNA theory an agent is commonly named "an actor".

Table 4. Social communications evaluation results for the academic group

Agent	1	2	3	4	5	6	7	8	9	10	11	12	13
1	0	0	0	0	0	0	5	0	10	0	0	15	5
2	−5	0	0	0	35	2	68	−13	25	0	0	53	0
3	0	0	0	0	0	0	0	5	0	0	10	0	0
4	4	4	4	0	4	4	4	4	4	4	4	4	8
5	13	25	2	0	0	2	25	13	25	10	4	54	10
6	5	5	5	5	5	0	5	5	5	5	5	8	5
7	15	50	9	1	41	8	0	29	50	6	16	57	11
8	0	0	0	0	0	0	0	0	0	0	0	0	0
9	20	18	−5	5	20	5	25	12	0	12	5	58	10
10	40	40	34	34	45	40	40	34	40	0	32	40	34
11	0	0	4	3	0	3	4	0	0	0	0	4	0
12	5	30	7	2	75	5	55	19	58	15	9	0	0
13	3	0	0	0	0	0	0	5	10	15	0	0	0

closure will be broken. In the social communication network (Fig. 2), some closed contours (paths) were detected. They are presented in Table 5, link weights and circularity attribute are also reflected there.

As one can see from Table 5, not all of the detected closed contours are of the circular type. For example, in the 9th contour the 3rd agent didn't put any mark to the fourth agent and didn't do it for the 5th agent as well, although from the topological point of view, this contour is a closed one. The same issue can

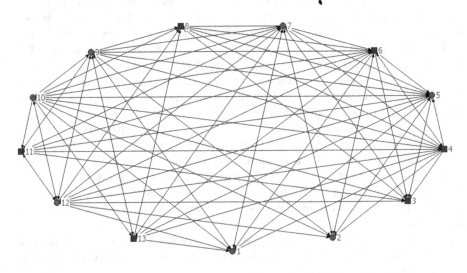

Fig. 2. Student communication sociogram (made in UCINET 6 for Windows).

Table 5. Closed contours examples

Number	Contour	Circularity	Weights (marks)	Autopoiesis
1	$3 \to 11 \to 4 \to 5 \to 3$	yes	$10 + 3 + 4 + 2 = 19$	yes
2	$3 \to 11 \to 6 \to 5 \to 3$	yes	$10 + 3 + 5 + 3 = 21$	yes
3	$3 \to 11 \to 7 \to 5 \to 3$	yes	$10 + 4 + 41 + 3 = 57$	yes
4	$3 \to 11 \to 12 \to 5 \to 3$	yes	$10 + 4 + 75 + 3 = 92$	yes
5	$3 \to 11 \to 4 \to 9 \to 3$	yes	$10 + 3 + 4 - 5$	no
6	$3 \to 11 \to 6 \to 9 \to 3$	yes	$10 + 3 + 5 - 5$	no
7	$3 \to 11 \to 7 \to 9 \to 3$	yes	$10 + 3 + 50 - 5$	no
8	$3 \to 11 \to 12 \to 9 \to 3$	yes	$10 + 3 + 58 - 5$	no
9	$3 \leftarrow 4 \to 5 \to 3$	no	$4 + 4 + 2 = 10$	no
10	$3 \leftarrow 6 \to 7 \to 9 \to 3$	no	$5 + 5 + 50 + 9 = 19$	no
11	$3 \leftarrow 7 \to 11 \to 4 \to 3$	no	$9 + 16 + 3 + 4 = 32$	no

be found in the 10th and 11th contours (according to Table 5 numeration): they are not circular too. The 5th, 6th, 7th, and 8th circular contours have positive and negative link weights (marks) that is why they can not be considered as autopoietic ones. Only the 1st, 2nd, 3rd, and the 4th contours are of autopoietic nature, because they are circular and have links with positive weights (marks). The set of autopoietic contours (from 1 to 4) forms the autopoietic pattern of the social network under research.

As was considered for the agents who integrate these contours in one pattern, they are the 9th and 11th ones. As was estimated with the help of UCINET 6.0, the 9th and 11th agents have the maximum values of the Freeman's centrality betweenness (11.943 and 10.167 respectively) in this network.

5   Conclusions

As the main result of this research, it was demonstrated that in economic and social networks on the basis of a consumption/communication/adjacency matrix the autopoietic patterns can be revealed (identified). According to the theory of autopoiesis, these patterns support the network self-reproduction. Every pattern is a set of closed circular contours with some corporate elements (nodes/agents). Usually, these "corporate" agents have high values of centrality measures.

We argue that in the set of closed contours it is very useful to identify the subset of circular contours as the specific class for network reproduction (autopoiesis) support.

In the network of economic communications, one link can reflect an autopoietic type communication and also an ordinary product shipping. Consumption matrix of an autopoietic pattern is the balanced one, and reproduction in this case can be described with the help of the Leontiefs intersectoral equilibrium.

Acknowledgement. The work was supported by the RFBR grant no. 15-06-04863 ("Mathematical models of local payment system lifecycles").

References

1. Luhmann, M.: Social systems. Sketch of the general theory. St. Petersburg: Science, p. 668 (2000)
2. Popkov, V.V.: The metaphysics of economics: what is a meaning worth? University of Athens, School of Philosophy, Zografos, p. 575 (2013)
3. Castells, M.: The Rise of the Network Society. The Information Age: Economy, Society and Culture, vol. I, p. 656. Blackwell, Cambridge (2000)
4. Maturana, H.R., Varela, F.J.: Autopoiesis and Cognition: The Realization of the Living. (Boston Studies in the Philosophy of Science), vol. 42. D. Reidel Publishing, Boston (1980)
5. Ignatova, M.A., Selezneva, N.A. Ulyanov, E.A.: Municipal Economics: Domestic Market Financial Network Model. Modern Problems of Science and Education, 2 (2014). http://www.science-education.ru/116-12901. Accessed 04 November 2014
6. Stephen, P., Borgatti, R.: Centrality and network flow. Soc. Netw. **27**, 55–71 (2005)
7. da Costa, L.F., Rodrigues, F.A., Travieso, G., Villas, P.R.: Characterization of complex networks: a survey of measurements. Adv. Phys. **56**(1), 167–242 (2007)
8. Opsahl, T., Agneessens, F., Skvoretz, J.: Node centrality in weighted networks: generalizing degree and shortest paths. Soc. Netw. **32**, 245–251 (2010)
9. Leontief, V.V.: Essays in Economics. Theories, Theorizing, Facts and Policies, p. 415. Politizdat, Moscow (1990)
10. Macal, C.M., North, M.J.: Tutorial on agent–based modelling and simulation. J. Simul. **4**, 151–162 (2010)
11. Borrill, P.L., Tesfatsion, L.: Agent–Based Modeling: The Right Mathematics for the Social Sciences. Working Paper 10023, July 2010
12. Zvereva, O.M., Berg, D.B.: Economic system agent–based communication model based on leontiefs intersecoral equilibrium. St.Peterssburg State Polytechnical University Journal. Computer Science. Telecommunications and Control Systems, 6(186), 77–86 (2013)
13. Carnegie Mellon Heinz School. Dr. David Krackhardt. Sample Questionnaires. http://www.andrew.cmu.edu/user/krack/documents/questionnaires/. Accessed 05 November 2014
14. Open electronic paper Forum.msk.ru. Total quality management. http://forum-msk.org/material/economic/627694.html. Accessed 06 November 2014
15. ISO 9001 Quality Management Systems. http://the9000store.com/what-is-iso-9001-quality-management-system.aspx. Accessed 07 November 2014
16. Robert, A., Riddle, H., Riddle, M: Introduction to social network methods. http://faculty.ucr.edu/~hanneman/nettext/C7_Connection.html. Accessed 04 November 2015
17. Grannovetter, M.G.: The strength of weak ties. AJS **78**(6), 1360–1380 (2001)

Text Mining and Natural Language Processing

A Heuristic Strategy for Extracting Terms from Scientific Texts

Elena I. Bolshakova[✉] and Natalia E. Efremova

Lomonosov Moscow State University,
National Research University Higher School of Economics, Moscow, Russia
eibolshakova@gmail.com, nvasil@list.ru

Abstract. The paper describes a strategy that applies heuristics to combine sets of terminological words and words combination pre-extracted from a scientific text by several term recognition procedures. Each procedure is based on a collection of lexico-syntactic patterns representing specific linguistic information about terms within scientific texts. Our strategy is aimed to improve the quality of automatic term extraction from a particular scientific text. The experiments have shown that the strategy gives 11–17 % increase of F-measure compared with the commonly-used methods of term extraction.

Keywords: Multiword terms · Automatic term extraction · Text variants of terms · Term occurrences in scientific text · Lexico-syntactic patterns

1 Introduction

Automatic extraction of terms from domain specific texts plays crucial role in various natural language processing (NLP) applications, such as compiling terminology dictionaries, constructing thesauri and ontologies, text abstracting, etc. The problem of automatic term recognition was studied over last two decades, and significant results were obtained – see, for example, [4, 5, 11].

Most terms, including scientific ones are multiword units, e.g., *nonlinear plan*, *coefficient adjustment learning*. In order to recognize them within NL texts, shallow syntactic analysis along with statistical and linguistics criteria are used, based on assumption that terms are frequently encountered within texts in specific grammatical forms. The applied extraction techniques do not guarantee extracted text units to be true terms (e.g., non-term phrase *general plan* may be extracted), so they are only *term candidates* and usually need to be confirmed by human experts [5].

In most modern applications of automatic term extraction (in particular, for thesaurus construction), large text collections and corpora are processed, and exploiting statistical criteria of term recognition along with even poor linguistic information (such as part of speech of words) gives acceptable quality of extraction measured by precision and recall. Statistical criteria include *tf.idf* measure (widely used in information retrieval [12]) and its numerous modifications, as well as their combinations (see, for example, [15]). Meanwhile, in such applied tasks as text abstracting and summarization, computer-aided writing and editing of specialized texts, recognition of terms is performed from a single text, and statistical measures becomes less significant because

© Springer International Publishing Switzerland 2015
M.Y. Khachay et al. (Eds.): AIST 2015, CCIS 542, pp. 297–307, 2015.
DOI: 10.1007/978-3-319-26123-2_29

of less volume of the text under processing. Moreover, contrast corpora needed to compute *tf.idf* value are not always available. Thus, more comprehensive linguistic information is to be applied for reliable term extraction based on shallow syntactic analysis. Besides grammatical patterns of multiword terms, linguistic information may comprise local context data similar to that described in [16] for extraction of terms (concepts) from highly specialized texts.

In contrast to corpus-based term extraction, we consider the task of term recognition in a single text, the task is necessary for several applications, including computer-aided construction of glossaries and subject indexes for text documents [1, 6]. We should note that task of term recognition somewhat differs from keywords extraction (e.g., [13]), since terms denote concepts of problem domain, while keywords may be non-terms (such as *trend monitoring* or *banking application*). We assume that consideration of various linguistic features of terms and their occurrences within texts facilitates their detection in texts and improve the quality of term extraction. Since the intensive use of terminological phrases with diverse structure is typical for scientific papers, in our study we consider term extraction from scientific texts.

Heterogeneous information about terms in Russian scientific texts was formalized and represented as linguistic patterns of term structures, text variants of terms, and terminological contexts. For purposes of formalization, we used LSPL (Lexico-Syntactic Pattern Language) [3] developed for specifying linguistics features of Russian phrases. LSPL programming tools were exploited as well to implement several term extraction procedures. Each procedure is based on a collection of LSPL patterns representing specific linguistic information about terms in scientific texts.

The term extraction procedures were studied experimentally, and analysis of the results gave us a strategy for improving the quality of automatic term extraction from a particular scientific text. The strategy implies heuristics on how to combine sets of term candidates pre-extracted by the procedures. The experiments have shown that our strategy gives 11–17 % increase of F-measure (the combined measure of precision and recall) compared with commonly-used methods for term extraction [7, 9, 17].

The objectives of this paper are:

- To clarify the categorization of extracted term candidates on the basis of linguistic information used for term recognition;
- To briefly describe a collection of LSPL patterns representing linguistic information of specific types;
- To overview the term extraction procedures developed for groups of patterns, as well as results of their experimental evaluation on Russian scientific texts;
- To describe our heuristics strategy for combining sets of term candidates extracted by the procedures.

2 Scientific Terms and Their Lexico-Syntactic Patterns

In order to reveal heterogeneous linguistic information useful for automatic term extraction, we have performed an empirical study of scientific texts in Russian (approx. 330 texts), as well as terminological dictionaries in several scientific fields (including computer science and physics) [2, 8]. Based on the study, we formalized revealed linguistics features of multi-word terms and their occurrences in texts on the basis of LSPL language [3] intended to support development of various applications for information extraction from Russian texts.

LSPL is a declarative formal language flexible enough to specify both lexical and syntactic features of phrases to be extracted from Russian texts. Phrases are specified in the form of *lexico-syntactic patterns*; elements of patterns include particular word forms, lexemes, arbitrary words of particular part of speech (POS), morphological attributes of words, conditions of grammatical agreement. The agreement conditions are especially important for describing Russian noun phrases. To specify complex phrases, auxiliary patterns can be defined and used within LSPL pattern.

Formalizing linguistics features of scientific terms gave us a representative set of LSPL patterns, which comprises 6 groups. Every group of patterns corresponds to specific linguistic information used to recognize term occurrences in texts. Therefore, our categorization of extracted terms reflects types of linguistic information represented in the patterns.

Examples of LSPL patterns of each group, as well as examples of terms or term contexts extracted by patterns of the corresponding groups are presented in Table 1. Let us consider the groups and types of extracted terms in more detail.

The first group of patterns specifies grammatical structures of one-, two- and tree-word terms frequently used in scientific texts. This linguistics information is commonly-used by most term extraction methods. We should note that each LSPL pattern fixes not only POS of constituent term words (N is noun, A is adjective), but also their morphological attributes (if necessary). In both patterns given in Table 1 grammatical agreement of adjectives and nouns are specified ($A = N$ and $A1 = A2 = N$). Symbol => denotes extraction of terms recognized by the patterns.

Regretfully, not only terms have the specified grammatical structures, in particular, collocations of common scientific lexicon may have the similar structure, e.g., *main problem, developed scheme*. Therefore, grammatical patterns of multiword terms are not enough reliable for term extraction, and hereinafter text units extracted by LSPL grammatical patterns we call **term candidates**.

The second and the third groups of patterns specify typical contexts of term occurrences. Primarily, we consider phrases for definition of new terms, which are often encountered in scientific papers, for example: *"A set of entities created by expansion process we call reference markers"*. Such new terms that are explicitly introduced by authors and then used within their texts are called **author's terms** [2].

Each LSPL pattern of the second group specifies typical one-sentence definition of author's term, the pattern includes both particular lexical units (verbs *call, define*, and so on) and special auxiliary patterns *Term* and *Defin*. The former describes all allowable grammatical patterns of terms (i.e. patterns of the first group), while the latter specifies structure of phrases explicating meaning of the defined term.

In the first example of definition pattern given in Table 1, *Term* is to be in nominative case ($c = nom$), and the *Defin* phrase is to be in instrumental case ($c = ins$); for Russian verb *называться* morphological attributes of time and person ($t = pres, p = 3$) and agreement in number with *Defin* phrase are specified as well. Each pattern of the

Table 1. Groups of LSPL patterns

Groups and Examples of Patterns	Examples of Terms and Contexts of Terms
1. Grammatical patterns of terms	
A N <A=N> => A N	*Rus. стерильный нейтрино –* *Eng. sterile neutrinos*
A1 A2 N <A1=A2=N> => A1 A2 N	*Rus. горячая темная материя –* *Eng. hot dark matter*
2. Definitions of authors' terms	
Defin <c=nom> V<называться, *t=pres, p=3> Term<c=ins>* *<V.n=Defin.n> => Term*	*Rus. ... яркое кольцо называется кольцом* *Эйнштейна – Eng. ... a bright ring is called* *Einstein ring*
Term "is" Defin => Term	*Rus. Вероятность есть степень* *возможности... – Eng. Probability is the* *measure of the likeliness...*
3. Contexts of introduction of term synonyms	
Term1 "("Term2")" *<Term1.c=Term2.c>* *=> Term1, Term2*	*Rus. двумерный электронный газ (ДЭГ) –* *Eng. two-dimensional electron gas (2DEG)*
4. Dictionary terms	
N1<адрес> [N2<возврат,c=gen> *| N2<результат,c=gen>* *| N2<точка,c=gen>* *N3<вход,c=gen>]*	*Rus. адрес, адрес возврата, адрес результата,* *адрес точки входа – Eng. address, return* *address, result address, entry point address*
5. Combinations of several terms	
A1 "и" A2 N <A1=A2=N> => *A1 N <A1=N>, A2 N <A2=N>*	*Rus. гравитационная и инертная масса =>* *гравитационная масса, инертная масса –* *Eng. gravitational and inertial mass =>* *gravitational mass, inertial mass*
N1 A N2<c=gen> <A=N2> => *N1 N2<c=gen>, A N2 <A=N2>*	*Rus. разрядность внутреннего регистра =>* *разрядность регистра, внутренний регистр –* *Eng. capacity of internal register => capacity of* *register, internal register*
6. Text variants of term	
A1 N <A1=N> => N, *A2 N <A2=N> <Syn(A1,A2)>*	*Rus. сильное взаимодействие =>* *взаимодействие, ядерное взаимодействие –* *Eng. strong force => force, nuclear force*

group includes the element => *Term* that denotes text item to be extracted from the recognized definition phrase.

The third group of LSPL patterns specify typical contexts encountered in scientific texts and intended to introduce **term synonyms** (including synonyms for author's terms), for example: ... *the generalized momentum, also known as the canonical momentum*.... Term synonyms include, in particular, acronyms, such as *CPU* for term *central processing unit*.

Besides grammatical structures of terms and contexts of term definitions, linguistic information useful for term recognition are represented in terminological dictionaries (given that they are available for text processing). Such dictionaries seldom accumulate a complete set of terms in particular scientific field but yet fix many stable terminological words and word combinations. As LSPL language proved to be convenient for describing entries of terminology dictionaries, we have created a group of patterns that specify terms from several text dictionaries in the fields of physics and computer science. The patterns contain particular lexemes with particular morphological attributes (if needed), for example, genitive case ($c = gen$) for Russian words *возврат, результат, точка, вход* in the example pattern from Table 1 (the forth group of patterns) is specified. This pattern specifies four terms, symbol | separate alternatives, optional elements are written in square brackets. Term occurrences recognized in text by such dictionary patterns are called ***dictionary terms***.

The last two groups of LSPL patterns describe general rules for text variants of terms that often occur in Russian scientific texts. Variation of terms and methods to detect term variants is relatively well studied for English and French texts [10, 14]. In our empirical study we have revealed analogous term variants for Russian [8]. Besides variations of single term – cf. the group of ***text variants*** in Table 1 – we additionally consider typical ***combinations*** of several multi-word terms and formalized their features as LSPL patterns – cf. the fifth group of patterns.

Rules of combining several multi-word terms take into account two different cases:

- combinations with coordinating conjunctions (*Rus. шина адреса, шина данных, шина управления => шина адреса, данных и управления* – *Eng. address bus, data bus, control bus => address, data, and control bus*);
- conjunctionless combinations (*Rus. поляризация волн, электромагнитные волны => поляризация электромагнитных волн* – *Eng. polarization waves, electromagnetic waves => polarization electromagnetic waves*).

In both cases within described combinations one or more multiword terms are broken (discontinuous) or truncated, and this is the real problem of their automatic recognition in texts. Any LSPL pattern of combination fixes its grammatical structure and also specifies (after sign =>) patterns of its constituent elements (i.e. grammatical structure of multi-word terms that compose this term combination).

Each pattern of the last group describes (in a similar manner) grammatical structure of the term being varied; grammatical structure of its possible text variants are specified after sign =>. In particular, if the structure of multiword term is *N1 N2<c=gen>*, its text variants comprises:

- insert (or deletion) of word from the term (e.g., *Rus. ввод данных => ввод* – *Eng. data input => input*);
- substitution of a synonym (in the given scientific domain) for constituent part of the term (e.g., *Rus. фрейм активации => запись активации* – *Eng. activation frame => activation record*);
- substitution of a word with the same root but another part of speech (e.g., *Rus. шина адреса => адресная шина* – *Eng. address bus => bus of address*).

We should note that to recognize synonymy variation of some constituent word of the term, certain dictionary of synonyms is to be incorporated into term extraction procedure. The element *<Syn(A1,A2)>* of the LSPL pattern given in Table 1 denotes necessary check-up of adjectives *A1* and *A2* in the incorporated synonymy dictionary.

3   Testing Term Extraction Procedures

For each described group of patterns, an automatic term extraction procedure was developed based on the LSPL programming tools.

Five developed procedures, namely procedures for extraction terms candidates, authors' terms, term synonyms, dictionary terms, and term combinations, are applied to a given text, and resulting sets of extracted terms are formed, along with recognized occurrences of the terms within the text. The procedures were tested with the aim to evaluate the quality of term extraction by traditional measures, i.e. recall and precision. For testing, a collection of Russian medium-sized texts (from 1597 to 4767 words) on computer science and physics were taken. Output results of the procedures were compared with sets of terms recognized and extracted by human experts.

The results of experimental evaluation of the procedures are shown in Table 2. Besides extraction of terms (in this case their presence in the given text was tested), we evaluate recognition of all occurrences of extracted terms within the text under processing. For example, let us consider a text fragment:

> The geodetic effect represents the effect of the curvature of spacetime, predicted by general relativity, on a vector carried along with an orbiting body. The geodetic effect was first predicted by Willem de Sitter in 1916.

The term *geodetic effect* can be extracted from it, and also two its occurrences can be recognized (occurrences are required for such applied task as subject index construction). For term synonyms and term combinations, recognition of their occurrences in the texts was not evaluated, since both occurrences of synonyms and occurrences of terms extracted from combinations are recognized as term candidates.

Recall of automatic extraction proved to be from 58 % (for term candidate recognition) to 94 % (for dictionary terms), while precision varies from 25 % (for term combination recognition) to 96 % (for authors' terms). For recognition of term occurrences, the values are 60–89 % and 49–78 % for recall and precision respectively.

As for the procedure for extraction of term variants, information about its performance is absent in Table 2, because the procedure is intended to merely reveal variants among terms yet extracted by other procedures.

Our analysis of detected cases of incompleteness and inaccuracy of term extraction from texts shows that restrictions of the applied linguistics criteria is evidently the main reason of the imperfect results. In particular, certain terms are not extracted because of their complex grammatical structure that is not represented in our LSPL patterns, however it is almost impossible to compile any exhaustive inventory of possible term structures. Some patterns of term definitions are ambiguous, and their addition to the

Table 2. Recall and precision of extraction procedures

Procedure and type of terms	Extraction of terms		Recognition of their occurrences	
	Recall	Precision	Recall	Precision
Term candidates	58 %	27 %	60 %	49 %
Authors' terms	92 %	96 %	74 %	78 %
Synonyms of terms	64 %	50 %	–	–
Dictionary terms	94 %	83 %	89 %	72 %
Term combinations	82 %	25 %	–	–

set of extraction patterns increases recall of extraction but simultaneously decreases precision. Moreover, patterns of term combinations, as well as patterns of term candidates fix only their grammatical structure, so many non-term word combinations (e.g., *important problem of astronomy*) match the patterns. Dictionary terms are not recognized in the cases, when they are broken within term combinations, whereas some text units are falsely recognized as dictionary terms, though they are only fragments of other terms or non-term collocations (such as mathematical term *series* in common scientific collocation *series of experiments*).

In general, using additional linguistics information facilitates more reliable term extraction based on another specific information, in particular, accounting for patterns of term combinations increases the recall of recognition of dictionary terms, while ac-counting for dictionary terms increases the precision of recognition of authors' terms.

Since linguistics features of terms used for their extraction by the tested procedures are not mutually exclusive, the resulting sets of extracted terms are intersected, for example, extracted term candidates include terms recognized by other procedures. Therefore, in order to accomplish more accurate and complete term detection and extraction, it is reasonable to combine the output sets of our term extraction procedures. It should be pointed out that simple union of the sets is not appropriate because it evidently gives increase of recall with simultaneous decrease of precision.

4 Strategy for Combining Sets of Extracted Terms

Based on the results of the described experiments, we derived certain heuristics on how to combine the sets of terms recognized by the extraction procedures from the given text. In order to improve the overall quality of term extraction, we elaborated a strategy that iteratively forms a resulting set of terms by applying these heuristics while selecting elements of the pre-extracted sets of terms.

Let us denote the output sets produced by the extraction procedures as TCAND (term candidates), AUTH (authors' terms), SYN (term synonyms), DICT (dictionary terms), COMB (conjunctionless term combinations), COMBCON (term combinations with conjunctions), TVAR (term variants).

The steps of our strategy are as follows.

$S:=\varnothing$

Step 1. $S1:=\text{AUTH} \cup \text{DICT}_{\text{not\_part\_TCAND}}$

Step 2. $S2:=\text{DICT}_{\text{from\_COMB}} \cup \text{COMB}_{\text{with\_DICT}}$
$S:=S1 \cup S2$

CL: **Step 3.** $S3:=\text{SYN}_{\text{for\_s}}$; $S:=S \cup S3$

Step 4. $S4:=\text{DICT}_{\text{from\_COMBCON}} \cup \text{TCAND}_{\text{from\_COMCON}}$; $S:=S \cup S4$

Step 5. $S5:=\text{DICT}_{\text{from\_COMB}} \cup \text{TCAND}_{\text{from\_COMBCON}} \cup \text{COMB}_{\text{without\_TCAND}}$
If $S3 \cup S4 \cup S5 \neq \varnothing$ then $S:=S \cup S3 \cup S4 \cup S5$; goto CL

Step 6. $S6:=\text{TVAR}_{\text{for\_s}}$
If $S6 \neq \varnothing$ then $S:=S \cup S6$; goto CL

Step 7. $S7:=\text{TCAND}_{\text{freq}>F}$
If $S7 \neq \varnothing$ then $S:=S \cup S7$; goto CL

Step 8. $S8:=\text{DICT}$
If $S8 \neq \varnothing$ then $S:=S \cup S8$; goto CL

The final set S of terms is formed incrementally. Initially S is empty, and whenever some element of abovemention sets of pre-extracted terms are included into S, the element is removed from the source set. $S1$–$S8$ denotes terms, selected in the corresponding steps.

First of all, we put into S terms recognized with a high degree of precision. In step 1 all author's terms (AUTH) are included into S, dictionary terms are included as well unless they are fragments of term candidates ($\text{DICT}_{\text{not\_part\_TCAND}}$). Thereby, we do not approve as actual term any text unit recognized as dictionary, if all its occurrences in the text are embedded into occurrences of some term candidate (such as pair *dark matter – hot dark matter*). However, in Step 2 we add to S those remaining dictionary terms that are constituents of conjunctionless combinations ($\text{DICT}_{\text{from\_COMB}}$), since this condition increases the likelihood they are really terms. For similar reasons, any conjunctionless combination is added to S provided that it includes some dictionary term as constituent ($\text{COMB}_{\text{with\_DICT}}$).

In steps 3–5 synonyms and combinations of terms are considered. We include into S synonyms of all terms that belong to actual S ($\text{SYN}_{\text{for\_s}}$). If any term combination with conjunction includes as constituent an element either from S, from DICT or from TCAND (as broken term), in Step 4 we add to S all constituents of this term combination (we denote them $\text{DICT}_{\text{from\_COMBCON}}$ and $\text{TCAND}_{\text{from\_COMBCON}}$). Similarly, in Step 5 we add to S all constituents of conjunctionless term combinations provided that they include as constituent (in broken form) any element either from S, from DICT or from TCAND ($\text{DICT}_{\text{from\_COMB}}$ and $\text{TCAND}_{\text{from\_COMB}}$). If there are no such conjunctionless combinations, we add to S whole conjunctionless combinations ($\text{COMB}_{\text{without\_TCAND}}$) instead of their constituent terms.

If the set S is extended in steps 3–5, these steps need to be repeated, otherwise the following steps are performed. Similarly, after each of the steps 6, 7, 8 in the case of extension of the set S, steps of our strategy are to be repeated from the step 3.

In step 6 term variants for all terms that belong to actual $S(TVAR_{for\_s})$ are added to S. Then we add to S all remaining term candidates ($TCAND_{freq>F}$) with frequency more than F, where F is computed as rounded weighted arithmetic average of term candidates frequencies. Thus, we exclude from consideration relatively rare term candidates.

In the final step 8 we include into S all dictionary terms that are not yet in S, and then if needed we repeat the Steps 3–7.

The described term extraction strategy produces a set of selected terms with higher degree of reliability. We have performed experiments to evaluate and compare recall and precision of the strategy and several commonly-used extraction methods, which we consider as baseline methods. For experiments, a collection of texts (approx. 33,000 words) in the same scientific fields (computer science and physics) was taken.

The baseline methods use information about frequencies of words and grammatical structures of terms. Method Mutual-Inf extracts two-word terms (possibly with prepositions) based solely on statistics of word occurrences and co-occurrences [17]. Method Mod-Mutual is a modification of the previous method, it additionally accounts for single word terms and POS of words [7]. Method C-Value recognizes multiword terms by using frequencies of words and information about embedded terms [9]. Method SP extracts multi-word terms according to their grammatical patterns.

The results of experimental evaluation of our strategy in comparison with the baseline method are shown in Table 3. The strategy shows significantly better performance in precision and compatible performance in recall compared with the method Mod-Mutual. In overall, for term extraction our strategy gives 17,6 % increase of F-measure (the combined measure of precision and recall) compared with the best baseline method, and 11,7 % increase of F-measure for recognition of term occurrences.

Table 3. Comparative evaluation of the proposed strategy

Method	Extraction of terms			Recognition of term occurrences		
	Recall	Precision	F-measure	Recall	Precision	F-measure
Mutual-Inf	27,3 %	13,0 %	17,6 %	24,4 %	20,4 %	22,2 %
Mod-Mutual	**54,1 %**	37,4 %	44,2 %	**69,2 %**	41,5 %	51,9 %
C-Value	35,5 %	4,9 %	8,6 %	21,3 %	5,9 %	9,3 %
SP	51,4 %	22,6 %	31,4 %	37,3 %	29,7 %	33,1 %
Strategy	53,6 %	**73,1 %**	**61,8 %**	68,1 %	**59,7 %**	**63,6 %**

5 Conclusion

Aiming to improve the overall quality of term extraction from a given scientific text, we propose a heuristic strategy based on various linguistics information including grammatical structures of multiword scientific terms, their text variants, and contexts of their usage. The information about various features of term occurrences within Russian scientific texts has been formalized and represented as a set of LSPL lexico-syntactic patterns. Several term extraction procedures have been implemented with the aid of LSPL programming tools, each procedure uses a particular group of extraction patterns. This made it possible to experimentally evaluate efficiency of the implemented extraction procedures and then to reveal certain heuristics on how to combine and select output sets of terms extracted by the procedures in order to produce a set of terms with higher degree of reliability. Experimental evaluation of the proposed heuristic strategy shows significant increase of F-measure in comparison with the commonly-used methods of term extraction.

Nevertheless, our heuristic strategy needs additional verification on texts of various scientific domains and sizes. We believe that further improvement of the strategy is feasible by extending and refining LSPL extraction patterns. Necessary experiments can be performed without reprogramming our extraction procedures (only the input sets of patterns should be changed).

The described extraction procedures and strategy are undoubtedly useful for various NLP applications with complex processing of terminological units, especially for computer-aided abstracting of scientific texts and for computer support of scientific writing. In fact, writing support involves checkups of term consistency and accuracy within a particular scientific document, as well as construction of a problem-oriented glossary and a subject index for the document, the work can be done only by means of automatic term extraction procedures similar to those described in our paper.

Acknowledgements. We would like to thank the anonymous reviewers of our paper for their helpful and constructive comments.

References

1. Arora, C., Sabetzadeh, M., Briand, L., Zimmer, F.: Improving requirements glossary construction via clustering: approach and industrial case studies. In: Proceedings of the 8th ACM/IEEE International Symposium on Empirical Software Engineering and Measurement. ACM, New York, NY (2014)
2. Bolshakova, E.I.: Recognition of author's scientific and technical terms. In: Gelbukh, A. (ed.) CICLing 2001. LNCS, vol. 2004, pp. 281–290. Springer, Heidelberg (2001)
3. Bolshakova, E., Efremova, N., Noskov, A.: LSPL-patterns as a tool for information extraction from natural language texts. In: Markov, K., Ryazanov, V., Velychko, V., Aslanyan, L. (eds.) New Trends in Classification and Data Mining, pp. 110–118. ITHEA, Sofia (2010)
4. Bosma, W., Vossen, P.: Bootstrapping language neutral term extraction. In: Proceedings of the 7th Language Resources and Evaluation Conference, pp. 2277–2282. LREC, Valetta (2010)
5. Castellvi, M., Bagot, R., Palatresi, J.: Automatic term detection: a review of current systems. In: Bourigault, D., Jacquemin, C., L'Homme, M.-C. (eds.) Recent Advances in Computational Terminology, pp. 53–87. John Benjamins, Amsterdam (2001)

6. Csomai, A., Mihalcea, R.: Investigations in unsupervised back-of-the-book indexing. In: Proceedings of the Florida Artificial Intelligence Research Society Conference, pp. 211–216 (2007)
7. Dobrov, B., Loukachevich, N., Syromiatnikov, S.: Forming base of terminological word combinations from problem oriented texts. In: Proceedings of the 5th Russian Scientific Conference "Digital Libraries: Perspective Methods and Technologies, Electronic Collections", pp. 201–210 (2003) (in Russian)
8. Efremova, N.E.: Methods and Programming Tools for Extraction of Terminological Information from Scientific and Technical Texts. PhD Thesis, Lomonosov Moscow State University (2013) (in Russian)
9. Frantzi, K., Ananiadou, S., Mima, H.: Automatic Recognition of Multi-Word Terms: The C-value/NC-value method. In: Nikolau, C. et al. (Eds.) International Journal on Digital Libraries, vol. 3(2), pp. 115–130 (2000)
10. Jacquemin, C., Tsoukermann, E.: NLP for term variant extraction: synergy between morphology, lexicon, and syntax. In: Strzalkowski, T. (ed.) Natural Language Information Retrieval, pp. 25–74. Kluwer Academic Publishers, Dordrecht (1999)
11. Korkontzelos, I., Ananiadou, S.: Term extraction. In: Oxford Handbook of Computational Linguistics (2nd Ed.). Oxford University Press, Oxford (2014)
12. Manning, C.D., Raghavan, P., Schütze, H.: Introduction to Information Retrieval. Cambridge University Press, New York (2008)
13. Matsuo, Y., Ishizuka, M.: Keyword extraction from a single document using word co-occurrence statistical information. Int. J. Artif. Intell. Tools 13(1), 157–169 (2004)
14. Nenadic, G., Ananiadou, S., McNaught, J.: Enhancing automatic term recognition through recognition of variation. In: Proceedings of 20th International Conference on Computational Linguistics COLING 2004, pp. 604–610. Morristown, NJ (2004)
15. Nokel, M.A., Bolshakova, E.I., Loukachevich, N.V.: Combining multiple features for single-word term extraction. In: Computational Linguistics and Intellectual Technologies: Papers from the Annual International Conference "Dialogue", vol. 1, no. 11 pp. 490–501. RGGU, Moscow (2012)
16. Paice, C.D., Jones P.A.: The identification of important concepts in highly structured technical papers. In: Korfhage, R., Rasmussen, E., Willett, P. (eds.) Proceedings of the 16th Annual International ACM-SIGIR Conference on Research and Development in Information Retrieval, pp. 69–78. ACM, Pittsburgh, PA (1993)
17. Smadja, F., McKeown, K.: Automatically extracting and representing collocations for language generation. In: Proceedings of the 28th Annual Meeting on Association for Computational Linguistics, pp. 252–259. ACL, Pittsburgh, PA (1990)

Text Analysis with Enhanced Annotated Suffix Trees: Algorithms and Implementation

Mikhail Dubov[(✉)]

Computer Science Faculty, National Research University Higher School of Economics,
Moscow, Russia
mdubov@hse.ru

Abstract. We present an improved implementation of the Annotated suffix tree method for text analysis (abbreviated as the AST-method). Annotated suffix trees are an extension of the original suffix tree data structure, with nodes labeled by occurrence frequencies for corresponding substrings in the input text collection. They have a range of interesting applications in text analysis, such as language-independent computation of a matching score for a keyphrase against some text collection. In our enhanced implementation, new algorithms and data structures (suffix arrays used instead of the traditional but heavyweight suffix trees) have enabled us to derive an implementation superior to the previous ones in terms of both memory consumption (10 times less memory) and runtime. We describe an open-source statistical text analysis software package, called "EAST", which implements this enhanced annotated suffix tree method. Besides, the EAST package includes an adaptation of a distributional synonym extraction algorithm that supports the Russian language and allows us to achieve better results in keyphrase matching.

Keywords: Text analysis · Algorithms on strings · Annotated suffix trees · Suffix arrays · Synonym extraction

1 Introduction

The annotated suffix tree (AST) method has first been introduced in 2006 by Pampapathi, Mirkin, and Levene [12] as a novel text classification tool with application to anti-spam e-mail filtering. It differs from the majority of text processing methodologies in that it considers the text as a sequence of letters, not as a set of words. While the latter approach, called the *bag-of-words* model, is much more commonly exploited both in literature and in practice (e.g. [13, pp. 167–200]), the former one has the advantage that it makes the text analysis more language-independent [10]. Indeed, with a letter-based approach there is usually no need to perform such preprocessing step as stemming since the performance of algorithms no longer depends on the uniformity of grammatical word forms. The annotated suffix tree method allows omitting the stop words collection and the filtering step as well since it is not sensitive to the "noise" in the text. Seemingly, the most significant feature of this approach is that it captures the sequential structure of the texts being analyzed [11].

© Springer International Publishing Switzerland 2015
M.Y. Khachay et al. (Eds.): AIST 2015, CCIS 542, pp. 308–319, 2015.
DOI: 10.1007/978-3-319-26123-2_30

The basic computation in the AST method is the matching (relevance) score for some keyphrase against the input text collection indexed by the AST. This "approximate" matching score, lying in the $[0; 1]$ interval, serves as a basis for several interesting applications. Among the most interesting examples in this field, let us mention the AST-based feature extraction and text classification algorithms presented by Pampapathi [12], and also the work by Mirkin, Chernyak, & Chugunova [10], where the AST method is exploited to build a keyphrase reference graph.

While the annotated suffix tree methodology is a very promising tool for a great variety of text analysis applications, it has proven to be not efficient enough in terms of running time and space consumption to be able to process large-scale text collections. In this paper, we describe an enhanced implementation of this method. This implementation uses suffix arrays instead of pure suffix trees as the main underlying data structure, which allows it to require up to 10 times less memory and be faster than any other previous implementation of the AST method known to us. We also discuss injecting one language-dependent feature into the AST method, namely the context-based extraction of synonyms, which is neccessary for more accurate keyphrase relevance score computation.

We structure this paper as follows: in Sect. 2, we describe the foundations of the original AST method. In Sect. 3, we review the suffix array data structure and show how it can substitute suffix trees in an AST method implementation. We devote Sect. 4 to a Python package called EAST that implements these algorithms and describe some of its features. Finally, we discuss some experimental results of our study obtained using this package in Sect. 5.

2 Annotated Suffix Trees

A *suffix tree* is considered to be one of the central data structures in the field of string processing algorithms [5, pp. 89–93]. Its importance stems from the fact that it establishes a linear time solution for one of the classical tasks in strings processing, namely the *exact pattern matching* problem, and serves as an efficient data structure for *full-text indexing*. Formally, the suffix tree data structure for a string S ($|S| = n$) is defined as a rooted directed tree encoding all the suffixes of that string. It is constructed in such a way that the concatenation of edge labels on every path from the root node to one of the leaves makes up one of the suffixes of that string, i.e. $S[i \ldots n]$. It is also required that each internal node has two or more children, and each edge is labeled with a non-empty substring of S [5, p. 90]. Suffix trees can be constructed in linear time and stored in linear space [15]. A generalized suffix tree is a suffix tree built for multiple strings; it can be constructed in linear time as well [5, p. 116].

While being linear in size and in construction runtime, suffix trees still suffer from extensive memory consumption: a typical implementation needs 30 bytes of memory on average per one input symbol. Besides, suffix trees have poor locality of memory reference, which leads to a loss of efficiency on modern cached processor architectures [1].

An *annotated suffix tree* is an extension of the original suffix tree data structure. In addition to the edge labels in suffix trees, annotated suffix trees have their nodes labeled with integer numbers indicating the number of entries of the substring on the path from the root to that node (as spelled out by the edge labels) in the original text collection. These node annotations can then be used to calculate the matching score (a real number in [0; 1]) of a given keyphrase against the text the annotated suffix tree has been built for [10]. In this way, annotated suffix trees provide an alternative solution to the *approximate pattern matching* problem.

For example, when built for a single string *"XABXAC"*, the annotated suffix tree gives us a relevance score of 0.25 for the string *"ABC"* and 0.11 for the string *"XYZ"* (which has only one common letter with *"XABXAC"*). When used to index the text of *"Alice in Wonderland"* by L. Carroll[1], the AST returns 0.32 as a matching score for *"Alice"* and 0.04 for *"Bob"* (as expected). This score can be thus seen as a measure of relevance of a keyphrase to some text collection (in our experience, values greater than 0.2 are usually strong indicators of relevance).

In the first publications [12], the annotated suffix tree has been naively depicted as a tree where each node v contains exactly one letter and stores the value of $f(v)$, which is the number of occurrences of a substring, obtained by the concatenation of all the symbols on the path from root to that node, in the original text collection (see Fig. 1). In this representation, the tree is actually not a suffix but a *keyword tree* (a *trie*), which requires quadratic space to store and therefore quadratic time to be constructed. It allows, however, to make a definition of the key phrase matching score that is easy to state and to interpret.

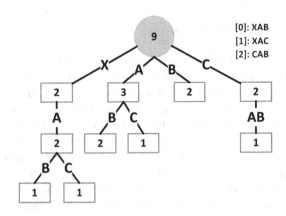

Fig. 1. Naive AST representation

To describe the matching score computation procedure for the naive AST representation, let us define the *conditional probability of a node* given its parent as

$$\hat{p}(v) = \frac{f(v)}{f(parent(v))}$$

---

[1] http://archive.org/stream/alicesadventures19033gut/19033.txt.

Where v is a node, $f(v)$ is the node's annotation (the frequency of the corresponding substring) and $parent(v)$ is the parent node of v. Imagine that we try to match some sequence of letters against the suffix tree just as in the exact pattern matching problem, starting from the root and proceeding to the child nodes. If, for some string s, we matched exactly k symbols in the tree, then

$$score_{seq}(s) = \frac{\sum_{i=1}^{k} \hat{p}(v_i)}{k},$$

Where v_i is the i-th node on the matching path starting at the root. Now we are ready to define the actual AST matching score:

Definition 1. *AST matching score* for a given keyphrase S against some text collection T is a value computed for $AST(T)$ as

$$SCORE(S) = \frac{\sum_{i=1}^{|S|} score_{seq}(S[i :])}{|S|}$$

In other words, we first match each suffix of the input keyphrase against the constructed annotated suffix tree in terms of the *conditional probabilities* of the nodes corresponding to its symbols and then produce the final AST matching score by averaging these suffix scores. This final matching score has can be interpreted as the *average conditional probability* of an occurrence of a single symbol of the input keyphrase in the text collection the AST has been built for.

Obviously, quadratic runtime and memory consumption of such a naive implementation of the AST methodology as described above are unsatisfactory for real applications. Optimized implementations are based on the actual suffix tree data structure (not on a keyword tree) with compacted edges and no "chain" nodes, i.e. having exactly one child (Fig. 2, top-level leaves for unique symbols omitted for simplicity).

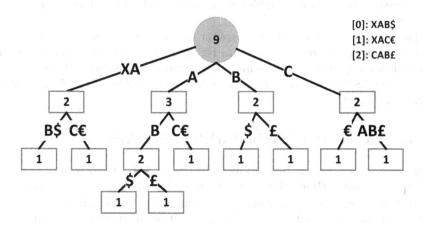

Fig. 2. Optimized AST representation

It has been shown by the author [3] that a very simple preprocessing of the text collection (appending unique termination characters to each text) makes it possible to annotate the tree with substring frequencies *after* the pure suffix tree was constructed for the preprocessed text collection (using some linear algorithm like the *Ukkonen's algorithm*). Indeed, with unique symbols appended, the number of occurrences for each text is guaranteed to be equal to 1, so that the leaf nodes can be annotated with 1. Then, in a bottom-up tree traversal, each inner node gets annotated with the sum of the annotations of its children. This allows to obtain a $O(n)$ algorithm for AST construction [3]:

Algorithm. **LinearASTConstruction(C)**
Input. String collection $C = \{S_1, \ldots, S_m\}$
Output. Generalized annotated suffix tree for C.

1. Construct $C' = \{S_1\$_1, \ldots, S_m\$_m\}$, where $\$_i$ are unique characters that do not appear in $S_1 \ldots S_m$.
2. Construct a generalized suffix tree T for collection C' using a linear-time algorithm (e.g. the *Ukkonen algorithm*).
3. **for** l in $leaves(T)$
4.   **do** set $f(l) \leftarrow 1$
5. Run a postfix depth-first tree traversal on the suffix tree T. For each inner node v, set $f(v) \leftarrow \sum_{u \in children(v)} f(u)$.

The compacted annotated suffix tree representation requires only a minor change in the formula for the helper score:

$$score_{seq}(s) = \frac{\sum_{i=1}^{k} \hat{p}(v_i) + l - k}{l},$$

Since for all *"chain"* nodes that had exactly one child in the naive representation and were removed, their conditional probability $\hat{p}(v) = 1$ (in the formula k is, as before, the number of nodes in the match and l is the number of symbols in the match).

These approximate matching scores for various keyphrases alone allow us to solve a number of interesting problems. For example, having an annotated suffix tree built for a spam corpus, one may split the analyzed e-mails into keyphrases and use the AST score to determine whether the e-mail text is likely to contain a spam message [12].

3   Enhanced AST Method with Suffix Arrays

Suffix arrays have first been introduced in the work by Manber and Myers [9] as a space efficient alternative to suffix trees. Technically, a *suffix array* for a string S of length n is an array of n integer numbers, enumerating the n suffixes of S in lexicographic order [5, p. 149]. Linear-time algorithms for suffix array construction are known as well [6], but tend to be more complicated than their suffix tree counterparts.

The potentiality of suffix arrays to serve as a space efficient replacement for suffix trees has naturally lead to an intense interest to the problem of effective porting of algorithms that use suffix trees as an auxiliary data structure to suffix arrays [1,7]. It is important to note that not only suffix arrays can yield a significant gain in space efficiency, but they also can speed up certain algorithms: modern processors make use of the fact that suffix arrays are a more compact data structure than suffix trees and utilize several low-level optimizations [1].

Abouelhoda, Kurtz, & Ohlebusch [1] have shown how to systematically replace every algorithm that uses suffix trees with another one based on suffix arrays by introducing the concept of an *lcp-interval tree*. The nodes of such a tree correspond to those of the original suffix tree, but the tree itself actually does not have to be stored in memory: it is sufficient to enhance the suffix array with two auxiliary arrays called *lcp-table* and *child-table*. It has been shown that it is possible to perform both bottom-up and top-down suffix tree traversals using these arrays without loss in asymptotic time complexity. Together with the original *suffix array*, these two auxiliary arrays (*lcp* and *child*) take no more than 10 bytes per symbol in the input text collection, when implemented accurately. Thus, using suffix arrays instead of suffix trees is at least twice as efficient in terms of memory usage (because of the additional arrays used to achieve this result, the actual data structure became known as an *enhanced suffix array*).

A sample enhanced suffix array for string *"XABXAC"* is presented in Table 1. The suffixes in the last column are not actually stored in memory. There are also *three child-tables* instead of one mentioned above; we keep them to be consistent with analogous illustrations in [1]. With certain implementation tricks, these three arrays can be compacted to one.

Table 1. Enhanced suffix array for string "XABXAC"

i	Suffix array	lcp-table	Child-table 1.	2.	3.	S[suff[i]:]
0	1	0		1	2	ABXAC
1	4	1				AC
2	2	0	1		3	BXAC
3	5	0			4	C
4	0	0				XABXAC
5	3	2				XAC

Let us briefly mention that enhanced suffix arrays are not the only way to solve the problem of extensive memory consumption by suffix trees. Another approach is to store suffix trees in external memory [2]. Besides, there are methods to compress suffix trees in the main memory [14]. These methods, however, do not yield a compact representation of the full-text index in memory and thus disable possible low-level optimizations implemented in modern processors.

In what follows, we will show how one can further enhance this data structure to simulate node annotations and use them to compute the AST matching score. Recall that the enhanced suffix array, built for a text collection of total length of n symbols, is a set of three arrays of length n. In the annotated suffix tree, however, the total number of nodes was larger than n and each node stored a substring frequency annotation.

In fact, it is easy to overcome this issue. It follows from the definition of a suffix tree that the number of internal nodes cannot exceed $(n-1)$. Recall that, to build an annotated suffix tree, we first made each text in the collection to end with a unique symbol and then annotated all the leaves with 1. But that means that we can omit storing these leaf annotations explicitly, because we always know that they are all equal to 1. We only need to store $(n-1)$ annotations for the root and internal nodes, and we can introduce yet another auxiliary *annotation-table* of length n for that purpose. Thus, this further enhanced data structure (that we call *"enhanced annotated suffix array"*) is a set of four arrays of length n: *suffix array*, *lcp-table*, *child-table*, and *annotation-table*.

Further, we need to define how exactly the frequency annotations of inner nodes of a suffix tree should be stored in this new *annotation-table* of length n. In the *virtual lcp-tree* model mentioned above, it is possible to restore the original suffix tree using the information from the *lcp-table*. The nodes of this *lcp-tree* have a one-to-one correspondence with inner nodes of a suffix tree and are represented using three parameters $\langle l, i, j \rangle$: the *lcp-value* l and the left and right boundaries of the *lcp-interval* (i, j). It can be shown (as done in [1]) that for each *lcp-interval* $v = \langle l, i, j \rangle$ there exists a unique index, $index(v) \in [0; n-1]$, which is equal to the smallest k, such that $k > i$ and $lcp[k] = l$. It is this mapping that we use to store the inner node frequency annotations.

In Table 2, there is an example of an enhanced annotated suffix array constructed for a single string "XABXAC" (the annotated suffix tree for the same string can be seen in Fig. 3, unique symbols omitted for simplicity). We need to store only 3 numbers in the *annotation-table* since the corresponding tree has only two inner nodes annotated with 2, plus the root (it's frequency annotation is 6).

Table 2. Enhanced annotated suffix array for string "XABXAC"

i	Suffix array	lcp-table	Child-table 1.	2.	3.	Annotation-table	S[suff[i]:]
0	1	0		1	2	6	ABXAC
1	4	1				2	AC
2	2	0	1		3		BXAC
3	5	0			4		C
4	0	0					XABXAC
5	3	2				2	XAC

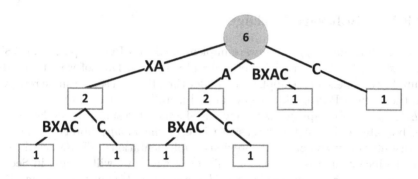

Fig. 3. Annotated suffix tree for string "XABXAC"

We are now ready to present a linear enhanced annotated suffix array construction algorithm. This algorithm has a structure which is very similar to that of **LinearASTConstruction**. First, it preprocesses the input text collection by appending unique symbols to each of its texts and builds and enhanced suffix array for the concatenation of these texts. In the last step, the algorithm simulates a bottom-up tree traversal to fill up the *annotation-table*.

Algorithm. LinearEASAConstruction(C)
Input. String collection $C = \{S_1, \ldots, S_m\}$
Output. Enhanced suffix array for C with substring frequency annotations.

1. Construct a string $S = S_1\$_1 + \cdots + S_m\$_m$, where $\$_i$ are unique symbols that do not appear in $S_1 \ldots S_m$ and "+" is string concatenation operator.
2. Construct a suffix array A for string S using a linear-time algorithm (e.g. the *Kärkkäinen-Sanders algorithm*) and two auxiliary arrays: *lcp-array* and *child-array*.
3. Simulate a postfix depth-first tree traversal on the suffix array A. At each of the *virtual inner nodes*, corresponding to an *lcp-interval* $v = \langle l, i, j \rangle$, where $i < j$, set $annotation[index(v)] = \sum_{u \in children(v)} annotation[index(u)] + \#(\langle l, i, j \rangle : i = j)$.

Since in the *virtual lcp-tree* model there is no processing of leaves, we have to take them into account by appending the number of *lcp-intervals* with $i = j$ at each virtual inner node in step 3. If the length of the i-th string in the collection $|S_i| = n_i$, then the runtime complexity of this algorithm is $\Theta(n_1 + \cdots + n_m)$.

The formula for the AST matching score stays unchanged with this new suffix array-based implementation: indeed, this data structure allows us to simulate not only postfix, but also prefix tree traversals [1], which is exactly what is needed for the score computation described in Sect. 2.

One last detail about the AST method implementation is that it is worth splitting every text in the input collection into a sequence of substrings, each consisting of k words, where k should be the expected number of words in the input keyphrases (usually $k = 3$ or $k = 4$). This approach has proven to provide more accurate matching scores in most practical applications [10].

4   EAST Software Package

The algorithms decribed above were implemented in a Python package EAST[2] (EAST stands for "Enhanced Annotated Suffix Trees"). This software tool is distributed as an open-source application under the MIT license. It can be retrieved from the Python Package Index system and installed via the *pip* tool.

EAST not only implements the enhanced annotated suffix arrays data structure, but also relies on the extensive usage of the *numpy* library to increase memory efficiency compared to previous implementations. This allowed us to get a significant increase in memory efficiency which we will discuss in Sect. 5. The underlying algorithms used in *EAST* are the *Kärkkäinen-Sanders algorithm* for suffix array construction [6] and the *Kasai algorithm* for lcp-table computation [7]. The package also includes the previous implementation of the AST method which uses the *Ukkonen algorithm* for suffix tree construction [15]. This implementation can be used as an alternative one.

The package not only can be used as a Python library, but also provides the user with a command line interface. For example, to produce a table of matching scores for a set of keyphrases against some text collection, one should run:

```
$ east keyphrases table <keyphrases_list.txt>
  <path/to/the/text/collection/>
```

An interesting feature of the *EAST* package is that it introduces *synonym extraction*, a language-dependent feature to enhance the AST method even more (AST is basically language-independent). Its purpose is to handle situations when one tries to match a keyphrase (say, *"plant taxonomy"*) against some text collection that may cover the topic corresponding to that keyphrase, but mostly using a different terminology (say, *"plant classification"*) and, as a result, the relevance score is unreasonably small.

The extensive body of literature concerning synonym extraction comprises two basic approaches: building synonymic thesauri based on some already existing *lexica* (dictionaries), and the analysis of word collocations in *corpora* (the so-called *distributional* approach) [16]. The former strategy usually includes the detailed analysis of the dictionary definitions structure, while the latter one is based on the assumption that similar words appear in similar contexts.

In our software, we use a distributional synonym extraction algorithm similar to that of Lin [8]. The advantage of this approach, which is of huge importance to us, is the fact that it extracts synonyms based on the same set of texts that is analyzed later using the AST method. Consequently, it is, on the one hand, capable of extracting domain-specific synonyms, and, on the other hand, does not extract redundant synonyms not present in the text collection, thus bringing no performance decrease to our algorithms.

We use the resulting set of synonyms as follows: for each input keyphrase query S against the constructed annotated suffix tree (array), we first find a set of synonymous queries $syn(S) = w_1, w_2, ..., w_k$. We then define the matching

---

[2] https://pypi.python.org/pypi/EAST.

score of S as $\max_{w \in syn(S)} SCORE(w)$, where $SCORE(w)$ is the matching score for keyphrase w as computed by the original AST method.

The distributional synonym extraction algorithm by Lin [8] employs the so-called dependency triples (w_1, r, w_2) that model semantic relationships between words in the analyzed texts and then allow to extract words that often appear in common dependency triples as synonyms. In his original paper, Lin uses a *broad-coverage parser* to extract dependency triples from English corpora. In our adaptation of that algorithm to the Russian language, we use the *Yandex Tomita parser*[3] to extract such triples based on a set of grammatical templates like *"adjective + substantive"* or *"verb + arverb"*.

5   Experimental Results

We have conducted a series of experiments based on the implementation described in Sect. 4. To compare our new suffix array-based AST implementation against the algorithms used previously, we have measured both their runtime and memory consumption. In our experiments, we used randomly generated text collections. Each text collection contained 100 random strings, with their lengths increasing from 10 to 1000 from iteration to iteration. At each iteration, the strings in the collection started with the same prefix and differed only in several ending symbols. Such collections not only allowed us to get close to the worst-case for the naive construction algorithm, but also represented a situation that often occurrs in natural languages, where lots of words with the same stem are not a rare case. Using this input, we have compared three algorithms:

1. "Naive" construction algorithm that appeared in the early works on the AST method. It has the worst-case complexity of $\Theta(n_1^2 + n_2^2 + \cdots + n_m^2)$, where n_i is the length of the i-th string in the collection and m is the collection size;
2. Linear algorithm from Sect. 2, producing the annotated suffix tree data structure in $\Theta(n_1 + n_2 + \cdots + n_m)$ in the worst case;
3. Linear algorithm for enhanced annotated suffix arrays construction from Sect. 3, running in $\Theta(n_1 + n_2 + \cdots + n_m)$ in the worst case.

The results are visualized in Fig. 4. They demonstrate a significant increase in memory efficiency in our new implementation. When compared to the previous suffix tree-based implementation, it uses up to 10 times less memory. Indeed, to construct a usual AST for 100 strings, each of length 500, the software needs 250 MB of memory, while the enhanced implementation requires only 20–25 MB. Such a drastic improvement can be partially explained by the usage of the *numpy* library that wasn't used in the previous implementations (they were based on heavy-weight Python dictionaries to store child nodes), but there is no doubt as well that it is due to the usage of suffix arrays (which in theory should have given us a 2–3x improvement). As for the runtime, our new algorithm shows only a modest improvement over the previous solution.

---

[3] http://api.yandex.ru/tomita.

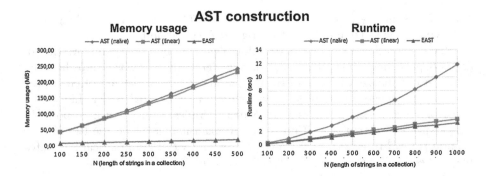

Fig. 4. Experimental results

Our adaptation of a distributional synonym extraction algorithm has shown relatively good results for Russian texts as well. When run over a set of articles from the newspaper "Izvestia" devoted to economics[4], it marked as synonyms, for example, the words *"head"* (*"глава"*) and *"CEO"* (*"гендиректор"*), but also wrongly (but rather expectedly) extracted such synonym pairs as *"high"* (*"высокий"*) and *"low"* (*"низкий"*). The low precision of this method, however, does not seem to be a big issue: in fact, wrong synonyms usually do not affect the resulting matching score for keyphrases consisting of multiple words (which is chosen as the maximal score among all the possible synonymous phrases).

6   Conclusion

We have developed and implemented a new approach to the construction of annotated suffix trees that uses enhanced suffix arrays as its main underlying data structure. As a result, we have an implementation of the AST method for text analysis which is superior to any other implementation known to us in terms of both memory usage and runtime efficiency.

The corresponding software it the first AST method implementation that has been published as an open-source Python package. It follows the best practices of software engineering, such as self-documented code or unit tests coverage, which makes it easy to integrate and to extend for other researchers and developers.

The EAST package has already been successfully integrated with an automatic Russian text processing system called *LM Monitor*[5] [4]. This system performs the collection and analysis of a Russian Newspaper Web Corpus (*RuNeWC*). The *EAST* package is used in the corpus browser to analyse the relevance of different keyphrases to the newsfeeds. The relevance scores are also used to build the so-called *keyphrase reference graphs* that visualize the relations between keyphrases (computed in a manner similar to *associative rules*). Our enhanced implementa-

---

[4] http://izvestia.ru/rubric/16.

[5] http://cs.hse.ru/vitext.

tion has proven to work fast and efficiently enough to be able to process large amounts of texts even with modest computational powers.

Acknowledgements. This research carried out in 2015 was supported by "The National Research University 'Higher School of Economics' Academic Fund Program" grant (№ 15-05-0041). The financial support from the Government of the Russian Federation within the framework of the implementation of the 5–100 Programme Roadmap of the National Research University – Higher School of Economics is acknowledged.

References

1. Abouelhoda, M.I., Kurtz, S., Ohlebusch, E.: Replacing suffix trees with enhanced suffix arrays. J. Discrete Algorithms **2**, 53–86 (2004)
2. Barsky, M., Stege, U., Thomo, A.: A survey of practical algorithms for suffix tree construction in external memory. Softw. Pract. Experience **40**(11), 965–988 (2010)
3. Dubov, M., Chernyak, E.: Annotated suffix trees: implementation details. Transactions of Scientific Conference on Analysis of Images, Social Networks and Texts (AIST), pp. 49–57. Springer, Switzerland (2013)
4. Dubov, M., Mirkin, B., Shal, A.: Automatic russian text processing system. Open Systems DBMS **22**(10), 15–17 (2014)
5. Gusfield, D.: Algorithms on Strings, Trees, and Sequences: Computer Science and Computational Biology. Cambridge University Press, Cambridge (1997)
6. Kinen, J, Sanders, P.: Simple Linear Work Suffix Array Construction. Automata, Languages and Programming. Lecture Notes in Computer Science, pp. 943–2719 (2003)
7. Kasai, T., Lee, G.H., Arimura, H., Arikawa, S., Park, K.: Linear-time longest-common-prefix computation in suffix arrays and its applications. In: Amir, A., Landau, G.M. (eds.) CPM 2001. LNCS, vol. 2089, pp. 181–192. Springer, Heidelberg (2001)
8. Lin, D.: Automatic retrieval and clustering of similar words. In: Proceedings of the 17th International Conference on Computational Linguistics, pp. 768–774 (1998)
9. Manber, U.: Suffix arrays: a new method for on-line string searches. SIAM J. Comput. **22**(5), 935–948 (1993)
10. Mirkin, B., Chernyak, E., Chugunova, O.: Method of annotated suffix tree for scoring the extent of presence of a string in text. Bus. Inf. **3**(21), 31–41 (2012)
11. Pampapathi, R.: Annotated suffix trees for text modelling and classification. Doctoral dissertation, Birkbeck College, University of London, Retrieved from CiteSeerX (2008)
12. Pampapathi, R., Mirkin, B., Levene, M.: A suffix tree approach to anti-spam email filtering. Mach. Learn. **65**(1), 309–338 (2006)
13. Perkins, J.: Python Text Processing with NLTK 2.0 Cookbook. Packt Publishing, Birmingham (2010)
14. Sadakane, K.: Compressed suffix trees with full functionality. Theory of Computing Systems **41**(4), 589–607 (2007)
15. Ukkonen, E.: On-Line Construction of Suffix Trees. Algorithmica **14**(3), 249–260 (1995)
16. Wang, T.: Extracting Synonyms from Dictionary Definitions. Retrieved from Focus on Research, Master dissertation, University of Toronto (2009)

Morphological Analyzer and Generator
for Russian and Ukrainian Languages

Mikhail Korobov[✉]

ScrapingHub, Inc., Yekaterinburg, Russia
kmike84@gmail.com

Abstract. pymorphy2 is a morphological analyzer and generator for
Russian and Ukrainian languages. It uses large efficiently encoded lexicons built from OpenCorpora and LanguageTool data. A set of linguisti-
cally motivated rules is developed to enable morphological analysis and
generation of out-of-vocabulary words observed in real-world documents.
For Russian pymorphy2 provides state-of-the-arts morphological analysis
quality. The analyzer is implemented in Python programming language
with optional C++ extensions. Emphasis is put on ease of use, documentation and extensibility. The package is distributed under a permissive
open-source license, encouraging its use in both academic and commercial setting.

Keywords: Morphological analyzer · Russian · Ukrainian · Morphological generator · Open source · OpenCorpora · LanguageTool · pymor-
phy2 · pymorphy

1 Introduction

Morphological analysis is an analysis of internal structure of words. For languages
with rich morphology like Russian or Ukrainian using the morphological analysis
it is possible to figure out if a word can be a noun or a verb, or if it can be
singular or plural. Morphological analysis is in an important step of natural
language processing pipelines for such languages.

Morphological generation is a process of building a word given its grammatical representation; this includes lemmatization, inflection and finding word
lexemes.

pymorphy2 is a morphological analyzer and generator for Russian and
Ukrainian language widely used in industry and in academia. It is being developed since 2012; Ukrainian support is a recent addition. The development of its
predecessor, pymorphy[1] started in 2009. The package is available[2] under a permissive license (MIT), and it uses open source permissively licensed dictionary
data.

---

[1] https://bitbucket.com/kmike/pymorphy.

[2] https://github.com/kmike/pymorphy2.

© Springer International Publishing Switzerland 2015
M.Y. Khachay et al. (Eds.): AIST 2015, CCIS 542, pp. 330–342, 2015.
DOI: 10.1007/978-3-319-26123-2_31

The rest of this paper is organized as follows. In Sect. 2 pymorphy2 software architecture and design principles are described. Section 3 explains how pymorphy2 uses lexicons and how analysis and morphological generation work for vocabulary words. In Sect. 4 methods used for out-of-vocabulary words are explained and compared with approaches used by other morphological analyzers. Section 5 is dedicated to a problem of selecting correct analysis from all possible analyses, and a role of morphological analyzer in this task. In Sect. 6 evaluation results are presented. Section 7 outlines a roadmap for future pymorphy2 improvements.

2   Software Architecture

pymorphy2 is implemented as a cross-platform Python[3] library, with a command-line utility and optional C++ extensions for faster analysis. Both Python 2.x and Python 3.x are supported. An extensive testing suite (600+ unit tests) ensures the code quality; test coverage is kept above 90%. There is online documentation[4] available.

When optional C++ extension is used (or when pymorphy2 is executed using PyPy[5] Python interpreter) the parsing speed is usually in tens of thousands of words per second; in some specific cases in can exceed 100000 words per second in a single thread. Without the extension parsing speed is in thousands of words per second. The memory consumption is about 15MB, or about 30MB if we account for Python interpreter itself.

Users are provided with a simple API for working with words, their analyses and grammatical tags. There are methods to analyze words, inflect and lemmatize them, build word lexemes, make words agree with a number, methods for working with tags, grammemes and dictionaries. Inherent complexity of working with natural languages is not hidden from the user. For example, to lemmatize the word correctly it is necessary to choose the correct analysis from a list of possible analyses; pymorphy2 provides $P(analysis|word)$ estimates and sorts the results accordingly, but requires user to choose the analysis explicitly before normalizing the word.

Analysis of vocabulary words and out-of-vocabulary words is unified. There is a configurable pipeline of "analyzer units"; it contains a unit for vocabulary words analysis and units (rules) for out-of-vocabulary words handling. Individual units can be customized or turned off; some rules are parametrized with language-specific data. Users can create their own analyzer units (rules). This all makes it possible to perform morphological analysis experiments without changing pymorphy2 source code, develop domain-specific morphology analysis pipelines and adapt pymorphy2 to work with languages other than Russian. The latter point is validated by introducing an experimental support for Ukrainian language.

---

[3] https://www.python.org/.

[4] http://pymorphy2.readthedocs.org.

[5] http://pypy.org/.

3   Analysis of Vocabulary Words

pymorphy2 relies on large lexicons for analysis of common words. For Russian it uses OpenCorpora [3] dictionary ($\sim 5 * 10^6$ word forms, $\sim 0.39 * 10^6$ lemmas) converted from OpenCorpora XML[6] format to a compact representation optimized for morphological analysis and generation tasks. End users don't have to compile the dictionaries themselves; pymorphy2 ships with prebuilt periodically updated dictionaries.

Any dictionary in OpenCorpora XML format can be used by pymorphy2. For Ukrainian there is such experimental dictionary ($\sim 2.5 * 10^6$ word forms) being developed[7] by Andriy Rysin, Dmitry Chaplinsky, Mariana Romanyshyn and other contributors; it is based on LanguageTool[8] data.

Source dictionary contains word forms with their tags, grouped by lexemes. For example, a lexeme for lemma "ёж" (a hedgehog) looks like this:

```
ёж        NOUN,anim,masc sing,nomn
ежа       NOUN,anim,masc sing,gent
ежу       NOUN,anim,masc sing,datv
...
ежами     NOUN,anim,masc plur,ablt
ежах      NOUN,anim,masc plur,loct
```

In source dictionaries there could also be links between lexemes. For example, lexemes for infinitive, verb, gerund and participle forms of the same lemma may be connected. Currently pymorphy2 joins connected lexemes into a single lexeme for most link types.

3.1   Morphological Analysis and Generation

Given a dictionary, to analyze a word means to find all possible grammatical tags for a word. Obtaining of a normal form (lemmatization) is finding the first word form in the lexeme. To inflect a word is to find another word form in the same lexeme with the requested grammemes.

As can be seen, all these tasks are simple. With an XML dictionary analysis of known words can be performed just by running queries on XML file.

The problem is that querying XML is O(N) with large constant factors, raw data takes quite a lot of memory, and the source dictionary is not well suited for morphological analysis and generation of out-of-vocabulary words.

To create a compact representation and enable fast access pymorphy2 encodes lexeme information: all words are stored in a DAFSA [5] using the dawgdic[9]

---

[6] http://opencorpora.org/?page=export.
[7] Conversion utilities: https://github.com/dchaplinsky/LT2OpenCorpora.
[8] https://languagetool.org/.
[9] https://code.google.com/p/dawgdic/.

C++ library [11] via Python wrapper[10]; information about word tags and lexemes is encoded as numbers. Storage scheme is close to the scheme described in aot.ru [10], but it is not quite the same.

Paradigms. Paradigm in pymorphy2 is an inflection pattern of a lexeme. It consists of $prefix_i, suffix_i, tag_i$ triples, one for each word form in a lexeme, such as that each word form i can be represented as $prefix_i + stem + suffix_i$ where $stem$ is the same for all words in a lexeme.

This representation allows us to factorize a lexeme into a stem and a paradigm.

Paradigm prefixes, suffixes and tags are encoded as numbers by pymorphy2; lexeme stems are discarded. It means that a paradigm is stored as an array of numbers (prefixes, suffixes and tags IDs), and lexemes are not stored explicitly - they are reconstructed on demand from word and paradigm information.

There are no paradigms provided in the source dictionary; pymorphy2 infers them from the lexemes. For Russian there are about 3200 paradigms inferred from 390000 lexemes.

Word Storage. Word forms with their analysis information are stored in a DAFSA. Other storage schemes were tried, including two tries scheme similar to described in [9] (but using double-array tries), and succinct (MARISA[11]) tries. For pymorphy2 data DAFSA provided the most compact representation, and at the same time it was the fastest and had the most flexible iteration support.

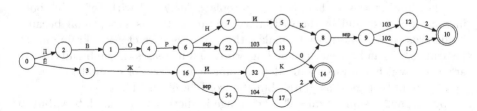

Fig. 1. DAFSA encoding example. Encoded (word, paradigmId, formIndex) triples: (двор, 103, 0); (ёж, 104, 0); (дворник, 101, 2); (дворник, 102, 2); (ёжик, 101, 2); (ёжик, 102, 2)

For each word form pymorphy2 stores (word, paradigmId, formIndex) triples:

- word form, as text;
- ID of its paradigm;
- word form index in the lexeme.

---

[10] https://github.com/kmike/DAWG.
[11] https://code.google.com/p/marisa-trie/.

DAFSA doesn't support attaching values to leaves; the information is encoded like the following: $< word > SEP < paradigmId >< formIndex >$ (see an example on Fig. 1)[12].

The storage is especially efficient because words with similar endings often have the same analyses, i.e. the same $(paradigmId, formIndex)$ pairs; this allows DAFSA to use fewer nodes/transitions to represent the data. DAFSA for Russian OpenCorpora dictionary ($5*10^6$ analyses, about $3*10^6$ unique word forms) enables fast lookups (hundreds thousand lookups/sec from Python) and takes less than 7MB of RAM; source XML file is about 400MB on disk.

To get all analyses of a $word$, DAFSA transitions for $word$ are followed, then a separator SEP is followed, and then the remaining subtree is traversed to get all possible $(paradigmId, formIndex)$ pairs.

Given $(paradigmId, formIndex)$ pair one can find the grammatical tag of a word: find a paradigm in paradigms array by $paradigmId$, get $(prefix_i, suffix_i, tag_i)$ triple from a paradigm by using $i := formIndex$. Given $(paradigmId, formIndex)$ pair and the $word$ itself it is possible to restore the lexeme and lemmatize or inflect the $word$ - from $word$, $prefix_i$ and $suffix_i$ we can get the stem, and given a stem and $(prefix_k, suffix_k, tag_k)$ it is possible to restore a full word for k-th word form.

3.2   Working with "ё" and "г" Characters Efficiently

The usage of "ё" letter is optional in Russian; in real texts it is often replaced with "е" letter. There rules for "г" / "ґ" substitutions are different in Ukrainian, but in practice there are real-world texts with "ґ" letters replaced with "г".

The simplest way to handle it is to replace "ё" / "ґ" with "е" / "г" both in the input text and in the dictionary. However, this is suboptimal because it discards useful information, makes the text less correct (in Ukrainian "г" instead of "ґ" can be seen as a spelling error) and increases the ambiguity: there are words which analysis should depend on е/ё and г/ґ. For example, the word "все" should be parsed as plural, but the word "всё" shouldn't.

pymorphy2 assumes that "ё" / "ґ" usage in dictionary is mandatory, but in the input text it is optional. For example, if a Russian input word contains "ё" letter then only analyses with this letter are returned; if there are "е" letters in the input word then possible analyses both for "е" and "ё" are returned.

An easy way to implement this would be to check each combination of е/ё and г/ґ replacement for the input word. It is not how pymorphy2 works. To do the task efficiently, pymorphy2 exploits DAFSA [5] dictionary structure: the result is built by traversing the word character graph and trying to follow "ё" transitions in addition to "е" transitions (for Russian) and "ґ" transitions in addition to "г" transitions (for Ukrainian).

---

[12] pymorphy2 encodes words to UTF-8 before putting them to DAFSA, so in practice there are more nodes than shown on Fig. 1. It is an implementation detail.

4 Analysis of Out-of-Vocabulary Words

It is not practical to try incorporate all the words in a lexicon - there is a long tail of rarely used words, new words appear; there is morphological derivation, loanwords, it is challenging to add all names, locations and special terms to the dictionary. Empirically, Zipf's Law seems to hold for natural languages [14]; one of the consequences is that even doubling the size of a lexicon could increase the coverage only slightly [6].

For languages without rich morphology it may be practical to assume that if word is not in a dictionary then it can be of any class from the open word classes, and then disambiguate the results on later processing stages, using e.g. a contextual POS tagger or a syntactic parser. For Slavic languages doing this on later stages is challenging because of large tagsets - for example, OpenCorpora [3] words have more than 4500 different tags. Morphological analyzer solves it by limiting the number of possible analyses based on word shape.

pymorphy2 uses a set of rules (analyzer units) to handle unknown words. Some of the rules are described in literature [4, 6, 8–10]; the resulting combination is novel. The order of in which the rules are applied is language-specific.

4.1 Common Prefixes Removal

There is a set of immutable prefixes which can be attached to words of open classes (nouns, verbs, adjectives, adverbs, participles, gerunds) without affecting the word grammatical properties. Examples of such prefixes for Russian: "не", "псевдо", "авиа"; pymorphy2 provides language-specific lists of these prefixes.

When a words starts with one of these prefixes, pymorphy2 removes the prefix, parses the reminder and re-attaches the prefix. A similar rule is described in [8]. Note that full analysis is performed on the reminder, so the reminder can be an out-of-vocabulary word itself. To speedup prefix matching built-in lists of prefixes are encoded to DAFSAs.

4.2 Words Ending with Other Dictionary Words

When all the following apply pymorphy2 assumes the whole word can be parsed the same way as the "suffix" word:

- a word being analyzed has another word from a dictionary as a suffix;
- the length of this "suffix" word is no less than 3;
- the length of the word without the "suffix" is no greater than 5;
- "suffix" word is of an open class (noun, verb, adjective, participle, gerund)

To search for suffixes pymorphy2 tries to consider 1st letter as a prefix, then two first letters as a prefix, etc., and lookups the reminder in a dictionary.

This rule is the same as described in [10]. A similar rule is described in [8], though its induction for concrete prefixes is different.

4.3   Endings Matching

In many languages, including Russian and Ukrainian, words with common endings often have the same grammatical form.

To exploit this, pymorphy2 first collects the information from the dictionary: for each word all endings of length 1 to 5 are extracted, and all possible analyses for these endings are stored. Then this *ending* → {*analyses*} mapping is cleaned up:

- only the most frequent analyses for each POS tag are kept;
- analyses from non-productive paradigms (currently these are paradigms which produced less than 3 lexemes in a dictionary) are discarded;
- rare endings (currently the ones which occur once) are also discarded.

The resulting mapping is encoded to DAFSA for fast lookups. Storage scheme is the following: $< ending > SEP < analysisInfo >$, where $analysisInfo$ consists of three 2-byte numbers: $(frequency, paradigmId, formIndex)$ - analysis frequency (a number of times a word with this ending had this analysis), ID of analysis paradigm and the form index inside the paradigm.

At prediction time pymorphy2 checks word endings from length 5 to 1, stopping at the first ending with some analyses found. To get possible analyses for a given ending pymorphy2 first follows all DAFSA transitions for the ending, then follows a separator, and then traverses the remaining subtree to get possible *analysisInfo* triples. The result is then sorted by analyses frequencies.

Recall that a *word* and a $(paradigmId, formIndex)$ pair is all what is needed to restore the lexeme and inflect the *word*. Lexemes are created on fly, so it doesn't matter *word* is not from the vocabulary as soon as we have $(paradigmId, formIndex)$ pair. It means morphological generation (lemmatization, inflection) works here.

Only analyses with open-class parts of speech (noun, verb, adjective, participle, gerund) are produced. Special care is taken to handle "ё" letter properly. Also, special care is required to handle paradigm prefixes properly - in fact, there are several *ending* → {*analyses*} DAFSAs built, one per each paradigm prefix.

This rule is based on [10]; similar approaches are also used in [4,9]. [8] uses similar rules, but derives them differently.

4.4   Words with a Hyphen

Unlike some other morphological analyzers, pymorphy2 opts to handle words with a hyphen.

In [7] it is argued that in most cases the parts of compound words should be handled as separate words if they are joined using a hyphen. A similar decision is made in OpenCorpora tokenization module [2]; it considers words like "Жан-Поль" as three tokens which should be analyzed separately and joined back at later processing stages. In both cases the decisions are not motivated by linguistic considerations; it is the technical difficulty which prevents analyzing and processing such words as single entities.

Currently pymorphy2 handles adverbs with a hyphen, particles separated by a hyphen and compound words with left and right parts separated by a hyphen.

Adverbs with a Hyphen. Russian words are parsed as adverbs if they

- start with a "по-" prefix;
- have total length greater than 5;
- can be parsed as a full singular adjective in dative case when "по-" is removed

Examples: "по-северному", "по-хорошему".

Particles Separated by a Hyphen. Though it is not clear if words with a particle separated by a hyphen (e.g. "смотри-ка" or "посмотрел-таки") should be handled as a single word or as two words, pymorphy2 supports parsing of such words. There are language-specific lists of common particles which can be attached, and if a word ends with one of these particles then it is parsed without the particle, and then the particle is re-attached to the result.

Compound Words with a Hyphen The main challenge in analysis of the compound words which parts are separated by a hyphen (like "человек-паук" and "Царь-пушка") is to figure out if the left part should be inflected together with the right part, or if it is a fixed prefix.

To do this, pymorphy2 parses left and right parts separately (they don't have to be dictionary words). Then it tries to find matching analyses. If there is a "left" analysis compatible with one of the "right" analyses then the resulting analysis is built where both word parts are inflected. After that, an analysis with a fixed left part is added to the result, regardless of whether a compatible "left" analysis was found or not. A similar method was used in [4].

Only words with a single hyphen are handled using heuristics described above. Words with multiple hyphens are likely represent different phenomena in Russian and Ukrainian languages; they could be interjections or phrases [13].

4.5   Other Tokens

Initial is an abbreviation of person's first or patronymic name. In most cases an initial is a single upper-cased character (language-specific). pymorphy2 parses such characters as fixed singular nouns, with variants for all possible gender and case combinations. For person first names (*Name*) two different lexemes are built for male and female names. For patronymic names (*Patr*) a single lexeme is returned. Unlike all other analyzer rules, detection of initials is case-sensitive. It is a way to decrease ambiguity.

The following tags are assigned to non-lexical tokens: *PNCT* for punctuation, *LATN* for tokens written in Latin alphabet, *NUMB, intg* for integer numbers, *NUMB, real* for floating-point numbers, *ROMN* for Roman numbers.

When analyzing the text, it is common to classify tokens during the tokenization step. The reason pymorphy2 handles non-lexical tokens during the morphological analysis step is that this allows users to use a simpler tokenizer when classifying tokenizer is not available; also, it means that information about all tokens is available in a common format.

4.6   Morphological Generation of Out of Vocabulary Words

Inflection is fully supported for out of vocabulary words. To achieve this pymorphy2 keeps track of the analyzer units (rules and their parameters) used to parse the word, requires each analyzer unit to provide a method for getting a lexeme, and calls this method for the last analyzer unit. To compute the lexeme analyzer unit can look at the analysis result, and it can ask previous analyzer units for the lexeme.

For example, Common Prefixes Removal analyzer removes the prefix from a word, then gets a lexeme from the previous analyzer, and then attaches the prefix to each word form in a lexeme to build a resulting lexeme.

5   Probability Estimation

Morphological analyzer may return multiple possible word parses. The problem of choosing the right analysis from a list of possible options is called disambiguation. Generally, to select the correct analysis it is required to take word context in account. Morphological analyzer takes individual words as an input, so it can't disambiguate the result robustly. However, it can provide an estimation for $P(analysis|word)$ conditional probability. Such probability estimations can be used in absence of a dedicated disambiguator to select the more probable analysis. In addition to that, these probabilities can be used on later stages of text analysis, for example by a disambiguator.

To estimate $P(analysis|word)$ conditional probability for Russian words pymorphy2 uses partially disambiguated OpenCorpora corpus [3] and assumes that $P(analysis|word) = P(tag|word)$. The conditional probability is estimated for words which have multiple analysis according to pymorphy2, but have occurrences with a single remaining analysis in the OpenCorpora corpus; the estimation is a maximum-likelihood estimation with Laplace (add-one) smoothing.

$$W_{disambiguated} := \{word : |tags_{corpus}(word)| = 1, word \in corpus\}$$

$$W_{ambiguous} := \{word : |tags_{pymorphy2}(word)| > 1, word \in W_{disambiguated}\}$$

$$B(word) = \max(|tags_{pymorphy2}(word)|, |tags_{corpus}(word)|)$$

$$\forall word \in W_{ambiguous},$$

$$\forall tag \in tags_{pymorphy2}(word) :$$

$$P_{MLE}(tag|word) = \frac{count(word, tag) + 1}{count(word) + B(word)} \qquad (1)$$

Counts are computed based on OpenCorpora corpus data; all words with a single remaining analysis are taken in account.

Once estimated, the result is stored on disk as a DAFSA; keys are

$$< word >:< tag >< NULL >< int(10^6 * P_{MLE}(tag|word)) >$$

For words without $P_{MLE}(tag|word)$ estimates the probabilities are assigned uniformly during the parsing.

For Ukrainian language probabilities are assigned uniformly because at the moment of writing there is no a freely available Ukrainian corpus similar to OpenCorpora.

6 Evaluation

Evaluating analysis quality of different morphological analyzers for Russian is not straightforward because most analyzers (as well as annotated corpora) use their own incompatible tagsets. And when a corpus and a dictionary have a compatible tagset it usually means that the dictionary was enhanced from the corpus, and it is a problem because quality numbers obtained on a corpus the dictionary was enhanced from shouldn't be relied on - they are too optimistic.

pymorphy2 analysis quality was compared to an analysis quality of a well-known morphological analyzer[13], Mystem 3.0 [9]. Testing corpus consists of 100 randomly selected sentences (1405 tokens) from OpenCorpora (microcorpus[14]) and 100 randomly selected sentences (1093 tokens) from ruscorpora.ru - 2498 manually disambiguated tokens in total.

Full details for this evaluation can be found online[15].

OpenCorpora (pymorphy2) tagset is not the same as ruscorpora.ru tagset, and ruscorpora.ru tagset differs from Mystem tagset. For evaluation purposes all tags were converted to Mystem format using a set of automatic rules. Quality was evaluated on full morphological tags, i.e. tags must match exactly to be considered correct, with a few exceptions related to tags conversion problems. All reported errors were checked manually to filter out false positives (Table 1).

Table 1. Errors

	pymorphy2	mystem 3.0
microcorpus	10	15
ruscorpora	9	8
Total	19	23

---

[13] https://tech.yandex.ru/mystem/.
[14] https://github.com/kmike/microcorpus.
[15] http://kmike.ru/links/aist2015/pymorphy2-mystem3.

Both pymorphy2 and Mystem made less than 1% errors (without disambiguation, i.e. in less than 1% cases the correct analysis was not in a set of analyses
returned by an analyzer). It should be noted that 9 out of 19 pymorphy2 errors
and 14 out of 23 Mystem errors were related to abbreviation handling. Mystem
handled first and last names better (1 mistake versus 4 for pymorphy2); pymor-
phy2 made less mistakes for "regular" words (4 versus 6 for mystem). Mystem
can't parse many hyphenated words as a single token; such words were not considered. Punctuation, numbers and non-Russian words were also removed from
the input.

It is hard to draw a quantitative conclusion because the corpus size is small.
Both analyzers has a similar analysis quality, and the resulting numbers depend
on evaluation minutiae: whether abbreviations are considered or not, should we
require hyphenated words to be parsed, do we require verb transitivity to be
predicted correctly, is it important to distinguish adverbs from parenthesis, etc.

Several human annotation errors were found by parsing OpenCorpora data
with mystem (1 error) and ruscorpora data with pymorphy2 (6 errors). OpenCorpora shares a dictionary with pymorphy2, and ruscorpora annotation is related
to mystem; this shows an utility of using cross-corpora tools to check the annotations.

The most sophisticated Russian morphological parser evaluation so far is [1];
it happened in 2010. Previous version of pymorphy2 (pymorphy) participated[16]
in tracks without disambiguation; it finished 1st on Full Morphology Analysis,
3rd on Lemmatization, 3rd on POS tagging and 5th on the Rare Words track.
pymorphy haven't participated in disambiguation tracks.

pymorphy used some pymorphy2 rules (not all) and a different dictionary
(extracted from [10] instead of [3]). Generally, pymorphy2 should work better
than pymorphy because of an improved dictionary and rules, but this has not
been not measured quantitatively yet.

7   Conclusion and Future Plans

Permissive open-source license (MIT) is used for pymorphy2. All the dictionaries
and corpora pymorphy2 depends on are also available under permissive open-
source licenses. This encourages usage and contributions. There are volunteers
working on Russian and Ukrainian dictionaries and corpora, related tools and
pymorphy2 itself.

Development of pymorphy2 is by no means finished. There are word classes
for which pymorphy2 analysis can be improved. Some of them: people last
and patronymic names, foreign people names, diminutive first names, locations,
uppercase and other abbreviations, some classes of hyphenated words, ordinal numbers (including ordinal numbers written in digit notation like "22-й").
According to [1], similar issues are common for Russian morphological analyzers.

Non-contextual $P(tag|word)$ estimates can be made better by transferring
some information about similar words and by improving the corpora.

---

[16] Anonymized results: http://ru-eval.ru/tables_index.html.

A better comparison between pymorphy, pymorphy2, Mystem and other morphological analyzers could require a robust tagset conversion library.

The support for Ukrainian is experimental. The dictionary requires work, pymorphy2 needs more Ukrainian-specific rules for handling of out of vocabulary words, and for better $P(tag|word)$ estimates an annotated Ukrainian corpus is needed: even a small corpus (or even a manually crafted frequency list) should fix a substantial amount of "obvious" errors.

There are plans to add Belarusian language support to pymorphy2 based on Belarusian N-korpus[17] grammar database.

Although pymorphy2 is already fast enough for many use cases (tens of thousands words per second in a single thread), there is a room for further speed improvements.

References

1. Astaf'eva, I., Bonch-Osmolovskaya, A., Garejshina, A., Ju, G., D'jachkov, V., Ionov, M., Koroleva, A., Kudrinsky, M., Lityagina, A., Luchina, E., Sidorova, E., Toldova, S., Lyashevskaya, O., Savchuk, S., Koval', S.: NLP evaluation: Russian morphological parsers. In: Kibrik, A. (ed.) Computational Linguistics and Intellectual Technologies: Papers from the Annual International Conference "Dialouge", vol. 1 (2010)
2. Bocharov, V.V., Granovsky, D.V., Surikov, A.V.: Probabilistic Tokenization model in the OpenCorpora project [Veroyatnastnaya model' tokenizacii v proekte Otkritiy Korpus]. In: New Information Technology in Automated Systems: Proceedings of the 15th Seminar [Noviye informacionnie tehnologii v avtomatizirovannih sistemah: materiali pyatnadcatogo nauchno-prakticheskogo seminara] (2012)
3. Bocharov, V.V., Alexeeva, S.V., Granovsky, D.V., Protopopova, E.V., Stepanova, M.E., Surikov, A.V.: Crowdsourcing morphological annotation. In: Selegey, V. (ed.) Computational Linguistics and Intellectual Technologies. Papers from the Annual International Conference "Dialogue", vol. 1 (2013)
4. Bolshakov, I.A., Bolshakova, E.I.: An automatic morphological classifier of noun phrases in Russian. In: Kibrik, A. (ed.) Computational Linguistics and Intellectual Technologies. Papers from the Annual International Conference "Dialogue", vol. 1 (2012)
5. Daciuk, J., Watson, B.W., Mihov, S., Watson, R.E.: Incremental construction of minimal acyclic finite-state automata. Comput. Linguist. **26**(1), 3–16 (2000)
6. Daciuk, J.: Treatment of unknown words. In: Boldt, O., Jürgensen, H. (eds.) WIA 1999. LNCS, vol. 2214, p. 71. Springer, Heidelberg (2001)
7. Krylov, S.A., Starostin, S.A.: Current morphological analysis and synthesis challanges in the STARLING system [Aktualniye zadachi morfologicheskogo analiza i sinteza v integrirovannoy informacionnoy srede STARLING]. In: Proceedings of the International Conference "Dialog 2003" (2003)
8. Mikheev, A.: Automatic rule induction for unknown word guessing. Comput. Linguist. **23**(3), 405–423 (1997)
9. Segalovich, I.: A fast morphological algorithm with unknown word guessing induced by a dictionary for a web search engine. In: Proceedings of MLMTA 2003, Las Vegas (2003)

---

[17] http://bnkorpus.info.

10. Sokirko, A.: Morphological modules on the web-site www.aot.ru [Morphologich-eskie Moduli na saite www.aot.ru]. In: Computational Linguistics and Intelligent Technologies: Proceedings of the International Conference "Dialog 2004" (2004)
11. Yata, S., Morita, K., Fuketa, M., Aoe, J.: Fast string matching with space-efficient word graphs. In: Innovations in Information Technology (Innovations 2008), Al Ain, United Arab Emirates, pp. 79–83, December 2008
12. Zaliznjak, A.A.: Grammaticeskij slovar' russkogo jazyka, Moscow, Russia (1977)
13. Zanegina, N.N.: Improvised-temporary-compounds as a new expressive mean in Russian. In: Kibrik, A. (ed.) Computational Linguistics and Intellectual Technologies. Papers from the Annual International Conference "Dialogue", vol. 1 (2012)
14. Zipf, G.K.: Selected Studies of the Principle of Relative Frequency in Language. Harvard University Press, Cambridge (1932)

Semantic Role Labeling for Russian Language Based on Russian FrameBank

Ilya Kuznetsov[✉]

Higher School of Economics, Moscow, Russia
iokuznetsov@hse.ru

Abstract. Semantic Role Labeling (SRL) is one of the major research areas in today's natural language processing. The task can be described as follows: given an input sentence, that refers to some situation, find the participants of this situation in text and assign them semantically motivated labels, or roles. Although the topic has become increasingly popular in the last decade, there have been only a few attempts to apply SRL to Russian language. We present a supervised semantic role labeling system for Russian based on FrameBank, an actively developing Russian SRL resource analogous to FrameNet and PropBank.

Keywords: Semantic role labeling · Semantic parsing · Russian · Framebank · Supervised

1 Introduction

Semantic Role Labeling (SRL), often referred to as shallow semantic parsing, is one of the most popular topics in modern natural language processing. The task at hand has been formulated by D. Gildea and D. Jurafsky [1] and can be described as follows. Given an input sentence, that describes some **situation** or event, find the participants of this situation and assign them semantically motivated labels, or roles. It is important to distinguish SRL from syntactic parsing which works on the grammar level, and from deep semantic analysis which may include complex reasoning. Consider the following example:

[He] bought [a pie] for [1 dollar]

SRL would result in a following analysis: given the situation of *buying* (envoked by the predicate *buy. V*) and the target sentence, [He] is the *Buyer*, [a pie] is *Goods* and [1 dollar] is the *Price* paid. Buyer, Goods and Price belong to the predefined set of semantic roles, and the task is to find the text spans (or syntactic nodes) that correspond to those roles. There are two major classes of SRL tasks regarding the role inventory used. PropBank-style SRL assigns predicate-specific argument labels (*Arg0, Arg1...*) as in the PropBank corpus [2], and FrameNet-style SRL uses a large set of semantically rich labels as in FrameNet [3] and as in our example above.

© Springer International Publishing Switzerland 2015
M.Y. Khachay et al. (Eds.): AIST 2015, CCIS 542, pp. 333–338, 2015.
DOI: 10.1007/978-3-319-26123-2_32

Since the SRL task has been set, it has attracted a lot of attention in the NLP research community. Most of the reported studies are dedicated to English SRL due to the lack of preprocessing tools and training data for other languages. The Russian FrameBank initiative aims to create a FrameNet-like resource which includes an SRL-annotated corpus. The corpus is still in development, but it is already possible to use it for preliminary experiments.

In this work we present a simple SRL system trained on a subset of the FrameBank corpus. The system shows promising results and can be used as baseline for future research. We also describe data filtering and preprocessing steps and a mapping procedure that maps span-based FrameBank annotation to the syntax tree nodes.

2   FrameBank-Based SRL for Russian

2.1   Source Data

In our study we use the developing Russian FrameBank corpus [4] as source data. The corpus consists of automatically tokenized and morphology-tagged sentences from the Russian National Corpus [5]. An extra annotation layer provides the SRL-related information. Each example sentence contains one or several tokens marked as **predicate** tokens. For each predicate the **construction name** is specified, which corresponds to the lexical sense. Additionally, each example sentence contains one or several tokens marked as **arguments**. For each token, a letter-based label is specified which uniquely refers to the role of this token within the construction at hand. A detailed explication information is also given. Most data is taken from modern literature and similar sources.

In order to be used as training data for our task, the FrameBank content has to be filtered out and preprocessed. According to the convention, each sentence in FrameBank must contain SRL markup only for a single construction. Due to markup errors, it is sometimes not the case. We have selected the sentences from the training data that contain at least and at most one token marked as verbal predicate and used additional heuristics to filter out spurious data. We have also restricted our scope to the constructions that occured more than 10 times in source data. That resulted in a reduced, but relatively clean source corpus consisting of 4852 sentences and 113897 tokens of which 7883 are semantic role fillers.

2.2   Preprocessing

The cleaned data has to be further processed with lower-level NLP tools. For this purpose we use the pipeline developed by S. Sharof [6] that includes a sentence tokenizer, a morphology analyser and a dependency parser trained on the SynTagRus corpus [7].

One additional issue we have to solve is that argument and predicate annotations in FrameBank (as well as in FrameNet) are span-based, i.e. not related to syntactical nodes, which makes extracting syntax-related feature problematic. We use the **span mapping** technique proposed by D. Bauer [8], who used it to create a dependency-parsed version of the FrameBank corpus. To determine the node that represents the argument marked up by a span, following steps are suggested. Let **yield** be the set of all children of a target node in the target dependency tree, including non-direct children. Then for each argument span in the sentence we look for the node whose yield is most similar to this span in terms of F1-measure. This node inherits all the SRL markup from the source span and only this node is used as classification instance. Since the yield of a predicate often includes the whole clause, in this case a simple character overlap heuristic is used instead.

2.3 SRL Engine

Even in narrow case of supervised semantic role labeling, there are several possible task interpretations and experimental setups, so it seems reasonable to describe our setting in more detail.

We use **predicate-specific roles** because they are easily accessible in our source data. The classification unit is **dependency node**, which is in line with modern SRL approaches. Argument detection and classification could be performed as two consecutive steps, but to keep the model simple we perform **joint detection and classification** of arguments. High-quality verb sense disambiguation and frame detection are crucial for SRL, but this is traditionally seen as word sense disambiguation task and is not included in SRL. Our system also **does not perform VSD** and works on pre-set verb senses. Finally, we **ignore unknown prediates**, which is left for further work.

The following procedure is used to train the SRL system. Sentences from the training data are grouped based on construction label (which is obtained from gold data). This results in relatively small subcorpora for each predicate sense. Each token in those subcorpora is considered as an instance, and the classification label is either the name of the arguments role (if the token has been decided to be representative for this role on the argument mapping stage) or *None* otherwise.

Given the sets of instances, we train one-vs-all SVM classifiers for each role and for the *None* class using the LinearSVC implementation from python library *scikit-learn* [9] with default parameter settings.

Our feature set is in line with traditional features used for semantic role labeling. The set of features we use can be naturally divided in two groups:

- Syntactic features (SYN)
  - **Path** - path in dependency tree from the target predicate to the instance node in terms of dependency edge labels and directions
  - **Voice** - voice of the target, active or passive
  - **Vform** - verb form of the target, participle, gerund, etc.
  - **P.lemma** - if instance token is a preposition, we use its lemma as feature

– Lexical features (LEX)
  • **Cluster** - lexical cluster the instances lemma belongs to.
  • **POS** - part of speech of the instance token.

One of the well known problems of SRL is lexical sparseness of argument nodes. To solve this problem, instead of using argument lemmas directly we use **cluster labels**. Clusters are obtained by applying the Chinese Whispers [10] graph clustering algorithm to the sets of top similar words according to a vector space model extracted from a large corpus [11]. This fast solution produces clusters of reasonable quality, but using another clustering procedure or a hand-built thesaurus (such as RuThes [12]) could also be beneficial for or task.

3   Evaluation and Results

The following evaluation procedure was used to measure the quality of our classifiers. For each predicate sense, a set of sentences containing markup with this sense was chosen. Then, instances contained in each sentence set are randomly split into training (80 % instances) and test (20 %) sets. The classifiers are trained on the training data and evaluated using the Precision, Recall, F1 and Accuracy measures. The measures are computed for each role classifier and the average is taken. This macro-averaging allows us to reduce the influence of the majority class (which is in all cases *None*, i.e. no role is assigned). The scores are then averaged across all the predicate senses. The following table summarizes evaluation results for the full feature set and syntax/lexical features alone. For all combinations averaged values are provided. We also provide value for majority baseline (which always assigns the *None* class) (Table 1).

Table 1. SRL results for different feature sets

FeatureSet	P	R	F1	Acc
SYN+LEX	**0.71**	**0.66**	**0.64**	0.94
SYN	0.68	0.63	**0.64**	0.94
LEX	0.40	0.40	0.40	0.91
Baseline	0.33	0.35	0.34	0.90

As we can see, using lexical features alone yields the lowest results. This may be due to the insufficient clustering quality and to the fact that the information about semantic roles is mostly encoded in syntax. Syntactical features are strong, partially because the syntax model we use (MaltParser trained on SynTagRus markup) contains information closely related to semantic roles, e.g. the completive dependencies (see [7] for details). At the same time, the evaluation shows that SRL systems benefits from lexical information too, and the joint setting yields best results.

4   Related Work

Compared to state of art in modern English-based SRL, our system is very simple. The studies reported in CONLL 2008 and 2009 shared tasks [13,14] showed that there are two major modeling features that are beneficial for SRL and that our system lacks. First, instead of assigning roles locally and independently (as we do), a global model can be used, that choses the best-scoring set of assignments given the whole sentence. Second, it has been shown that both syntax and semantic parsing benefit from joint modeling, when the syntax parser and semantic role labeler are trained simultaneously. A detailed review of state-of-art methods in English-based SRL can be found in [15].

To the best of our knowledge, the only study that reported results for Russian is [16]. They approach the task from the semi-supervised perspective and use a hand-built dictionary to annotate the SynTagRus corpus with semantic roles. However, due to the large differences in experimental setup, it is hard to compare their results to the ones obtained during our experiment.

5   Conclusion

In this paper we have reported a supervised SRL model trained on subset of the developing FrameBank corpus. Our simple model yields promising results, and it can be improved in several ways.

First, we plan to investigate different lexical models that supply cluster data to the SRL engine.

Second, our classification model assumes independence between role assignments within one sentence. One universal property of semantic roles is that given a predicate and a sentence containing it, each role can be assigned only once. Adding a mechanism that enforces this restriction could improve the system's performance. ILP optimization could be used for this task.

Third, syntax parsing for Russian still has place for improvement and it would be interesting to train a joint model. However, a corpus annotated with both syntax trees and SRL information is needed for this task, which is not available.

Finally, we hope that the cleanup techniques and support tools developed during our study will be useful for the FrameBank project and will help to speed up the development and publishing of this resource.

References

1. Gildea, D., Jurafsky, D.: Automatic labeling of semantic roles. Comput. Linguist. **28**(3), 245–288 (2002)
2. Palmer, M., Kingsbury, P., Gildea, D.: The proposition bank: an annotated corpus of semantic roles. Comput. Linguist. **31**, 71–105 (2005)
3. Baker, C.F., Fillmore, C.J., Lowe, J.B.: The berkeley framenet project. In: Proceedings of the 36th Annual Meeting of the Association for Computational Linguistics and 17th International Conference on Computational Linguistics, ACL 1998, vol. 1, pp. 86–90. Association for Computational Linguistics, Stroudsburg (1998)

4. Lyashevskaya, O.: Bank of russian constructions and valencies. In: Proceedings of the Seventh Conference on International Language Resources and Evaluation (LREC 2010). European Languages Resources Association (ELRA) (2010)
5. Russian national corpus. http://ruscorpora.ru. Accessed 20 January 2014
6. Sharoff, S., Nivre, J.: The proper place of men and machines in language technology: processing russian without any linguistic knowledge. Russian State University of Humanities, Moscow (2011)
7. Boguslavsky, I., Chardin, I., Grigorieva, S., Grigoriev, N., Iomdin, L., Kreidlin, L., Frid, N.: Development of a dependency treebank for russian and its possible applications in nlp. In: Proceedings of the 3rd International Conference on Language Resources and Evaluation, Las Palmas, Gran Canaria (2002)
8. Bauer, D., Fürstenau, H., Rambow, O.: The dependency-parsed framenet corpus. In: Proceedings of the Eighth International Conference on Language Resources and Evaluation (LREC-2012). European Language Resources Association (ELRA) (2012)
9. Scikit-learn. http://scikit-learn.org/stable/index.html. Accessed 20 January 2014
10. Biemann, C.: Chinese whispers: an efficient graph clustering algorithm and its application to natural language processing problems. In: Proceedings of the First Workshop on Graph Based Methods for Natural Language Processing, TextGraphs-1, pp. 73–80. Association for Computational Linguistics, Stroudsburg (2006)
11. Semantic calculator: find semantically related words in russian based on their typical contexts. http://ling.go.mail.ru/quazy-synonyms. Accessed 20 January 2014
12. Loukachevitch, N., Dobrov, B., Chetviorkin, I.: Ruthes-lite, a publicly available version of thesaurus of russian language ruthes. In: Computational Linguistics and Intellectual Technologies. Proceedings of International Conference Dialog, vol. 13(20), pp. 340–350. Russian State University of Humanities, Moscow (2014)
13. Surdeanu, M., Johansson, R., Meyers, A., Màrquez, L., Nivre, J.: The conll 2008 shared task on joint parsing of syntactic and semantic dependencies. In: CoNLL 2008: Proceedings of the Twelfth Conference on Computational Natural Language Learning, pp. 159–177. Coling 2008 Organizing Committee Manchester, August 2008
14. Hajič, J., Ciaramita, M., Johansson, R., Kawahara, D., Martí, M.A., Màrquez, L., Meyers, A., Nivre, J., Padó, S., Štěpánek, J., Straňák, P., Surdeanu, M., Xue, N., Zhang, Y.: The conll-2009 shared task: syntactic and semantic dependencies in multiple languages. In: Proceedings of the Thirteenth Conference on Computational Natural Language Learning (CoNLL 2009): Shared Task, pp. 1–18. ACL, Boulder, June 2009
15. Das, D.: Statistical models for frame-semantic parsing. In: Proceedings Frame Semantics in NLP: A Workshop in Honor of Chuck Fillmore (19292014). Association for Computational Linguistics, June 2014
16. Shelmanov, A., Smirnov, I.: Methods for semantic role labeling of russian texts. In: Computational Linguistics and Intellectual Technologies. Proceedings of International Conference Dialog, vol. 13(20), pp. 607–620. Russian State University of Humanities, Moscow (2014)

Supervised Approach to Finding Most Frequent Senses in Russian

Natalia Loukachevitch[1](✉) and Ilia Chetviorkin[2]

[1] Research Computing Center of Lomonosov Moscow State University,
Moscow, Russia
louk_nat@mail.ru
[2] Lomonosov Moscow State University, Moscow, Russia
ilia2010@yandex.ru

Abstract. The paper describes a supervised approach for the detection
of the most frequent sense on the basis of RuThes thesaurus, which
is a large linguistic ontology for Russian. Due to the large number of
monosemous multiword expressions and the set of RuThes relations it is
possible to calculate several context features for ambiguous words and
to study their contribution in a supervised model for detecting frequent
senses.

Keywords: Lexical senses · Automatic sense disambiguation · The most
frequent sense

1 Introduction

The most frequent sense (MFS) is a useful heuristic in lexical sense disambiguation when the training or context data are insufficient. In sense-disambiguation
(WSD) evaluations the first sense labeling is presented as an important baseline
[1], which is difficult to overcome for many WSD systems [2].

Usually MFS is calculated on the basis of a large sense-tagged corpus such as
SemCor, which is labelled with WordNet senses [3]. However, the creation of such
corpora is a very labour-consuming task. Besides Princeton WordNet [4], only
for several other national wordnets such corpora are labeled [5]. In addition, the
MFS of a given word may vary according to the domain, therefore the automatic
processing of documents in a specific domain can require reestimation of MFS
on the domain-specific text collection. The distributions of lexical senses also
can depend on time [6].

One of the prominent approaches in the task of automatic calculation of the
most frequent sense [7–9] is to use distributional vectors to compare contexts of
an ambiguous word with sense-related words [9, 10]. In such experiments mainly
WordNet-like resources are studied. In [7] Macquarie Thesaurus serves as a basis
for the predominant sense identification.

In this paper we present our experiments on MFS identification extracting several types of context features and combining them with machine learning methods. The experiments are based on newly-published Thesaurus of

© Springer International Publishing Switzerland 2015
M.Y. Khachay et al. (Eds.): AIST 2015, CCIS 542, pp. 339–349, 2015.
DOI: 10.1007/978-3-319-26123-2_33

Russian language RuThes-lite, which has been developed since 1994 and was applied in a number of tasks of natural language processing and information retrieval [11].

This paper is organized as follows. Section 2 compares our study with related works. In Sect. 3, we describe the main principles of RuThes-lite linguistic ontology construction. Section 4 is devoted to the manual analysis of the sense distribution described in RuThes, which is performed on the basis of Russian news flow provided by Yandex news service (news.yandex.ru). Section 5 describes the experiments on supervised prediction of the most frequent sense of an ambiguous word.

2   Related Work

It was found in various studies that the most frequent sense is a strong baseline for many NLP tasks. For instance, only 5 systems of the 26 submitted to the Senseval-3 English all words task outperformed the reported 62.5 % MFS baseline [12].

However, it is very difficult to create sense-labeled corpora to determine MFS, therefore techniques for automatic MFS revealing were proposed. McCarthy et al. [8,9] describe an automatic technique for ranking of word senses on the basis of comparison of a given word with distributionally similar words. The distributional similarity is calculated using syntactic (or linear) contexts and automatic thesaurus construction method described in [13]. WordNet similarity measures are used to compare the word senses and distributional neighbors. McCarthy et al. [9] report that 56.3 % of noun SemCor MFS (random baseline – 32 %), 45.6 % verb MFS (random baseline – 27.1 %) were correctly identified with the proposed technique.

In [10] the problem of domain specific sense distributions is studied. The authors form samples of ambiguous words having a sense in one of two domains: SPORTS and FINANCE. To obtain the distribution of senses for chosen words, the random sentences mentioning the target words in domain-specific text collections are extracted and annotated.

Lau et al. [14] propose to use topic models for identification of the predominant sense. They train a single topic model per target lemma. To compute the similarity between a sense and a topic, glosses are converted into a multinomial distribution over words, and then the Jensen Shannon divergence between the multinomial distribution of the gloss and the topic is calculated. Their model was primarily directed to sense clustering and in the MFS task their results were worse than results obtained in [9]. Mohammad and Hirst [7] describe an approach for acquiring predominant senses from corpora on the basis of the category information in the Macquarie Thesaurus.

A separate direction in WSD research is automatic extraction of contexts for ambiguous words based on so called "monosemous relatives" [15–17] that are related words having only a unique sense. It was supposed that extracted sentences mentioning monosemous relatives are useful for lexical disambiguation. These approaches at first determine monosemous related words for a given ambiguous word, then extract contexts where the relatives were mentioned, and use these contexts as automatically annotated data to train WSD classifiers.

In our case we use monosemous relatives in another way: to determine the most frequent senses of ambiguous words. We conduct our research for Russian and this is the first study on MFS prediction for Russian.

3   RuThes Linguistic Ontology

One of popular resources used for natural language processing and information-retrieval applications is WordNet thesaurus [4]. Several WordNet-like projects were also initiated for Russian [18–20]. However, at present there is no large enough and qualitative Russian wordnet. But another large resource for natural language processing – RuThes thesaurus, having some other principles of its construction, has been created and published. The first publicly available version of RuThes (RuThes-lite) contains 96,800 unique words and expressions and is available from http://www.labinform.ru/ruthes/index.htm.

RuThes Thesaurus of Russian language is a linguistic ontology for natural language processing, i.e. an ontology where the majority of concepts are introduced on the basis of actual language expressions. RuThes is a hierarchical network of concepts. Each concept has a name, relations with other concepts, a set of language expressions (words, phrases, terms) whose senses correspond to the concept, so called ontological synonyms.

In contrast to WordNet-like resources, ontological synonyms in RuThes can comprise words belonging to different parts of speech (*stabilization, stabilize, stabilized*); language expressions relating to different linguistic styles, genres; idioms and even free multiword expressions (for example, synonymous to single words).

So, a row of ontological synonyms can include quite a large number of words and phrases. For instance, the concept ДУШЕВНОЕ СТРАДАНИЕ (*wound in the soul*) has more than 20 text entries in Russian (several English translations may be as follows: *wound, emotional wound, pain in the soul* etc.). Besides, in RuThes introduction of concepts based on multiword expressions is not restricted and even encouraged if this concept adds some new information to knowledge described in RuThes. For example, such concept as ЗАСНУТЬ ЗА РУЛЕМ (*falling asleep at the wheel*) is introduced because it denotes a specific important situation in road traffic, has an "interesting" text entry заснуть во времядвижения (*falling asleep while driving*). Also this concept has an "interesting" relation, which cannot inferred from the phrase component structure, to concept ДОРОЖНО-ТРАНСПОРТНОЕ ПРОИСШЕСТВИЕ (*road accident*) [11].

So RuThes principles of construction give the possibility to introduce more multiword expressions in comparison with WordNet-like resources.

An ambiguous word is assigned to several concepts - this is the same approach to represent the lexical ambiguity as in WordNet-like resourcs. For instance, the Russian word картина (picture) has 6 senses in RuThes and attributed to 6 concepts:

1. ФИЛЬМ (*moving picture*)
2. ПРОИЗВЕДЕНИЕ ЖИВОПИСИ (*piece of painting*)
3. КАРТИНА (ОПИСАНИЕ) (*picture as description*)
4. КАРТИНА СПЕКТАКЛЯ (*scene as a part of a play*)
5. ЗРЕЛИЩЕ (ВИД) (*sight, view*)
6. КАРТИНА ПОЛОЖЕНИЯ, СОСТОЯНИЯ (*picture as circumstances*)

The relations in RuThes are only conceptual, not lexical (as antonyms or derivational links in wordnets). The main idea behind the RuThes set of conceptual relations is to describe the most essential, reliable relations of concepts, which are relevant to various contexts of concept mentioning. The set of conceptual relations includes the class-subclass relation, the part-whole relation, the external ontological dependence, and the symmetric association.

Thus, RuThes has considerable similarities with WordNet including concepts based on senses of real text units, representation of lexical senses, detailed coverage of word senses. At the same time the differences include attribution of different parts of speech to the same concepts, formulating names of concepts, attention to multiword expressions, a set of conceptual relations. A more detailed description of RuThes and RuThes-based applications can be found in [11].

4   Manual Analysis of Sense Distribution

To check the coverage of lexical senses described in RuThes we decided to verify their usage in a text collection. At this moment we do not have the possibility to create a sense-tagged corpus based on RuThes senses. In addition, as it was indicated in [5], in sense-labeling most time and efforts are spent on adding new word senses to a source resource. Another problem of a sense-labeled corpus is that it fixes the described sets of senses, and it is impossible to automatically update them for a new version of a thesaurus.

To verify the coverage of lexical senses described in RuThes the most important issue is to check that at least frequent senses have been already described. With this aim it is not necessary to label all senses of a word in a large text collection, it is enough to check out senses in an randomly selected sample of word usages in contemporary texts as it was made in [10]. In addition, from this analysis we obtain manual estimation of MFS.

We decided to check RuThes senses on news texts and articles through Yandex news service (http://news.yandex.ru/). We based our evaluation on a news collection because news reports and articles are one of the most popular documents for natural language processing, such as categorization, clustering, information extraction, sentiment analysis. Besides, the news collection comprises a lot of other text genres as legal regulations or literature pieces. Finally, this collection contains recently appeared senses, which can be absent in any fixed collection such as, for example, Russian national corpus (http://ruscorpora.ru/en/index.html) and dictionaries.

Yandex.news service (news.yandex.ru) collects news from more than 4,000 sources (including main Russian newspapers), receiving more than 100,000 news during a day. The news flow from different sources is automatically clustered into sets of similar news. When searching in the service, retrieval results are also clustered. Usually three sentences from the cluster documents (snippets) are shown to the user.

For a given ambiguous word, linguists analyzed snippets in Yandex news service, which returns the most recent news reports and newspaper articles containing the word. Considering several dozens of different usages of the word in news, the linguists estimated the distribution of senses of the word, which later would allow defining its most frequent sense. In news snippets, repetitions of the same sentences can be frequently met – such repetitions were dismissed from the analysis. Table 1 presents the results of the analysis for Russian ambiguous words провести (*provesti*), картина (*kartina*), and стрелка (*strelka*). The sense distributions for these three words have quite different behavior. Word провести has a single predominant sense; word картина has two main senses with approximately similar frequencies.

Table 1. Sense distribution of several Russian ambiguous words in the news flow (20 different contexts in current news flow were analyzed)

Word	Names of concept corresponding to senses	Number of usages met in contexts
Провести (*provesti*) 9 senses	ПРОВЕСТИ, ОРГАНИЗОВАТЬ *to organize*	19
	ПРОЛОЖИТЬ ЛИНИЮ, ПУТЬ *to build road, pipe*	1
картина (*kartina*) 6 senses	ПРОИЗВЕДЕНИЕ ЖИВОПИСИ *piece of painting*	10
	ФИЛЬМ *moving picture*	10
стрелка (*strelka*) 7 senses	СТРЕЛКА РЕК *river spit*	8
	СТРЕЛКА ПРИБОРА *pointer of the device*	6
	ЗНАК СТРЕЛКИ *arrow sign*	4
	ЖЕЛЕЗНОДОРОЖНАЯ СТРЕЛКА *raolroad pointer*	1
	СТРЕЛКА НА ЧАСАХ *clock hand*	1

Because of insufficient amount of data under consideration the experts could designate several senses as the most frequent ones if they saw that the difference in the frequencies does not allow them to decide what a sense is more frequent as for example for word картина two main senses were revealed: ФИЛЬМ (*moving picture*) and ПРОИЗВЕДЕНИЕ ЖИВОПИСИ (*piece of painting*) (Table 1).

In total, around 3,000 thousand ambiguous words with three and more senses described in RuThes (11,450 senses altogether) were analyzed in such a manner. As a result of such work, about 650 senses (5.7 %) were added or corrected. So the coverage of senses in RuThes was enough qualitative and improved after the analysis.

Certainly, the distribution of word senses in news service search results can be quite dependent on the current news flow; in addition, the subjectivity of individual expertise can appear. Therefore for 400 words the secondary labelling was implemented, which allows us to estimate inter-annotator (and inter-time) agreement. 200 words from these words had three senses described in RuThes, other 200 words had four and more described senses.

Table 2 demonstrates that for 88 % of words experts agreed or partially agreed on MFS for the analyzed words. The partial agreement means in this case that experts agreed on prominent frequency of at least one sense of a word and indicated other different senses as also prominent. For example, for word картина the first expert indicated two main senses (*moving picture* and *piece of painting*) with equal frequencies. The second expert revealed the *piece of painting* sense as much more frequent than other senses. Therefore we have here partial agreement between experts and suppose that the most frequent sense of a word is the *piece of painting* sense.

5   Supervised Estimation of Most Frequent Sense

In the previous section we showed that the manual analysis, which was performed for 3,000 ambiguous words, revealed the most frequent senses with precision more than 80 %. The analysis was implemented only for ambiguous words with three and more senses described in RuThes; but RuThes contains about 6,500 words with two senses, which were not analyzed manually. In addition, MFS can change in different domains; natural language processing of documents in a specific domain can require re-estimation of MFS on the domain collection.

Therefore we propose a method for supervised estimation of MFS based on several features calculated on the basis of a target text collection. To our knowledge, this is the first attempt to apply a supervised approach to MFS estimation. In addition, in contrast to previous works our method of MFS estimation is essentially based on unambiguous text entries of RuThes, especially on multiword expressions, which were carefully collected from many sources.

Automatic estimation of the most frequent sense was fulfilled on a news collection of 2 million documents. Computing features for the supervised method we used several context types of a word: *the same sentence context, the neighbor sentence context, full document context.*

Table 2. The agreement in manual estimation of the most frequent senses for ambiguous words described in RuThes

Number of words analyzed by two experts	400
Number of words for that experts agreed on MFS	216
Number of words for that experts partially agreed on MFS	125
Number of words for that experts did not agreed on MFS	49

Table 3. Document frequencies of monosemous relatives of word kartina

Monosemous relatives of word картина	sense of картина	document frequency
Фильм (*film*)	*1. moving picture*	45285
мультфильм (*cartoon*)	*1. moving picture*	4097
документальный фильм (*documentary film*)	*1. moving picture*	3516
живопись (*painting*)	*2. piece of painting*	3200
съемка фильма (*shooting a film*)	*1. moving picture*	2445
кинофильм (*movie*)	*1. moving picture*	1955
произведение искусства (*art work*)	*2. piece of painting*	1850
художественный фильм (*fiction movie*)	*1. moving picture*	1391
изобразительное искусство (*visual art*)	*2. piece of painting*	1102
режиссер картины (*director of the movie*)	*1. moving picture*	978
общая картина (*general picture*)	*6. picture as circumstances*	932

From the thesaurus we utilize several types of conceptual contexts of ambiguous word w:

1. one-step context of word w attached to concept C (ThesConw1) that comprises other words and expressions attached to the same concept C and concepts directly related to C as described in the thesaurus, that is the one-step thesaurus context of a word includes its synonyms, direct hyponyms, hypernyms, parts and wholes;
2. two and three-step contexts of word w attached to C (ThesConw2 (3)) comprising words and expressions from the concepts located at the distance of maximum 2 (3) relations to the initial concept C (including C); the path between concepts can consist of relations of any types,
3. one-step thesaurus context including only unambiguous words and expressions: UniThesConw1. From these text and thesaurus contexts we generate the following features for ambiguous word w and its senses Cw:

From these text and thesaurus contexts we generate the following features for ambiguous word w and its senses Cw:

1. the overall collection frequency of expressions from UniThesConw1 – here we estimate how often unambiguous relatives of W were met in the collection - Freqdoc1 and logarithm of this value logFreqdoc; Table 3 depicts frequencies of monosemous relatives of word картина in the source collection,

Table 4. Accuracy of MFS prediction for single features and the supervised algorithm for Etalon1

Feature	Accuracy
Freqdoc1	42.4 %
FreqdocW1	46.4 %
FreqSentWsum1	41.2 %
FreqSentWmax1	43.3 %
FreqSentWsum2	**48.2 %**
FreqSentWmax2	**48.2 %**
FreqSentWsum3	47.0 %
FreqSentWmax3	47.6 %
FreqNearWsum1	43.0 %
FreqNearWmax1	44.2 %
FreqNearWsum2	39.7 %
FreqNearWmax2	46.7 %
FreqNearWsum3	38.8 %
FreqNearWmax3	43.3 %
Supervised algorithm	**50.6 %**
Random	23.5 %

2. the frequency of expressions from UniThesConw1 in texts where w was mentioned - FreqdocW1,
3. the overall frequency and the maximum frequency of words and expressions from ThesConwi co-occurred with W in the same sentences – FreqSentWmaxi and FreqSentWsumi (i = 1, 2, 3),
4. the overall frequency and the maximum frequency of words and expressions from ThesConwi occurred in the neighbor sentences with w – FreqNearWmaxi and FreqNearWsumi (i = 1, 2, 3).

All real-valued features are normalized by dividing them by their maximal value.

We conducted our experiments on two sets of ambiguous words with three and more senses. The first set consisted of 330 words of 400 words that were analyzed by two linguists. They agreed with each other on one or two the most frequent senses (further Etalon1). We used this set to train machine learning models. Then we apply the trained model to the second set of ambiguous words – 2532 words, for which only one expert provided MFS (further Etalon2). Both sets include words of three parts of speech: nouns, verbs and adjectives.

Table 4 presents accuracy results of MFS detection for single features. One can see that many single features provide a quite high level of accuracy.

To combine the features regression-oriented methods implemented in WEKA machine learning package were utilized. The best quality of classification using labelled data was shown by the ensemble of three classifiers: Logistic Regression, LogitBoost and Random Forest. Every classifier ranged word senses according to the probability of this sense to be the most frequent one. We averaged probabilities of MFS generated by these methods and obtained 50.6 % accuracy of MFS prediction. The random baseline for this set is very low -23.5% (Table 4). Our estimation is based on ten-fold cross validation.

To check the robustness of the obtained supervised model we transferred it to the Etalon2 set. Table 5 describes the accuracy results for the best single features and the supervised method for Etalon2. The average level of results is higher than on the Etalon1 set, because Etalon2 contains the larger share of 3-sense words.

We can see that simple context features give results comparable with those described in [8,9], which have similar levels of random baselines. At this moment machine-learning combination of features did not demonstrate the significant growth in accuracy but the machine-learning framework allows adding distributional features utilized in the above-mentioned works.

Table 5. Accuracy of MFS prediction for words from Etalon2 including accuracy of the best single features and accuracy of the supervised algorithm trained on Etalon1

Feature	Accuracy
FreqSentWsum1	53.7 %
FreqSentWsum2	**57.4 %**
FreqSentWmax2	53.7 %
FreqSentWsum3	54.6 %
FreqNearWsum2	53.7 %
Supervised algorithm trained on Etalon1	**57.8 %**
Random	33.4 %

6   Conclusion

In this paper we describe a supervised approach to detecting the most frequent senses of ambiguous words on the basis of thesaurus of Russian language RuThes. The approach is considerably based on monosemous relatives of ambiguous words, in particular multiword expressions, described in RuThes. To check the proposed approach two linguists manually estimated most frequent senses for 3,000 ambiguous words described in RuThes with three and more senses.

Our approach demonstrates the quality, which is quite comparable to the state-of-art distributional approaches, but our approach is based on simpler context features. We found that some simple features (such as frequency of 2-step monosemous relatives of a word in sentences with this word –FreqSentWsum2) provide high level of prediction of the most frequent sense.

We believe that in combination with other distributional features of words proposed in previous works it is possible to achieve better results in future experiments on MFS prediction.

References

1. Agirre, E., Màrquez, L., Wicentowski, R. (eds.): Proceedings of the 4th International Workshop on Semantic Evaluations (SemEval). Association for Computational Linguistics, Prague (2007)
2. Navigli, R.: Word sense disambiguation: a survey. ACM Comput. Surv. (CSUR), **41**(2), 10, 1–69 (2009)
3. Landes, S., Leacock, C., Tengi, R.: Building semantic concordances. In: Fellbaum, C. (ed.) WordNet: An Electronic Lexical Database. The MIT Press, Cambridge (Mass) (1998)
4. Fellbaum, C.: WordNet: An Electronic Lexical Database. MIT Press, Cambridge (1998)
5. Petrolito, T., Bond, F.: A survey of wordnet annotated corpora. In: Proceedings of Global WordNet Conference, GWC-2014, pp. 236–245 (2014)
6. Mitra, S., Mitra, R., Riedl, M., Biemann, C., Mukherjee, A., Goyal, P.: That's sick dude!: automatic identification of word sense change across different timescales. In: Proceedings of ACL-2014 (2014)
7. Mohammad, S., Hirst, G.: Determining word sense dominance using a thesaurus. In: Proceedings of EACL-2006, pp. 121–128 (2006)
8. McCarthy, D., Koeling, R., Weeds, J., Carroll, J.: Finding predominant word senses in untagged text. In: Proceedings of ACL-2004 (2004)
9. McCarthy, D., Koeling, R., Weeds, J., Carroll, J.: Unsupervised acquisition of predominant word senses. Comput. Linguist. **33**(4), 553–590 (2007)
10. Koeling, R., McCarthy, D., Carroll, J.: Domain-specific sense distributions and predominant sense acquisition. In: Proceedings EMNLP-2005, Vancouver, pp. 419–426 (2005)
11. Loukachevitch, N., Dobrov, B.: RuThes linguistic ontology vs. Russian wordnets. In: Proceedings of Global WordNet Conference GWC-2014 (2014)
12. Snyder, B., Palmer, M.: The english all-words task. In: Mihalcea, R., Chklowski, T. (eds.) Proceedings of SENSEVAL-3: Third International Workshop on Evaluating Word Sense Disambiguating Systems, pp. 41–43 (2004)
13. Lin, D.: Automatic retrieval and clustering of similar words. In: Proceedings of the 17th International Conference on Computational linguistics COLING-1998, pp. 768–774 (1998)
14. Lau, J.H., Cook, P., McCarthy, D., Gella, S., Baldwin, T.: Learning word sense distributions, detecting unattested senses and identifying novel senses using topic models. In: Proceedings of ACL-2014, pp. 259–270 (2014)
15. Agirre, E., Lacalle, O.L.: Publicly available topic signatures for all wordnet nominal senses. In: Proceedings of LREC-2004 (2004)
16. Leacock, C., Miller, G., Chodorow, M.: Using corpus statistics and wordnet relations for sense identification. Comput. Linguist. **24**(1), 147–165 (1998)
17. Mihalcea, R.: Bootstrapping large sense tagged corpora. In: Proceedings of LREC-2002 (2002)

18. Azarowa, I.: RussNet as a computer Lexicon for Russian. In: Proceedings of the Intelligent Information systems IIS-2008, pp. 341–350 (2008)
19. Balkova, V., Suhonogov, A., Yablonsky, S.: Some issues in the construction of a Russian wordnet grid. In: Proceedings of the Forth International WordNet Conference, Szeged, pp. 44–55 (2008)
20. Braslavski, P., Ustalov, D., Mukhin, M.: A spinning wheel for YARN: user interface for a crowdsourced thesaurus. In: Proceedings of EACL-2014, Sweden (2014)

FrameBank: A Database of Russian Lexical Constructions

Olga Lyashevskaya[1,2] and Egor Kashkin[2(✉)]

[1] National Research University Higher School of Economics, Moscow, Russia
olesar@yandex.ru
[2] V.V. Vinogradov Russian Language Institute of RAS, Moscow, Russia
egorkashkin@rambler.ru

Abstract. Russian FrameBank is a bank of annotated samples from the Russian National Corpus which documents the use of lexical constructions (e.g. argument constructions of verbs and nouns). FrameBank belongs to FrameNet-oriented resources, but unlike Berkeley FrameNet it focuses more on the morphosyntactic and semantic features of individual lexemes rather than the generalized frames, following the theoretical approaches of Construction Grammar (C. Fillmore, A. Goldberg, etc.) and of Moscow Semantic School (J.D. Apresjan, E.V. Paducheva, etc.).

Keywords: Russian · Construction Grammar · Frames · Corpus linguistics · Morphosyntax · Semantic roles · Polysemy

1 Background

FrameBank[1] is an open access database which consists of a dictionary of Russian lexical constructions and a corpus of their uses tagged with a FrameNet-like annotation scheme [1–3]. The examples are randomly taken from the Russian National Corpus [4]. At present the dictionary provides data for ca. 4000 target verbs, adjectives, and nouns, and the corpus part includes ca. 50000 annotated examples.

The project under discussion started in 2011. The ideology of FrameBank has obviously been inspired by Berkeley FrameNet [5], but there are some crucial differences in how these two resources are organized. Firstly, FrameBank is more focused on morphosyntactic patterns than FrameNet. This is determined by the grammatical properties of Russian (which are not relevant in English), where different case structures often help to profile the situation differently. Secondly, the target entries in FrameNet are extralinguistic situations – frames, which are further linked to a list of semantically related verbs (e.g., the frame of Motion embraces such lexical units as *to come, to go, to fly, to float, to glide, to blow*, etc.). On the contrary, FrameBank has

---

The work was partly supported by the Russian Foundation for the Humanities, grant No.13-04-12020, and by the Russian Basic Research Foundation, grant No. 15-07-09306.
[1] www.framebank.ru.

M.Y. Khachay et al. (Eds.): AIST 2015, CCIS 542, pp. 350–360, 2015.
DOI: 10.1007/978-3-319-26123-2_34

particular lexical items as target entries, providing data on their morphosyntactic patterns and on the frames corresponding to different meanings of a lexeme.

The theoretical basis of FrameBank includes Construction Grammar (C. Fillmore, A. Goldberg, etc.) as well as some approaches developed in the Moscow Semantic School (J.D. Apresjan, E.V. Paducheva et al.) with its attention to the differences between close synonyms and to the interaction between lexical and grammatical features of lexical items. There is another resource developed within the Moscow Semantic School – namely, the Lexicographer database [6]. However, it does not seem to equally embrace all the main semantic classes of Russian verbs and all the possible constructions of the verbs it includes. Neither is it directly linked to a set of corpus examples, which is one of the main features of FrameBank.

The paper is structured as follows. After outlining how the dictionary of constructions is designed, we discuss the annotation scheme and some theoretical issues it raises. Further, we consider two databases included in FrameBank: the graph of semantic roles and the graph of formal and semantic shifts between constructions. The graph of semantic roles presents our own inventory, which correlates with the semantic classification of verbs and forms a hierarchy in order to support flexible search options. The other graph shows both formal changes of verbal constructions (omission of a participant, change of a morphosyntactic pattern, diathetic alternations etc.) and their semantic changes (metaphor, metonymy, and also some shifts which have not been discussed so widely, like specialization or rebranding). FrameBank also provides quantitative data on the frequency of semantic roles and semantic shifts, which could be used in the automatic annotation of texts (e.g. for the tasks of semantic role labelling). Finally, we outline some future steps in developing FrameBank.

2   Dictionary of Construction Patterns

We will discuss the architecture of FrameBank using the example of verbs, which form the core of the database. Information about each lexical construction is stored as a construction template, which includes:

1. the syntactic rank of the element (Subject, Object, Predicate, Peripheral, Clause);
2. the morphosyntactic features of the element[2] (including POS, case and preposition marking);
3. its status: lexical constant vs. variable;
4. the semantic roles of the argument (e.g., Agent, Patient, Instrument);
5. the lexical-semantic class of the element (e.g., human, animate, abstract entity, means of transport, etc.);
6. the morphosyntactic features of the target lexical unit itself (e.g. impersonal, passive participle, etc.);
7. one or several examples.

Figure 1 shows a sample pattern in the dictionary.

---

[2] This part was originally based on [7].

ID230. Cx name: *Pjatno vystupilo na rubaške* ['a stain appeared on the short']. Cx Pattern: Snom V na + Sloc.									
Cx Item ID	Pl	Letter	Head	Phrase	Explication	Syntactic Rank	Lexico-semantic constraints	Status [obligatory / optional]
2077	1	X	Snom [Nominative case]	NPnom	Theme	Subject	natural object	Oblig.
2078	2	–	*vystupit'* ['to appear; lit. to step forward']	–	to appear	Predicate	–	Oblig.
2079	3	Y	na + Sloc [preposition na 'on' + Locative case]	na + NPloc	Location	Peripheral	space and place	Oblig.

Lexical Index of target words

Index of Morphosyntactic Items

Fig. 1. The template of the construction *Pjatno*[Noun.Nom] *vystupilo*[Verb] *na rubaške* [PREP + Noun.Loc] 'a stain appeared on the short'.

Each verb is followed in the database by a list of lexical constructions in which it serves as a target word (each construction is named by a mnemonic sentence label). Lexical constructions are grouped in clusters usually corresponding to a particular lexical meaning; the constructions belonging to one cluster differ in the number of explicit arguments and in their morphosyntactic marking. Figure 2 shows two groups of LexCxs of the verb *vystupit'* 'to step forward' which correspond to the frame of motion and the frame of coming into existence, respectively.

Target Lexeme: *vystupit'*

1. 'to step forward'

ID220. <Snom V> *Vystupilo srazu pjat' soldat* 'Five soldiers stepped forward at once'

ID221. <Snom V PR_from+S>. *Iz stroja vystupil čelovek* 'A man stepped forward from the line'

ID222. <Snom V PR_to+S> *On vystupil na seredinu komnaty* 'He stepped forward to the center of the room'

...

5. 'to appear (about blood, tears, stains, etc.)'

ID 230. <Snom V na.PR+Sloc> *Pjatno vystupilo na rubaške* 'A stain appeared on the short'

ID 231. <Snom V na.PR+Sloc u.PR+Sgen> *Sljozy vystupili u nee na glazax* lit. 'Tears appeared on the eyes at her'

ID 232. <Snom V u.PR+Sgen ot.PR+Sgen> *U nee ot smexa vystupili sljozy* lit. 'Tears appeared at her from laughing'

Fig. 2. The passport of the lexeme *vystupit'*

3   Corpus Annotation

The dictionary of constructions is supplemented by examples tagged manually. The examples are randomly selected from the Russian National Corpus, each target lexical unit is illustrated by up to 100 sentences with their pre- and post-context. Each example is annotated by one of the annotators in the online FrameBank Markup environment, and then is checked and corrected by the editor. An example is matched to a suitable construction pattern, which includes establishing correspondences between their elements and assigning morphosyntactic and semantic features of the arguments in a particular example. If an example does not fit any of the existing patterns, an annotator should add a new item into the dictionary of constructions (this is often the case for colloquial constructions, for the on-going changes in the semantics of verbs, and for idiomatic expressions). Note that the participants of a frame are annotated irrespective of their syntactic relation to the predicate (this distinguishes FrameBank from the treebanks like SynTagRus or Prague Dependency Treebank). For example, if we annotate the verb *vyslušat'* 'listen to somebody' and come across sentence (1), we will mark the NPs 'Andropov' and 'the marshal' as the participants of the frame referred to by the verb *vyslušal* 'listened' (the fact that they are not syntactically related to the predicate will also be mentioned in the annotation).

(1)   *Andropov prin'al maršala v svojem rabočem kabin'et'e, **vyslušal** i ob'eščal razobrat's'a v etoj probl'em'e* 'Andropov received the marshal in his office, **listened** to him and promised to examine the problem'

The annotators of FrameBank also mark non-standard types of constructions or non-standard variants of argument realization, such as passive, imperative, participial or converbal constructions, constructions with infinitives, control, genitive of negation. The annotation takes into account not only construction arguments and the properties of the predicate, but also adjuncts and modal particles. More details on the annotation procedure can be found in the full version of the manual for annotators, which is available online[3].

4   Semantic Roles

As has already been mentioned, construction patterns in FrameBank contain information on the semantic roles of the participants. The inventory of semantic roles may have quite different volume and structure depending on the particular research task and theoretical framework (see, for example, [8: 587–588, 9, 10, 11: 125–126, 12: 370–377]). The most important principles governing the inventory of semantic roles in FrameBank are as follows:

- the inventory should be hierarchical in order to support flexible search options (it may be reduced to 5–10 basic roles, or enlarged to several dozen labels);

---

[3] http://framebank.wikispaces.com/.

- the roles should correlate with the semantic classification of verbs (what follows from this is that traditionally "broad" roles such as Agent or Patient should get different labels in different semantic classes, cf. Agent in destruction vs. speech vs. motion);
- the scope of a semantic role is defined in accordance with the Prototype Theory: for instance, the prototype of Patient is a participant changing under the physical

Table 1. Frequency of semantic roles in FrameBank (top-15).

Semantic role	Number of construction patterns	Example	Number of predicates in the dictionary
Agent	4787	*Prodav'ec r'ežet syr* 'The seller is cutting cheese'	1824
Patient	3086	*Prodav'ec r'ežet syr* 'The seller is cutting cheese'	1498
Theme	1591	*Na polu l'ežal č'elovek* 'There was a man lying on the floor'	1004
Subject of motion	1520	*My jed'em v Moskvu* 'We are going to Moscow'	515
Speaker	1304	*On govorit pravdu* 'He is telling the truth'	749
Patient of motion	1049	*Mal'čik v'el sl'epogo za ruku* 'The boy led a blind man by the hand'	358
Point of destination	921	*My jed'em v Moskvu* 'We are going to Moscow'	657
Place	903	*Na polu l'ežal č'elovek* 'There was a man lying on the floor'	738
Message	776	*On skazal, čto rabotajet nad knigoj* 'He said that he was working on a book'	454
Effector	643	*V'et'er povalil d'er'evo* 'The wind threw down a tree'	565
Subject of psychological state	643	*On toskujet po druz'jam* 'He misses his friends'	526
Mental content	637	*My sčitali jego opasnym č'elov'ekom* 'We considered him a dangerous person'	438
Content of action	634	*Potrudit'es' vstat', požalujsta!* 'Be so kind to stand up, please!'	526
Result	633	*Mama svarila sup* 'Mother has cooked soup'	445
Reason	616	*Komandira b'espokoilo, jesli razv'edčiki dolgo n'e vozvraščalis'* 'The commander was worried if the scouts didn't return long'	501

influence of an Agent; peripheral examples (Patient of a non-physical process, Patient which is not changing, Patient created as a result of a physical action) get specific labels (Theme, Result, etc.) and are considered as specific types of Patient.

The detailed list of semantic roles currently contains 91 items classified into seven domains (those of Agent, Possessives, Patient, Addressee, Experiencer, Instrument, Settings), which are further subdivided into smaller units. Initially, we intended to use a list of semantic roles suggested in [12: 370–377]. However, we had to work out some of its parts in further detail in order to be in line with our theoretical principles. For instance, the inventory suggested by J.D. Apresjan includes the role of Experiencer without any further semantic specification. To achieve our goals, we considered Experiencer not as a single semantic role, but as a domain including Subject of Perception ('see', 'hear'), Subject of Mental State ('think', 'understand'), Subject of Psychological State ('love', 'be afraid'), Subject of Physiological State ('feel pain', 'have a buzzing in one's ears'), Subject of Physiological Response ('tremble with cold', 'feel sick'), and Subject of Psychological Response ('laugh', 'cry (burst into tears)'). Similarly, the role of Agent is defined in our inventory as an active (proto-typically animate) participant of a situation, intentionally changing something in the world. This role is typically assigned to verbs of physical impact, eating and drinking, creation, causation of motion, while more specific verbs which are less closer to the prototype of Agent receive their own semantic roles (Speaker, Subject of motion, Subject of social relationship, etc.).

It should also be noted that the principles of FrameBank annotation allow marking double roles (following the ideas of [11: 140]). Thus, examples like *kormit' r'eb'enka s ložečki* '**to feed** a child with a spoon' or *myt's'a pod kranom* '**to wash oneself** under a tap' contain instrumental participants, which at the same time have locative properties (which influences their morphosyntactic marking). Therefore, these participants receive a double role Instrument and Place in our annotation scheme.

FrameBank also provides frequency data about semantic roles in lexical constructions. Table 1 shows the top-15 roles (the calculation is based on the number of construction patterns with this role; the data on the other roles are left out of this paper due to size limits). These data supplemented with the morphosyntactic patterns may be useful for the tasks of semantic role labelling [13, 14], see [15] for a case study based on FrameBank.

5 Non-core Elements

Along with marking the arguments of target lexical units, the annotation of examples in FrameBank covers their adjuncts (non-obligatory valencies), see, for instance, [8: 72–79] on the theoretical foundations of the distinction between arguments and adjuncts. This provides large amounts of empirical evidence for discussing the restrictions imposed on the combinability of adjuncts with different types of predicates (cf. a traditional view touched upon in [8: 75] and stating that arguments are specific for each verb, while adjuncts are compatible with various verbs). Table 2 contains statistical data on co-occurrence of verbs and adjuncts depending on the semantic classes of both.

Table 2. Co-occurrence of verbs and adjuncts.

verb class	time	place	degree	manner	usualness	reason	duration	simultaneity	sequence	purpose	precision	frequency	comparison	speed
motion	426	172	133	219	148	93	107	62	58	64	26	38	74	82
speech	120	131	44	139	68	56	31	18	24	18	22	38	11	9
physical impact	70	89	44	80	36	28	33	20	21	19	4	24	12	7
emotion	69	44	224	21	41	39	21	11	11	6	29	3	8	
mental	79	79	40	36	31	28	32	18	17	12	21	5	6	
social interaction	71	64	58	41	43	35	14	17	9	11	13	10	6	4
start of existence	90	104	5	24	27	29	8	28	17	18	4	13	5	2
possessive	62	69	10	20	12	14	7	14	2	25	4	5	2	3
psychical	33	22	71	8	16	22	14	16	6	5	9	2	3	6
sound	17	42	2	16	9	2	2	5	6	2	7	2	2	
physiology	23	21	18	12	10	6	6	4	5		2	2		2
change of state	17	8	12	15	15	10	8	6	5	4		6	4	4
end of existence	19	6	10	2	10	8	7	8	10		4	5	1	8
SUM	1096	851	671	633	466	370	290	227	191	184	145	153	134	127

As can be seen in Table 2, the ratio of co-occurrence is much higher than average for verbs of emotion and the psychical sphere with adjuncts of degree, for verbs of motion with adjuncts of time, speed and comparison, for verbs of speech with adjuncts of manner and place, for verbs expressing start of existence or possessive relations with adjuncts of place (the overrepresented combinations are marked in bold). On the contrary, the ratio of co-occurrence is lower than average for verbs of motion and adjuncts of place, degree, reason and precision, for verbs of speech and adjuncts of degree and time, for verbs of physical impact and adjuncts of time, etc. (the underrepresented combinations are marked with a gray background; the least represented cases are on a dark-gray background). Interestingly, adjuncts referring to usualness, frequency, simultaneity and sequence do not tend to favor any particular verb class. Nevertheless, the data of FrameBank show that the combinability of adjuncts is not arbitrary: the choice of an adjunct with a particular semantics is to some extent predetermined by the semantic class of a verb.

6 Construction Grapher

Another component of FrameBank is the graph of lexical constructions. It documents the systematic relations between constructions. First, it systematizes semantic shifts in verbal lexemes (metaphor, metonymy and some more complex relations). Second, the graph represents formal changes in argument structure, such as omission of a participant, diathetic alternations (cf. [8]), the inheritance of a pattern from another verb etc. The semantic part of the project is inspired by FrameNet grapher as well as by E. Rakhilina's research database on Russian polysemous adjectives and adverbs (see [16] and references therein). The formal part is guided by E. Paducheva and G. Kustova's theoretical and empirical analysis of polysemy in Russian verbs [6, 8, 17].

The types of formal and semantic changes are represented below (for the previous stage of its discussion see [3]). The figures in brackets after the name of a shift indicate the number of its occurrences in the database. Sometimes a construction undergoes more than one formal or semantic change, in such cases all changes are counted. For each verb in the database we construct a graph showing the formal and semantic changes undergone by its constructions. These graphs are tied into a larger graph of lexical constructions, since some edges of the latter establish linkages between different verbs, consider "Inheritance of a pattern" below. A case study of how the construction grapher works can be found in [18].

6.1   Formal Changes

1. Morphosyntactic alternation (1796): *Vy* **govorit***'e* <u>*pravdu*</u> 'You **are telling** <u>the truth</u>' ↔ *Papa* **govorit**, <u>*čto bojat's'a n'ečego*</u> 'Father **says** <u>that there is nothing to be afraid of</u>' ↔ <u>*"Moemu drugu groz'at n'eprijatnosti"*</u>, – **govoril** *on* '<u>"My friend is facing troubles"</u>, – he **said**'. This formal change is bidirectional (as well as all the changes marked with the left-right arrow), as we assume all the morphosyntactic variants to have equal status in the graph, instead of choosing the primary one, which would often be not quite evident.

2. Focus shift between participants (1230): *Žuravli* **l'et'at** <u>*s vostoka*</u> 'The cranes **are flying** <u>from the east</u>' ↔ *Lastočki* **l'et'at** <u>*na jug*</u> 'The swallows **are flying** <u>to the south</u>' ↔ <u>*Nad gorami*</u> **letit** *or'el* 'An eagle **is flying** <u>over the mountains</u>' In particular, this change is typical of motion verbs. We treat all the constructions with a mover + one locative participant as basic and formally interrelated by means of a focus shift, instead of deriving them from constructions like *Pticy* **l'et'at** *s vostoka na jug nad gorami* 'The birds **are flying** from the east to the south over the mountains', as the latter ones are quite rare in our corpus data and do not seem to be natural for human language.

3. Diathetic alternation (407): <u>*Korma lodki*</u> **ušla** *v vodu* '<u>The stern of the boat</u> **plunged** (**lit.: went**) into water' → <u>*Lodka*</u> **ušla** *v vodu* <u>*kormoj*</u> 'lit.: <u>The boat</u> **went** into water <u>with its stern</u>'.

4. Omission of a participant belonging to a definite class (335): *On* **rastvor'aet** *sahar* <u>*v vod'e*</u> 'He **is dissolving** sugar <u>in water</u>' → *On* **rastvor'aet** *sahar* 'He **is dissolving** sugar'.

5. Omission of a participant which is deictically or situationally defined (875): *Avtobus* **prišel** <u>*na stanciju*</u> 'The bus **arrived** <u>at the station</u>' → *Begite, avtobus* **prišel**! 'Hurry up, the bus **has arrived**!'

6. Omission of an indefinite (or unimportant) participant (1152): *Korabl'* **plyv'et** <u>*iz gavani*</u> 'The ship **is sailing** <u>from the harbour</u>' → *Korabl'* *m'edl'enno* **plyv'et** 'The ship **is sailing** slowly'.

7. Addition of a participant (2269): *Lastočki* **l'et'at** 'The swallows **are flying**' → *Lastočki* **l'et'at** <u>*za kormom*</u> 'The swallows **are flying** <u>to find some food</u>' This formal shift usually involves adding peripheral participants like Goal, Reason,

Method, etc. Omission is in its turn marked when there is a core participant of a frame missing in a derived construction (e.g., Instrument in the frames of destruction or any kind of locative participant in the frames of motion).

8. Hybrid of two constructions (91): *Ptica* **prygala** <u>*po trav'e*</u> 'A bird **jumped** <u>on the grass</u>', *Ptica* **prygala** <u>*p'er'ed domom*</u> 'A bird **jumped** <u>in front of the house</u>' → *Ptica* **prygala** <u>*po trav'e p'er'ed domom*</u> 'A bird **jumped** <u>on the grass in front of the house</u>'.

9. Inheritance of a pattern (706): <u>*'"Sl'edujt'e za mnoj"*</u>, *– skazal oficiant* '<u>"Follow me"</u>, – **said** the waiter' → <u>*'"Sl'edujt'e za mnoj"*</u>, *– brosil oficiant* '<u>"Follow me"</u>, – **dropped** the waiter' The annotation of such examples sheds light on the most productive sources of inherited morphosyntactic patterns. These are the verbs *govorit'* 'to speak, to say' (66 constructions acquiring its pattern), *nakazat'* 'to punish' (32 cases), *bol'et'* 'to be ill' (21 cases), *bit'* 'to beat' and *udarit'* 'to hit once' (total 20 cases), *dat'* 'to give' (14 cases), *byt'* 'to be' (12 cases). The position of *govorit'* at the top of the list can be explained by the high productivity of metaphors referring to speech, as well as by the frequent occurrence of metonymic contexts which describe expressing emotions, cf. *"Vot eto fokus!" – udivils'a on* 'lit.: "What a trick!", he **was surprised**' In this example the verb *udivit's'a* 'to be surprised' not only denotes the emotional state of the experiencer, but also indicates that he is saying something. The latter part of meaning is supported by the use of direct speech inherited from verbs like 'to say'. In the case of *bol'et'* 'to be ill', the number of inherited patterns is high, as this semantic domain is inherently metaphorical: according to the cross-linguistic data analyzed in [19], most pain sensations are described with verbs borrowed from other domains (burning, cutting and breaking, sound, etc.), rather than with specific pain expressions. This semantic shift tends to be accompanied with morphosyntactic changes which make source verbs more "similar" to verbs of pain in their construction patterns (see [19] for details). The case of the verb *nakazat'* 'to punish' is a bit different. Many verbs become embedded into a construction with the preposition *za* + NPacc describing Motivation. This argument is typical of *nakazat'* and occurs with other verbs when they denote an action evaluated as punishment, cf. *ar'estovat'* <u>*za ubijstvo*</u> '**to arrest** <u>for murder</u>', *iskl'učit' iz komandy* <u>*za opozdanije*</u> '**to expel** from the team <u>for being late</u>', *S'erg'ej byl ostanovl'en policijej* <u>*za to, čto projehal na krasnyj signal sv'etofora*</u> 'Sergej **was stopped** by the police <u>for running a red light</u>'.

6.2   Semantic Changes

1. Metonymy: an associated participant (517): <u>*Voda* **zam'erzla**</u> '<u>The water</u> **has frozen**' → <u>*Prud* **zam'erz**</u> '<u>The pond</u> **has frozen up**'.

2. Metonymy caused by diathetic alternations (432): *Pojezd jed'et v gorod* 'The train **is going** to the city' → *Ja jedu v gorod* <u>*pojezdom*</u> '**I am going** to the city <u>by train</u>'.

3. Metonymy: an associated domain (726): *Vasilij* **int'er'esujets'a** <u>*russkoj lit'eraturoj*</u> '*Vasilij* **is interested** <u>in Russian literature</u>' → *Vasilij* **int'er'esujets'a**, <u>*vo skol'ko prihodit pojezd*</u> '*Vasilij* **wonders (lit.: is interested)** <u>when the train arrives</u>' Here in the first example the verb *int'er'esovat's'a* 'to be interested in sth.' describes the

mental state of the experiencer, while in the second example it shifts to expressing the speech of a person aiming at find something out.

4. Metaphor (5498): *Mat' budit syna* 'Mother **is waking** her son' → *Tišina budit vospominanija* 'Silence **evokes (lit.: wakes)** memories'.

5. Rebranding (146): a semantic shift where the derived meaning is an implicature from the source meaning [16], e.g. *Smotri: zv'er' podhodit* 'Look: a beast **is approaching**' → *Eto pal'to t'eb'e podhodit* 'This coat **suits (lit.: approaches)** you' In this example the idea of something approaching, conveyed in the direct use, implies meeting some standard as a figurative meaning. However, these two domains are not adjacent and therefore are not related metonymically. Neither is there a direct metaphoric relation which could be established between these two meanings.

6. Idiomatization (89): *On ulybnuls'a i prot'anul ruku* 'He smiled and **stretched** his hand' → *Vy tak nogi prot'an'et'e* 'You'll turn up your toes (lit.: **stretch** your legs)'.

7. Specialization (94): *Po utram on pjet čaj* 'He **drinks** tea in the morning' → *On pjet* 'He **drinks** (abuses alcohol)'.

8. Semantic bleaching (46): *javl'at's'a* '**to be** (lit.: to come, to appear)'; *obratit'* vni-*manije* '**to pay (lit.: to turn)** attention'.

7 Future Prospects

In the previous sections we have discussed the main parts of FrameBank: the dictionary of construction patterns, the annotation of constructions in corpus examples, the graphs of semantic roles and of shifts between constructions. Since FrameBank is an ongoing project, its development entails many further goals and challenges. The first task is to work out a graph of frames which could tie the constructions from the dictionary to the ontological classification of the lexicon. Although this graph may be to a great extent based on the broad inventory of semantic roles already existing in the database, it will sometimes require a more fine-grained semantic specification of the verbal ontology. The second task is to enlarge the database with constructions of nouns, adjectives, and adverbs which are now on the periphery of our research. It will also be promising to add full-text annotation, as this would allow studying the distribution and interaction of constructions in paragraphs and large texts.

References

1. Lyashevskaya, O.N., Kuznetsova, J.L.: Russian FrameNet: constructing a corpus–based dictionary of constructions [Russkij Frejmnet: k zadache sozdanija korpusnogo slovarja konstruktsij] (in Russian). In: Computational Linguistics and Intellectual Technologies: Proceedings of the International Conference "Dialog" [Komp'juternaja lingvistika i intelleltual'nye tehnologii: po materialam ezhegodnoj Mezhdunarodnoj konferentsii "Dialog"], vol. 8, pp. 306–312. RSUH, Moscow (2009)
2. Lyashevskaya, O.: Bank of Russian Constructions and Valencies. In: Proceedings of the Seventh conference on International Language Resources and Evaluation (LREC 2010), pp. 1802–1805. ELRA, Valletta (2010)

3. Kashkin, E.V., Lyashevskaya, O.N.: Semantic roles and construction net in Russian FrameBank [Semanticheskie roli i set' konstrukcij v sisteme FrameBank] (in Russian). In: Computational Linguistics and Intellectual Technologies. Proceedings of International Conference "Dialog", vol. 12-1, pp. 297–311. RSUH, Moscow (2013)

4. Russian National Corpus. http://ruscorpora.ru

5. FrameNet. http://framenet.icsi.berkeley.edu

6. Lexicographer database. http://lexicograph.ruslang.ru

7. Apresjan, J.D., Pall, E.: Russian verb – Hungarian verb. Government and combinability [Russkij glagol – vengerskij glagol. Upravlenie i sochetaemost'] (in Russian). Tankyonvkiado, Budapest (1982)

8. Paducheva, E.V.: Dynamic patterns in lexical semantics [Dinamicheskie modeli v semantike leksiki] (in Russian). Jazyki slavjanskoj kul'tury, Moscow (2004)

9. Fillmore, C.J.: The case for case. In: Bach, E., Harms, R.T. (eds.) Universals in Linguistic Theory, pp. 1–88. Holt, Rinehart and Winston, New York (1968)

10. Dowty, D.R.: Thematic proto roles and argument selection. Language 67, 547–619 (1991)

11. Apresjan, J.D.: Selected papers, vol. 1, Lexical Semantics [Izbrannye trudy, tom I. Leksicheskaja semantika] (in Russian). Jazyki Russkoj Kul'tury, Vostochnaja Literatura, Moscow (1995)

12. Apresjan, J.D., Boguslavskij, I.M., Iomdin, L.L., Sannikov, V.Z.: Theoretical issues of Russian syntax: the interrelation between grammar and vocabulary [Teoreticheskie problemy russkogo sintaksisa: vzaimodejstvie grammatiki i slovar'a] (in Russian). Jazyki slavjanskih kul'tur, Moscow (2010)

13. Màrquez, L., Carreras, X., Litkowski, K.C., Stevenson, S.: Semantic role labeling: an introduction to the special issue. Comput. Linguist. 34–2, 145–159 (2008)

14. Palmer, M.S., Wu, S., Titov, I.: Semantic role labeling tutorial. NAACL 2013 tutorials. http://naacl2013.naacl.org/Documents/semantic-role-labeling-part-3-naacl-2013-tutorial.pdf

15. Lyashevskaya, O.N., Kashkin, E.V.: Evaluation of frame-semantic role labeling in a case-marking language. In: Computational linguistics and intellectual technologies. Proceedings of International Conference "Dialog", vol. 13-1, pp. 362–379. RSUH, Moscow (2014)

16. Rakhilina, E.V., Reznikova, T.I., Karpova, O.S.: Semantic shifts in attributive constructions: metaphor, metonymy, and rebranding [Semanticheskie perehody v atribultivnyh konstrukcijah: metafora, metonimija i rebrending] (in Russian). In: Rakhilina, E.V. (ed.) Linguistics of Constructions [Lingvistika konstrukcij], pp. 398–455. Azbukovnik, Moscow (2010)

17. Kustova, G.I.: Types of figurative meanings and mechanisms of linguistic broadening [Tipy proizvodnyh zhachenij i mehanizmy jazykovogo rasshirenija] (in Russian). Jazyki slavjanskoj kul'tury, Moscow (2004)

18. Kashkin, E.V., Lyashevskaya, O.N.: Construction Grapher in Russian FrameBank: semantic and formal kinship map [Semanticheskie i formal'nye sv'azi russkih glagol'nyh konstrukcij: grafovoe predstavl'enije v sisteme FrameBank] (in Russian). In: Proceedings of the Kazan School on computational and cognitive linguistics TEL-2014 [Trudy Kazanskoj shkoly po kompjuternoj i kognitivnoj lingvistike TEL-2014], pp. 165–170. Fen, Kazan (2014)

19. Bricyn, V.M., Rakhilina, E.V., Reznikova, T.I., Yavorska, G.M. (eds.): The concept of pain from a typological point of view [Kontsept bol' v tipologicheskom osveschenii] (in Russian). Vidavnicjij Dim Dmitra Burago, Kiev (2009)

TagBag: Annotating a Foreign Language Lexical Resource with Pictures

Dmitry Ustalov[1,2,3](✉)

[1] N.N. Krasovskii Institute of Mathematics and Mechanics,
Ural Branch of the Russian Academy of Sciences, Ekaterinburg, Russia
[2] Ural Federal University, Ekaterinburg, Russia
[3] NLPub, Ekaterinburg, Russia
dau@imm.uran.ru

Abstract. Such forms of art as photography or drawing may serve as a uniform language, which represents things that we can either see or imagine. Hence, it is reasonable to use such pictures in order to connect nouns of the natural languages by their meanings. In this paper a study of mapping noun images from an annotated collection to the word senses of a foreign language lexical resource through the usage of a bilingual dictionary has been conducted. In this study, the English-Russian dictionary by V.K. Mueller has been used to enhance the Yet Another RussNet synsets with Flickr photos.

Keywords: Multimedia search · Bilingual dictionary · Image database · Lexical ontology · Natural language processing

1 Introduction

The problem of mapping pictures to word senses is important in various information retrieval and natural language processing tasks such as multimedia search, quality assessment, automatic text illustrating, etc. Modern search engines, including Yandex, Google, Bing and Yahoo! provide the corresponding possibilities and tools to find images by a search query, although refinement and alignment of the search results are still handled by a particular user or application.

English—among other resource-rich languages—has developed annotated image collections that are widely used for approaching computer vision and especially object detection tasks. However, those are rarely translated into Russian, thus it is rational to reuse their deliverables for processing Russian.

The work, as described in this paper, makes the following contributions: (1) it presents a survey on annotated image collections and those mapping to word senses, and (2) proposes and evaluates TAGBAG, an unsupervised approach to mapping pictures to the—hopefully—correspondent word senses.

The rest of this paper is organized as follows. Section 2 is devoted to the survey on the related work. Section 3 presents and exemplifies TAGBAG. Section 4 is dedicated to evaluation of the present approach. Section 5 interprets and explains the obtained results. Section 6 concludes with final remarks and directions for the future work.

© Springer International Publishing Switzerland 2015
M.Y. Khachay et al. (Eds.): AIST 2015, CCIS 542, pp. 361–369, 2015.
DOI: 10.1007/978-3-319-26123-2_35

2   Related Work

ImageNet is a project aimed at annotation of the Princeton WordNet noun synsets with images [1]. There is a study of Gelfenbein et al. to apply basic machine translation methods to produce a Russian WordNet [2] through the usage of a bilingual English-to-Russian dictionary, a Russian synonymy dictionary, and an English frequency dictionary. Unfortunately, intersection between ImageNet and this resource is negligible, suggesting to finding another approach.

ImageCLEF is an evaluation forum for the cross–language annotation and retrieval of images [3], which also aggregates various related tools and resources. It is aimed primarily at the problems of image indexing and visual feature extraction. Nevertheless, its output can be applied in creation of gold standards.

Joshi et al. use two large image databases, namely Terra Galleria and The Art Museum Image Consortium, which contain a lot of high quality photos with annotations, but these resources are highly expensive for academic use [4].

Mihalcea & Leong presented a PicNet illustrated dictionary [5]. Using such a knowledge base makes it possible to draw pictorial representations of simple sentences to make the cross-language communication less complicated. PicNet is a proprietary resource, which is unavailable for general public.

There are several studies of cross-language ontology mapping and information retrieval. Reiter et al. proposed the PanImages method for disambiguating word senses when searching images for a given single word query [6]. Trojahn et al. developed a framework for multilingual ontology mapping that operates with the Semantic Web ontologies and represents the deterministic mappings in the $\mathcal{SHOIN}^{(\mathcal{D})}$ description logic formalism [7].

Disambiguation of Flickr tags is a recently developed topic, e.g. Stampouli et al. presented a Semantic Web-based method to do so with the Wikipedia data [8]. However, mapping pictures to word senses—which is the inverse problem—requires another set of approaches.

Recent works in the field of automatic text-to-picture synthesis use Flickr as an annotated image collection (see Jiang et al. [9] and Li & Zhuge [10] for details). The target language of the systems described in these works is English, and these works do not consider multilingual extensions.

Flickr has a convenient API[1], which is freely available for research applications. It also has license information attached to the photos allowing one to use only pictures under the Creative Commons (CC) terms. Therefore, Flickr is a viable choice of an annotated image collection.

3   TagBag

Given an annotated image $I \in \mathcal{I}$ and a bilingual dictionary $\mathcal{B}$, the **bag** vector is getting constructed (the used notation is present at Table 1). The proposed algorithm is named TagBag since it has two sequential phases: phase "Tag" (lines 3–9) and phase "Bag" (lines 10–14).

---

[1] https://www.flickr.com/services/api/flickr.photos.search.html

Algorithm. TAGBAG

1: **function** TAGBAG$(I, f \mid \mathcal{B}) \rightarrow \textbf{\textit{bag}}$	
2: $\textbf{\textit{bag}} \leftarrow \emptyset$	▷ initialize a bag-of-words vector for I
3: **for all** $t \in I$ **do**	▷ iterate over all tags of the image I
4: **for all** $s \in \mathcal{B}(t)$ **do**	▷ iterate over all senses of the tag t
5: **for all** $w \in s$ **do**	▷ iterate over all translations of the sense s
6: $\textbf{\textit{bag}}_w \leftarrow \textbf{\textit{bag}}_w + 1$	▷ increase the frequency of the word w
7: **end for**	
8: **end for**	
9: **end for**	
10: **for all** $w \in \textbf{\textit{bag}}$ **do**	▷ iterate over the **bag** dimensions
11: **if** $\textbf{\textit{bag}}_w < f$ **then**	▷ detect the low frequency dimensions
12: $\textbf{\textit{bag}} \leftarrow \textbf{\textit{bag}} - w$	▷ remove the w dimension
13: **end if**	
14: **end for**	
15: **return** $\textbf{\textit{bag}}$	
16: **end function**	

Table 1. Notation

Object	Notation	Example
Image collection	$\mathcal{I}$	$\{\{\text{cat, kitten}\}, \{\text{nature, mountain, kodak}\}\}$
Bilingual dictionary	$\mathcal{B}$	$\{\{\text{crane} \rightarrow \{\{\text{кран}\}, \{\text{журавль, журавлиные}\}\}\}\}$
Bag-of-words vector	$\textbf{\textit{bag}}$	$(\text{people, march, event})$
Synset	s	$\{\text{quick, rapid, fast}\}$

The purpose of the TAGBAG algorithm is to (1) construct a bag-of-words vector that counts all the foreign language translations for all the tags attached to an image, and then (2) prune that vector by removing the coordinates of low frequency words using the defined cut-off value $f \in \mathbb{N}$.

"Tag." In this phase, image tags should be counted and the bag-of-words vector is getting populated. Given an image I, each sense s of each its tag t is iterated (lines 3–4, 8–9). Then, each translation w for each sense s is added to the correspondent coordinate of the previously initialized **bag** vector (lines 5–7). It should be noted that a word w may appear several times and each appearance is considered independently, the appropriate coordinate gets increased according to the number the foreign word appeared (line 6).

"Bag." In this phase, the bag-of-words vector is getting pruned in order to reduce the resulting vector space and decrease the computational complexity. Given a vector **bag**, all its coordinates that are below the f frequency cut-off value are getting removed (lines 10–14), and the modified vector is returned from the TAGBAG function, i. e. attached to the image I (line 15).

The resulting vector **bag** can now be used in obtaining a mapping I_s that annotates a synset s of a foreign language lexical resource S with the image I:

$$I_s = \underset{s' \in S}{argmax}\ sim(\textbf{bag}, s').$$

Figure 1 shows three particularly good examples of the TAGBAG results: a) дорога (*food*), путь (*way*), b) человек (*human*), душа (*soul*), c) еда (*food*), питание (*nutrition*), продовольствие (*provisions*), etc. These pictures originally have tags in English, although they have been successfully mapped to the Yet Another RussNet synsets in Russian.

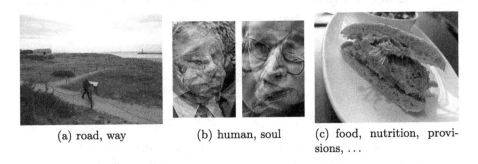

| (a) road, way | (b) human, soul | (c) food, nutrition, provisions, ... |

Fig. 1. Examples of the TAGBAG results

4   Evaluation

In order to evaluate the proposed TAGBAG approach, it has been applied to the task of mapping the Flickr photos to the synsets of the Yet Another RussNet electronic thesaurus for Russian (see [11] for more information about the lexical resource used). The synsets have been obtained in the CSV format[2], the number of noun synsets is 11 666.

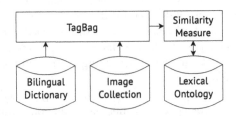

Fig. 2. The experimental setup used to evaluate TAGBAG

---

[2] http://nlpub.ru/YARN/Format.

4.1   Experimental Setup

The experimental setup is depicted at Fig. 2, where the derived English-Russian dictionary by V.K. Mueller is used as the bilingual dictionary, Flickr is used as the image collection, Yet Another RussNet acts as the lexical ontology. Two similarity measures have also been used: cosine similarity and the Jaccard index.

Then, the top 1500 English nouns list[3] has been used to query the Flickr API and retrieve images with their metadata including the licensing information and the assigned tags. Only publicly available images have been fetched. This resulted in 152 230 records, each of which represents metadata of the particular image including its title, license, ownership and tag information.

4.2   Deriving a Bilingual Dictionary

The English-Russian dictionary by V.K. Mueller is one of the well-known bilingual dictionaries available[4] in the dict format, which is sufficiently easy to parse. In order to produce the bilingual dictionary in the abovementioned notation of $\mathcal{B}$, it has been processed and cleansed, resulting[5] in 132 810 lines of the form of (en, sid, ru), where en is the word in English, sid is the sense identifier, and ru is the Russian word corresponding to the sid sense.

4.3   Questionnaire

The questionnaire is composed of single-answer questions representing the synset-image mappings produced by TAGBAG (Fig. 3). The questions are composed as follows: (1) the Flickr photo search results for the first 60 top English nouns have been obtained, (2) the mapping with highest similarity has been selected for each word, and (3) the corresponding synonyms and image URI have been extracted. The separate questionnaire has been created for each similarity measure.

4.4   Results

The used implementation of the TAGBAG approach is made in the Ruby programming language. The source code under the MIT license with all the used datasets under the CC BY-SA license are available for download in the form of tarball: http://ustalov.imm.uran.ru/pub/tagbag-aist.tar.gz.

> – synonym1, synonym2, . . . , synonymN
> ```
> https://www.flickr.com/photos/<user_id>/<photo_id>
> [ ] Yes
> [ ] No
> ```

Fig. 3. Format of a question in the questionnaire

---

[3] http://www.talkenglish.com/Vocabulary/Top-1500-Nouns.aspx.

[4] http://mueller-dic.chat.ru/.

[5] http://ustalov.imm.uran.ru/pub/mueller.tar.gz.

Table 2. Evaluation results

	Answers	Accuracy	Agreement
Cosine similarity	53	.623	.566
Jaccard index	51	.647	.567

Each questionnaire has been filled in by three independent annotators on Google Forms and then the obtained answers have been aggregated with majority voting, which is highly efficient for small number of annotators per question [12]. Accuracy, which is the rate of positive answers provided by the annotators, has been computed (Table 2). Since the answers are of nominal scale, the Fleiss' κ has been additionally computed to estimate the inter-annotator agreement [13].

5   Discussion

Both similarity measures show the similar results with good inter-annotator agreement suggesting that the proposed unsupervised approach provides quite reliable output despite of its simplicity. However, it is interesting to analyze its performance more thoroughly and discuss the most controversial mappings.

5.1   Controversial Mappings

Most image URIs in both datasets are equal. Particularly, the Jaccard index for these datasets is $\frac{43}{61} = .705$. Only 13 of these common 43 mappings are wrong. There are three sources of errors in the present results:

- *sloppy image tags*, when a user provides wrong, ambiguous or spamming tags for a photo (7 of 13),
- *actual mapping errors*, when incorrect mapping has been obtained due to the lack of the relevant concepts in the lexical ontology (3 of 13).
- *inaccurate photos* uploaded using batch upload (3 of 13).

Some results can be cleanly explained by the fact that the Flickr users often use batch photo upload feature allowing them to assign the same set of tags to all the photos being uploaded. For instance, a user can travel across the city and shoot a lot of different objects including his/her lunch, people walking around, moving cars, lovely kittens, and so on. Due to the lack of time all these photos can be uploaded into Flickr with such common tags as *people, travel, vacation* bringing errors to pure linguistic algorithms like that of TAGBAG.

5.2   Threshold

The cut-off value $f \in \mathbb{N}$ described in the Sect. 3 is the only input parameter affecting the behaviour of TAGBAG. During the evaluation it was manually set to $f = 2$ as a trade-off between the performance and the number of vector space dimensions. However, other options have also been tried. In case when $f = 1$ the computational complexity and the number of vector space dimensions are rapidly growing which results in poor performance. However, when $f > 2$, the number of vector dimensions tends to be zero for the most images that makes practical use of the present approach impossible.

5.3   Similarity Measures

The mean value of word count per synset is 3.09 for the cosine similarity dataset and 3.22 for the Jaccard index dataset. As according to Fig. 4, it seems that the higher the similarity measure is, the more likely the mapping is correct. Testing of such a hypothesis requires a quite larger dataset that covers wide range of similarity values and hence was not possible in this study.

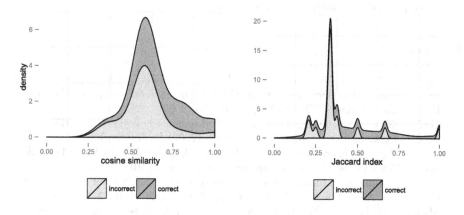

Fig. 4. Stacked densities of the similarity measures

6   Conclusion

TAGBAG is an unsupervised approach for mapping pictures to the correspondent synsets'. It demonstrates performance suitable to practical use, evidence of which has been proven through the conducted evaluation. The used datasets and software are available on the Internet for general public.

There are several directions for future work. Firstly, using visual saliency and the similar computer vision features to filter out inaccurate photos may improve the overall performance of the present approach [14]. Secondly, a more large-scale gamified evaluation may be useful to assess a much larger dataset [15]. Thirdly, it may be useful to evaluate TAGBAG on another lexical resource, such as RuThes [16]. Fourthly, the problem of sloppy tags could be probably addressed with stop words or more sophisticated information retrieval techniques [17]. Finally, there exist efficient algorithms to enhance and disambiguate a bilingual dictionary, and using of them may positively affect the performance [18].

Attribution. The picture at Fig. 1(a) is attributed to Joan Grífols[6] under the CC BY-NC-ND license, the picture at Fig. 1(b) is attributed to Pekka Nikrus[7] under the CC BY-NC-SA license, and the picture at Fig. 1(c) is attributed to Lori Greig[8] under the CC BY license.

Acknowledgements. This work is supported by the Russian Foundation for the Humanities, project no. 13-04-12020 "New Open Electronic Thesaurus for Russian", and by the Program of Government of the Russian Federation 02.A03.21.0006 on 27.08.2013. The URAN supercomputer located at the N.N. Krasovskii Institute of Mathematics and Mechanics of the Ural Branch of the Russian Academy of Sciences has been used to obtain the image collection. The author is grateful to those annotators who participated in the evaluation. He is also grateful to the anonymous referees who offered very useful comments on the present paper.

References

1. Deng, J., Dong, W., Socher, R., Li, L.J., Li, K., Fei-Fei, L.: ImageNet: a large-scale hierarchical image database. In: IEEE Conference on Computer Vision and Pattern Recognition, CVPR 2009, pp. 248–255 (2009)
2. Gelfenbein, I., et al.: Avtomaticheskij perevod semanticheskoj seti WORDNET na russkij yazyk. In: Proceedings of Dialog 2003 (2003) (in Russian)
3. Müller, H., Clough, P., Deselaers, T., Caputo, B. (eds.): ImageCLEF. The Information Retrieval Series, vol. 32. Springer, Heidelberg (2010)
4. Joshi, D., Wang, J.Z., Li, J.: The story picturing engine–a system for automatic text illustration. ACM Trans. Multimedia Comput. Commun. Appl. **2**, 68–89 (2006)
5. Mihalcea, R., Leong, C.W.: Toward communicating simple sentences using pictorial representations. Mach. Trans. **22**, 153–173 (2008)
6. Reiter, K., Soderland, S., Etzioni, O.: Cross-lingual image search on the web. In: Proceedings of the Workshop on Cross-Lingual Information Access (20th International Joint Conference on Artificial Intelligence) (2007)
7. Trojahn, C., Quaresma, P., Vieira, R.: A framework for multilingual ontology mapping. In: Proceedings of the Sixth International Conference on Language Resources and Evaluation, LREC 2008, Marrakech. European Language Resources Association (2008)
8. Stampouli, A., Giannakidou, E., Vakali, A.: Tag disambiguation through flickr and wikipedia. In: Yoshikawa, M., Meng, X., Yumoto, T., Ma, Q., Sun, L., Watanabe, C. (eds.) DASFAA 2010. LNCS, vol. 6193, pp. 252–263. Springer, Heidelberg (2010)
9. Jiang, Y., Liu, J., Lu, H.: Chat with illustration. Multimedia Syst. 1–12 (2014). http://link.springer.com/article/10.1007/s00530-014-0371-3
10. Li, W., Zhuge, H.: Summarising news with texts and pictures. In: 10th International Conference on Semantics, Knowledge and Grids (SKG), pp. 100–107 (2014)

---

[6] https://www.flickr.com/photos/33818912@N00/16224853830.

[7] https://www.flickr.com/photos/16391511@N00/16380713555.

[8] https://www.flickr.com/photos/39585662@N00/16204920679.

11. Braslavski, P., Ustalov, D., Mukhin, M.: A spinning wheel for YARN: user interface for a crowdsourced thesaurus. In: Proceedings of the Demonstrations at the 14th Conference of the European Chapter of the Association for Computational Linguistics, Gothenburg, pp. 101–104. Association for Computational Linguistics (2014)
12. Karger, D.R., Oh, S., Shah, D.: Budget-optimal task allocation for reliable crowdsourcing systems. Oper. Res. **62**, 1–24 (2014)
13. Fleiss, J.L., Levin, B., Paik, M.C.: Statistical Methods for Rates and Proportions, 3rd edn. Wiley, Hoboken (2003)
14. Cheng, M.M., Zhang, G.X., Mitra, N.J., Huang, X., Hu, S.M.: Global contrast based salient region detection. In: 2011 IEEE Conference on Computer Vision and Pattern Recognition (CVPR), pp. 409–416 (2011)
15. von Ahn, L., Dabbish, L.: Labeling images with a computer game. In: Proceedings of the SIGCHI Conference on Human Factors in Computing Systems, CHI 2004, pp. 319–326. ACM, New York (2004)
16. Loukachevitch, N.: Thesauri for Information Retrieval Tasks. MSU, Moscow (2011)
17. Ntoulas, A., Najork, M., Manasse, M., Fetterly, D.: Detecting spam web pages through content analysis. In: Proceedings of the 15th International Conference on World Wide Web, WWW 2006, pp. 83–92. ACM, New York (2006)
18. Flati, T., Navigli, R.: The CQC algorithm: cycling in graphs to semantically enrich and enhance a bilingual dictionary. J. Artif. Int. Res. **43**, 135–171 (2012)

BigARTM: Open Source Library for Regularized Multimodal Topic Modeling of Large Collections

Konstantin Vorontsov[1]($\boxtimes$), Oleksandr Frei[2], Murat Apishev[3], Peter Romov[1], and Marina Dudarenko[3]

[1] Yandex, Moscow Institute of Physics and Technology, Moscow, Russia
voron@forecsys.ru, peter@romov.ru
[2] Schlumberger Information Solutions, Oslo, Norway
oleksandr.frei@gmail.com
[3] Lomonosov Moscow State University, Moscow, Russia
great-mel@yandex.ru, m.dudarenko@gmail.com

Abstract. Probabilistic topic modeling of text collections is a powerful tool for statistical text analysis. In this paper we announce the BigARTM open source project (http://bigartm.org) for regularized multimodal topic modeling of large collections. Several experiments on Wikipedia corpus show that BigARTM performs faster and gives better perplexity comparing to other popular packages, such as Vowpal Wabbit and Gensim. We also demonstrate several unique BigARTM features, such as additive combination of regularizers, topic sparsing and decorrelation, multimodal and multilanguage modeling, which are not available in the other software packages for topic modeling.

Keywords: Probabilistic topic modeling · Probabilistic latent sematic analysis · Latent dirichlet allocation · Additive regularization of topic models · Stochastic matrix factorization · EM-algorithm · BigARTM

1 Introduction

Topic modeling is a rapidly developing branch of statistical text analysis [1]. Topic model reveals a hidden thematic structure of a text collection and finds a compressed representation of each document in terms of its topics. Practical applications of topic models include many areas, such as information retrieval for long-text queries, classification, categorization, summarization and segmentation of texts. Topic models are increasingly used for non-textual and heterogeneous data including signals, images, video and networks. More ideas, models and applications are outlined in the survey [4].

From a statistical point of view, a probabilistic topic model (PTM) defines each topic by a multinomial distribution over words, and then describes each document with a multinomial distribution over topics. From an optimizational point of view, topic modeling can be considered as a special case of approximate stochastic matrix factorization. To learn a factorized representation of a text

© Springer International Publishing Switzerland 2015
M.Y. Khachay et al. (Eds.): AIST 2015, CCIS 542, pp. 370–381, 2015.
DOI: 10.1007/978-3-319-26123-2_36

collection is an ill-posed problem, which has an infinite set of solutions. A typical approach in this case is to apply regularization techniques, which impose problem-specific constrains and ultimately lead to a better solution.

Modern literature on topic modeling offers hundreds of models adapted to different situations. Nevertheless, most of these models are too difficult for practitioners to quickly understand, adapt and embed into applications. This leads to a common practice of tasting only the basic out-of-date models such as *Probabilistic Latent Semantic Analysis*, PLSA [6] and *Latent Dirichlet Allocation*, LDA [3]. Most practical inconveniences are rooted in Bayesian learning, which is the dominating approach in topic modeling. Bayesian inference of topic models requires a laborious mathematical work, which prevents flexible unification, modification, selection, and combination of topic models.

In this paper we announce **the BigARTM open source project** for regularized multimodal topic modeling of large collections, http://bigartm.org. The theory behind BigARTM is based on a non-Bayesian multicriteria approach — *Additive Regularization of Topic Models*, ARTM [11]. In ARTM a topic model is learned by maximizing a weighted sum of the log-likelihood and additional regularization criteria. The optimization problem is solved by a general regularized expectation-maximization (EM) algorithm, which can be applied to an arbitrary combination of regularization criteria. Many known Bayesian topic models were revisited in terms of ARTM in [12,13]. Compared to the Bayesian approach, ARTM makes it easier to design, infer and combine topic models, thus reducing the barrier for entering into topic modeling research field.

BigARTM source code is released under the New BSD License, which permits free commercial and non-commercial usage. The core of the library is written in C++ and is exposed via two equally rich APIs for C++ and Python. The library is cross-platform and can be built for Linux, Windows and OS X in both 32 and 64 bit configuration. In our experiments on Wikipedia corpus BigARTM performs better than Vowpal Wabbit LDA and Gensim libraries in terms of perplexity and runtime. Comparing to the other libraries BigARTM offers several additional features, such as regularization and multi-modal topic modeling.

The rest of the paper is organized as follows. In Sect. 2 we introduce a multimodal topic model for documents with metadata. In Sect. 3 we generalize the fast online algorithm [5] to multimodal ARTM. In Sect. 4 we describe parallel architecture and implementation details of the BigARTM library. In Sect. 5 we report results of our experiments on large datasets. In Sect. 6 we discuss advantages, limitations and open problems of BigARTM.

2 Multimodal Regularized Topic Model

Let D denote a finite set (collection) of texts and W^1 denote a finite set (vocabulary) of all terms from these texts. Each term can represent a single word or a key phrase. A document can contain not only words, but also terms of other modalities. Each modality is defined by a finite set (vocabulary) of terms W^m, $m = 1, \ldots, M$. Examples of not-word modalities are: authors, class or category labels, date-time stamps, references to/from other documents, entities mentioned in texts, objects found in the images associated with the documents, users that read or downloaded documents, advertising banners, etc.

Assume that each term occurrence in each document refers to some latent topic from a finite set of topics T. Text collection is considered to be a sample of triples (w_i, d_i, t_i), $i = 1, \ldots, n$, drawn independently from a discrete distribution $p(w, d, t)$ over the finite space $W \times D \times T$, where $W = W^1 \sqcup \cdots \sqcup W^m$ is a disjoint union of the vocabularies across all modalities. Terms w_i and documents d_i are observable variables, while topics t_i are latent variables.

Following the idea of Correspondence LDA [2] and Dependency LDA [9] we introduce a topic model for each modality:

$$p(w \mid d) = \sum_{t \in T} p(w \mid t)\, p(t \mid d) = \sum_{t \in T} \phi_{wt}\theta_{td}, \quad d \in D,\ w \in W^m,\ m = 1, \ldots, M.$$

The parameters $\theta_{td} = p(t \mid d)$ and $\phi_{wt} = p(w \mid t)$ form matrices $\Theta = (\theta_{td})_{T \times D}$ of *topic probabilities for the documents*, and $\Phi^m = (\phi_{wt})_{W^m \times T}$ of *term probabilities for the topics*. The matrices Φ^m, if stacked vertically, form a $W \times T$-matrix Φ. Matrices Φ^m and Θ are *stochastic*, that is, their vector-columns represent discrete distributions. Usually $|T|$ is much smaller than $|D|$ and $|W|$.

To learn parameters Φ^m, Θ from the multimodal text collection we maximize the log-likelihood for each m-th modality:

$$\mathscr{L}_m(\Phi^m, \Theta) = \sum_{d \in D} \sum_{w \in W^m} n_{dw} \ln p(w \mid d) \to \max_{\Phi^m, \Theta},$$

where n_{dw} is the number of occurrences of the term $w \in W^m$ in the document d. Note that topic distributions of documents Θ are common for all modalities. Following the ARTM approach, we add a regularization penalty term $R(\Phi, \Theta)$ and solve a constrained multicriteria optimization problem via scalarization:

$$\sum_{m=1}^{M} \tau_m \mathscr{L}_m(\Phi^m, \Theta) + R(\Phi, \Theta) \to \max_{\Phi, \Theta}; \tag{1}$$

$$\sum_{w \in W^m} \phi_{wt} = 1,\ \phi_{wt} \geqslant 0; \qquad \sum_{t \in T} \theta_{td} = 1,\ \theta_{td} \geqslant 0. \tag{2}$$

The local maximum (Φ, Θ) of the problem (1), (2) satisfies the following system of equations with auxiliary variables $p_{tdw} = p(t \mid d, w)$:

$$p_{tdw} = \underset{t \in T}{\mathrm{norm}}\big(\phi_{wt}\theta_{td}\big); \tag{3}$$

$$\phi_{wt} = \underset{w \in W^m}{\mathrm{norm}}\left(n_{wt} + \phi_{wt}\frac{\partial R}{\partial \phi_{wt}}\right); \quad n_{wt} = \sum_{d \in D} n_{dw} p_{tdw}; \tag{4}$$

$$\theta_{td} = \underset{t \in T}{\text{norm}} \left(n_{td} + \theta_{td} \frac{\partial R}{\partial \theta_{td}} \right); \quad n_{td} = \sum_{w \in d} \tau_{m(w)} n_{dw} p_{tdw}; \quad (5)$$

where operator $\underset{t \in T}{\text{norm}} \, x_t = \frac{\max\{x_t, 0\}}{\sum_{s \in T} \max\{x_s, 0\}}$ transforms a vector $(x_t)_{t \in T}$ to a discrete distribution; $m(w)$ is the modality of the term w, so that $w \in W^{m(w)}$.

The system of Eqs. (3)–(5) follows from Karush–Kuhn–Tucker conditions. It can be solved by various numerical methods. Particularly, the simple-iteration method is equivalent to the EM algorithm, which is typically used in practice. For single modality ($M = 1$) it gives the regularized EM algorithm proposed in [11]. With no regularization ($R = 0$) it corresponds to PLSA [6].

Many Bayesian topic models can be considered as special cases of ARTM with different regularizers R, as shown in [12,13]. For example, LDA [3] corresponds to the entropy smoothing regularizer.

Due to the unified framework of additive regularization BigARTM can build topic models for various applications simply by choosing a suitable combination of regularizers from a build-in user extendable library.

3   Online Topic Modeling

Following the idea of Online LDA [5] we split the collection D into batches D_b, $b = 1, \ldots, B$, and organize EM iterations so that each document vector θ_d is iterated until convergence at a constant matrix Φ, see Algorithms 1 and 2. Matrix Φ is updated rarely, after all documents from the batch are processed. For a large collection matrix Φ often stabilizes after small initial part of the collection. Therefore a single pass through the collection might be sufficient to learn a topic model.

Algorithm 1 does not specify how often to synchronize Φ matrix at steps 5–8. It can be done after every batch or less frequently (for instance if $\frac{\partial R}{\partial \phi_{wt}}$ takes long time to evaluate). This flexibility is especially important for concurrent implementation of the algorithm, where multiple batches are processed in parallel. In this case synchronization can be triggered when a fixed number of documents had been processed since the last synchronization.

The online reorganization of the EM iterations is not necessarily associated with Bayesian inference used in [5]. Different topic models, from PLSA to multimodal and regularized models, can be learned by the above online EM algorithm.

4   BigARTM Architecture

The main goal for BigARTM architecture is to ensure a constant memory usage regardless of the collection size. For this reason each D_b batch is stored on disk in a separate file, and only a limited number of batches is loaded into the main memory at any given time. The entire Θ matrix is never stored in the memory. As a result, the memory usage stays constant regardless of the size of the collection.

Algorithm 1. Online EM-algorithm for multimodal ARTM

Input: collection $\{D_b : b = 1, \ldots, B\}$, discounting factor $\rho \in (0, 1]$;
Output: matrix Φ;

1 initialize ϕ_{wt} for all $w \in W$ and $t \in T$;
2 $n_{wt} := 0$, $\tilde{n}_{wt} := 0$ for all $w \in W$ and $t \in T$;
3 **for all** *batches* D_b, $b = 1, \ldots, B$
4 $\quad$ $(\tilde{n}_{wt}) := (\tilde{n}_{wt}) + \mathsf{ProcessBatch}(D_b, \phi_{wt})$;
5 $\quad$ **if** *(synchronize)* **then**
6 $\quad\quad$ $n_{wt} := \rho n_{wt} + \tilde{n}_{dw}$ for all $w \in W$ and $t \in T$;
7 $\quad\quad$ $\phi_{wt} := \underset{w \in W^m}{\text{norm}} \left(n_{wt} + \phi_{wt} \frac{\partial R}{\partial \phi_{wt}} \right)$ for all $w \in W^m$, $m = 1, \ldots, M$ and $t \in T$;
8 $\quad\quad$ $\tilde{n}_{wt} := 0$ for all $w \in W$ and $t \in T$;

Algorithm 2. $\mathsf{ProcessBatch}(D_b, \phi_{wt})$

Input: batch D_b, matrix ϕ_{wt};
Output: matrix $(\tilde{n}_{wt})$;

1 $\tilde{n}_{wt} := 0$ for all $w \in W$ and $t \in T$;
2 **for all** $d \in D_b$
3 $\quad$ initialize $\theta_{td} := \frac{1}{|T|}$ for all $t \in T$;
4 $\quad$ **repeat**
5 $\quad\quad$ $p_{tdw} := \underset{t \in T}{\text{norm}} \left(\phi_{wt} \theta_{td} \right)$ for all $t \in T$;
6 $\quad\quad$ $n_{td} := \sum_{w \in d} \tau_{m(w)} n_{dw} p_{tdw}$ for all $t \in T$;
7 $\quad\quad$ $\theta_{td} := \underset{t \in T}{\text{norm}} \left(n_{td} + \theta_{td} \frac{\partial R}{\partial \theta_{td}} \right)$ for all $t \in T$;
8 $\quad$ **until** θ_d *converges*;
9 $\quad$ increment $\tilde{n}_{wt}$ by $n_{dw} p_{tdw}$ for all $w \in d$ and $t \in T$;

Concurrency. An general rule of concurrency design is to express parallelism at the highest possible level. For this reason BigARTM implements a concurrent processing of the batches and keeps a single-threaded code for the $\mathsf{ProcessBatch}$ (D_b, ϕ_{wt}) routine.

To split collection into batches and process them concurrently is a common approach, introduced in AD-LDA algorithm [8], and then further developed in PLDA [15] and PLDA+ [7] algorithms. These algorithms require all concurrent workers to become idle before an update of the Φ matrix. Such synchronization step adds a large overhead in the online algorithm where Φ matrix is updated multiple times on each iteration. An alternative architecture without the synchronization step is described in [10], however it mostly targets a distributed cluster environment. In our work we develop an efficient single-node architecture where all workers benefit from the shared memory space.

To run multiple $\mathsf{ProcessBatch}$ in parallel the inputs and outputs of this routine are stored in two separate in-memory queues, locked for push and pop operations with spin locks (Fig. 1). This approach does not add any noticeable

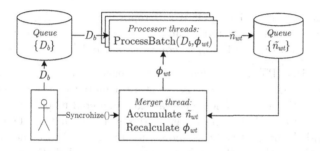

Fig. 1. Diagram of key BigARTM components

synchronization overhead because both queues only store smart pointers to the actual data objects, so push and pop operations does not involve copying or relocating big objects in the memory.

Smart pointers are also essential for lifecycle of the Φ matrix. This matrix is *read* by all processors threads, and can be *written* at any time by the merger thread. To update Φ without pausing all processor threads we keep two copies — an *active* Φ and a *background* Φ matrices. The active matrix is read-only, and is used by the processor threads. The background matrix is being built in a background by the merger thread at steps 6 and 7 of Algorithm 1, and once it is ready merger thread marks it as active. Before processing a new batch the processor thread gets the current active matrix from the merger thread. This object is passed via shared smart pointer to ensure that processor thread can keep ownership of its Φ matrix until the batch is fully processed. As a result, all processor threads keep running concurrently with the update of Φ matrix.

Note that all processor threads share the same Φ matrix, which means that memory usage stays at constant level regardless of how many cores are used for computation. Using memory for two copies of the Φ matrix in our opinion gives a reasonable usage balance between memory and CPU resources. An alternative solution with only one Φ matrix is also possible, but it would require a heavy usage of atomic CPU instructions. Such operations are very efficient, but still come at a considerable synchronization cost[1], and using them for all reads and writes of the Φ matrix would cause a significant performance degradation for merger and processor threads. Besides, an arbitrary overlap between reads and writes of the Φ matrix eliminates any possibility of producing a deterministic result. The design with two copies of the Φ matrix gives much more control over this and in certain cases allows BigARTM to behave in a fully deterministic way.

The design with two Φ matrices only supports a single merger thread, and we believe it should handle all $\tilde{n}_{wt}$ updates coming from many threads. This is a reasonable assumption because merging at step 6 takes only about $O(|W| \cdot |T|)$ operations to execute, while ProcessBatch takes $O(n|T|I)$ operations, where n is the number of non-zero entries in the batch, I is the average number of inner iterations in ProcessBatch routine. The ratio $n/|W|$ is typically from 100 to 1000

---

[1] http://stackoverflow.com/questions/2538070/atomic-operation-cost

(based on datasets in UCI Bag-Of-Words repository), and I is $10 \ldots 20$, so the ratio safely exceeds the expected number of cores (up to 32 physical CPU cores in modern workstations, and even 60 cores of the Intel Xeon Phi co-processors).

Data Layout. BigARTM uses dense single-precision matrices to represent Φ and Θ. Together with the Φ matrix we store a global dictionary of all terms $w \in W$. This dictionary is implemented as std::unordered_map that maps a string representation of $w \in W$ into its integer index in the Φ matrix. This dictionary can be extended automatically as more and more batches came through the system. To achieve this each batch D_b contains a local dictionary W_b, listing all terms that occur in the batch. The n_{dw} elements of the batch are stored as a sparse CSR matrix (Compressed Sparse Raw format), where each row correspond to a document $d \in D_b$, and terms w run over a local batch dictionary W_b.

For performance reasons Φ matrix is stored in column-major order, and Θ in row-major order. This layout ensures that $\sum_t \Phi_{wt} \theta_{td}$ sum runs on contiguous memory blocks. In both matrices all values smaller than 10^{-16} are always replaced with zero to avoid performance issues with denormalized numbers[2].

Programming Interface. All functionality of BigARTM is expressed in a set of extern C methods. To input and output complex data structures the API uses Google Protocol Buffers[3]. This approach makes it easy to integrate BigARTM into any research or production environment, as almost every modern language has an implementation of Google Protocol Buffers and a way of calling extern C code (ctypes module for Python, loadlibrary for Matlab, PInvoke for C#, etc.).

On top of the extern C API BigARTM already has convenient wrappers in C++ and Python. We are also planning to implement a Java wrapper in the near future. In addition to the APIs the library also has a simple CLI interface.

BigARTM has built-in libraries of regularizers and quality measures that can be extended in current implementation only through project recompilation.

Basic Tools. A careful selection of the programming tools is important for any software project. This is especially true for BigARTM as its code is written in C++, a language that by itself offers less functionality comparing to Python, .NET Framework or Java. To mitigate this we use various parts of the Boost C++ Libraries, Google Protocol Buffers for data serialization, ZeroMQ library for network communication, and several other libraries.

BigARTM uses CMake as a cross-platform build system, and it successfully builds on Windows, Linux and OS X in 32 and 64 bit configurations. Building the library require a recent C++ compiler with C++11 support (GNU GCC 4.6.3, clang 3.4 or Visual Studio 2012 or newer), and Boost Libraries 1.46.1 or newer. All the other third-parties are included in BigARTM repository.

We also use free online services to store source code (https://github.com), to host online documentation (https://readthedocs.org) and to run automated continuous integration builds (http://travis-ci.org).

---

[2] http://en.wikipedia.org/wiki/Denormal_number#Performance_issues.

[3] http://code.google.com/p/protobuf/.

5 Experiments

In this section we evaluate the runtime performance and the algorithmic quality of BigARTM against two popular software packages — Gensim [14] and Vowpal Wabbit[4]. We also demonstrate some of the unique BigARTM features, such as combining regularizers and multi-language topic modeling via multimodality, which are not available in the other software packages.

All three libraries (VW.LDA, Gensim and BigARTM) work out-of-core, e. g. they are designed to process data that is too large to fit into a computer's main memory at one time. This allowed us to benchmark on a fairly large collection — 3.7 million articles from the English Wikipedia[5]. The conversion to bag-of-words was done with gensim.make_wikicorpus script[6], which excludes all non-article pages (such as category, file, template, user pages, etc.), and also pages that contain less than 50 words. The dictionary is formed by all words that occur in at least 20 documents, but no more than in 10 % documents in the collection. The resulting dictionary was caped at $|W| = 100\,000$ most frequent words.

Both Gensim and VW.LDA represents the resulting topic model as Dirichlet distribution over Φ and Θ matrices: $\boldsymbol{\theta}_d \sim \mathrm{Dir}(\boldsymbol{\gamma}_d)$ and $\boldsymbol{\phi}_t \sim \mathrm{Dir}(\boldsymbol{\lambda}_t)$. On contrary, BigARTM outputs a non-probabilistic matrices Φ and Θ. To compare the perplexity we take the mean or the mode of the posterior distributions:

$$\phi_{wt}^{\mathrm{mean}} = \operatorname*{norm}_{w \in W} \lambda_{wt}, \qquad\qquad \theta_{td}^{\mathrm{mean}} = \operatorname*{norm}_{t \in T} \gamma_{td};$$

$$\phi_{wt}^{\mathrm{mode}} = \operatorname*{norm}_{w \in W}(\lambda_{wt} - 1), \qquad\qquad \theta_{td}^{\mathrm{mode}} = \operatorname*{norm}_{t \in T}(\gamma_{td} - 1).$$

The perplexity measure is defined as

$$\mathscr{P}(D, p) = \exp\left(-\frac{1}{n} \sum_{d \in D} \sum_{w \in d} n_{dw} \ln p(w \,|\, d)\right). \tag{6}$$

Comparison to Existing Software Packages. The *Vowpal Wabbit (VW)* is a library of online algorithms that cover a wide range of machine learning problems. For topic modeling VW has the VW.LDA algorithm, based on the Online Variational Bayes LDA [5]. VW.LDA is neither multi-core nor distributed, but an effective single-threaded implementation in C++ made it one of the fastest tools for topic modeling.

The *Gensim* library specifically targets the area of topic modeling and matrix factorization. It has two LDA implementations — LdaModel and LdaMulticore, both based on the same algorithm as VW.LDA (Online Variational Bayes LDA [5]). Gensim is entirely written in Python. Its high performance is achieved through the usage of NumPy library, built over low-level BLAS libraries (such

---

[4] https://github.com/JohnLangford/vowpal_wabbit/.
[5] http://dumps.wikimedia.org/enwiki/20141208/.
[6] https://github.com/piskvorky/gensim/tree/develop/gensim/scripts/.

Table 1. The comparison of BigARTM with VW.LDA and Gensim. *Train time* is the time for model training, *inference* is the time for calculation of θ_d of 100 000 held-out documents, *perplexity* is calculated according to (6) on held-out documents.

Library	Procs	Train Time	Inference Time	Perplexity Mode	Mean
BigARTM	1	35 min	72 s	4000	
LdaModel	1	369 min	395 s	4213	4161
VW.LDA	1	73 min	120 s	4061	4108
BigARTM	4	9 min	20 s	4061	
LdaMulticore	4	60 min	222 s	4055	4111
BigARTM	8	4.5 min	14 s	4304	
LdaMulticore	8	57 min	224 s	4379	4455

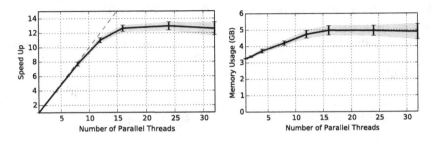

Fig. 2. Running BigARTM in parallel: speed up (left) and memory usage (right)

as Intel MKL, ATLAS, or OpenBLAS). In LdaModel all batches are processed sequentially, and the concurrency happens entirely within NumPy. In LdaMulticore the workflow is similar to BigARTM — several batches are processed concurrently, and there is a single aggregation thread that asynchronously merges the results.

Each run in our experiment performs one pass over the Wikipedia corpus and produces a model with $|T| = 100$ topics. The runtime is reported for an Intel-based CPU with 16 physical cores with hyper-threading. The collection was split into batches with 10000 documents each (`chunksize` in Gensim, `minibatch` in VW.LDA). The update rule in online algorithm used $\rho = (b + \tau_0)^{-0.5}$, where b is the number of batches processed so far, and τ_0 is an a constant offset parameter introduced in [5], in our experiment $\tau_0 = 64$. Updates were performed after each batch in non-parallel runs, and after P batches when running in P threads. LDA priors were fixed as $\alpha = 0.1$, $\beta = 0.1$, so that $\theta_d \sim \mathrm{Dir}(\alpha)$, $\phi_t \sim \mathrm{Dir}(\beta)$.

Table 1 compares the performance of VW.LDA, Gensim LdaModel, Gensim LdaMulticore, and BigARTM. Figure 2 shows BigARTM speedup and memory consumption depending on the number of CPU threads for Amazon AWS c3.8xlarge with 32 virtual cores, Gensim 0.10.3 under Python 2.7.

Table 2. Comparison of LDA and ARTM models. Quality measures: $\mathcal{P}_{10\,k}$, $\mathcal{P}_{100\,k}$ — hold-out perplexity on $10\,\mathrm{K}$ and $100\,\mathrm{K}$ documents sets, $\mathcal{S}_\Phi$, $\mathcal{S}_\Theta$ — sparsity of Φ and Θ matrices (in %), $\mathcal{K}_s$, $\mathcal{K}_p$, $\mathcal{K}_c$ — average topic kernel size, purity and contrast respectively.

Model	$\mathcal{P}_{10k}$	$\mathcal{P}_{100k}$	$\mathcal{S}_\Phi$	$\mathcal{S}_\Theta$	$\mathcal{K}_s$	$\mathcal{K}_p$	$\mathcal{K}_c$
LDA	3436	3801	0.0	0.0	873	0.533	0.507
ARTM	3577	3947	96.3	80.9	1079	0.785	0.731

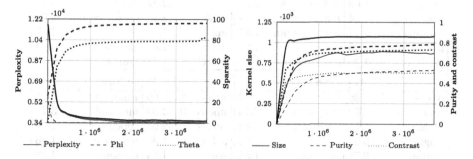

Fig. 3. Comparison of LDA (thin) and ARTM (bold) models. The number of processed documents is shown along the X axis.

Experiments with Combination of Regularizers. BigARTM has a built-in library of regularizers, which can be used in any combination. In the following experiment we combine three regularizers: sparsing of ϕ_t distributions, sparsing of θ_d distributions, and pairwise decorrelation of ϕ_t distributions. This combination helps to improve several quality measures without significant loss of perplexity, according to experiments on the offline implementation of ARTM [13]. The goal of our experiment is to show that this remains true for the online implementation in BigARTM. We use the following built-in quality measures: the hold-out perplexity, the sparsity of Φ and Θ matrices, and the characteristics of topic lexical kernels (size, purity, and contrast) averaged across all topics.

Table 2 compares the results of additive combination of regularizers (ARTM) and the usual LDA model. Figure 3 presents quality measures as functions of the number of processed documents. The left chart shows perplexity and sparsity of Φ, Θ matrices, and the right chart shows average lexical kernel measures.

Experiments on Multi-language Wikipedia. To show how BigARTM works with multimodal datasets we prepared a text corpus containing all English and Russian Wikipedia articles with mutual interlanguage links. We represent each linked pair of articles as a single multi-language document with two modalities, one modality for each language. That is how our multi-language collection acts as a multimodal document collection.

The dump of Russian articles[7] had been processed following the same technique as we previously used in experiments on English Wikipedia. Russian words

---

[7] http://dumps.wikimedia.org/ruwiki/20141203/.

Table 3. Top 10 words with $p(w \mid t)$ probabilities (in %) from two-language topic model, based on Russian and English Wikipedia articles with mutual interlanguage links.

Topic 68				Topic 79			
research	4.56	институт	6.03	goals	4.48	матч	6.02
technology	3.14	университет	3.35	league	3.99	игрок	5.56
engineering	2.63	программа	3.17	club	3.76	сборная	4.51
institute	2.37	учебный	2.75	season	3.49	фк	3.25
science	1.97	технический	2.70	scored	2.72	против	3.20
program	1.60	технология	2.30	cup	2.57	клуб	3.14
education	1.44	научный	1.76	goal	2.48	футболист	2.67
campus	1.43	исследование	1.67	apps	1.74	гол	2.65
management	1.38	наука	1.64	debut	1.69	забивать	2.53
programs	1.36	образование	1.47	match	1.67	команда	2.14
Topic 88				**Topic 251**			
opera	7.36	опера	7.82	windows	8.00	windows	6.05
conductor	1.69	оперный	3.13	microsoft	4.03	microsoft	3.76
orchestra	1.14	дирижер	2.82	server	2.93	версия	1.86
wagner	0.97	певец	1.65	software	1.38	приложение	1.86
soprano	0.78	певица	1.51	user	1.03	сервер	1.63
performance	0.78	театр	1.14	security	0.92	server	1.54
mozart	0.74	партия	1.05	mitchell	0.82	программный	1.08
sang	0.70	сопрано	0.97	oracle	0.82	пользователь	1.04
singing	0.69	вагнер	0.90	enterprise	0.78	обеспечение	1.02
operas	0.68	оркестр	0.82	users	0.78	система	0.96

were lemmatized with Yandex MyStem 3.0[8]. To further reduce the dictionary we only keep words that appear in no less than 20 documents, but no more than in 10 % of documents in the collection. The resulting collection contains 216175 pairs of Russian–English articles, with combined dictionary of 196749 words (43 % Russian, 57 % English words).

We build multi-language model with 400 topics. They cover a wide range of themes such as science, architecture, history, culture, technologies, army, different countries. All 400 topics were reviewed by an independent assessor, and he successfully interpreted all except four topics.

Table 3 shows top 10 words for four randomly selected topics. Top words in these topics are clearly consistent between Russian and English languages. The Russian part of last topic contains some English words such as "Windows" or "Server" because it is common to use them in Russian texts without translation.

6   Conclusions

BigARTM in an open source project for parallel online topic modeling of large text collections. It provides a high flexibility for various applications due to

---

[8] https://tech.yandex.ru/mystem/.

multimodality and additive combinations of regularizers. BigARTM architecture has a rich potential. Current components can be reused in a distributed solution that runs on cluster. Further improvement of single-node can be achieved by offloading batch processing into GPU.

Acknowledgements. The work was supported by the Russian Foundation for Basic Research grants 14-07-00847, 14-07-00908, 14-07-31176 and by Skolkovo Institute of Science and Technology (project 081-R).

References

1. Blei, D.M.: Probabilistic topic models. Commun. ACM **55**(4), 77–84 (2012)
2. Blei, D.M., Jordan, M.I.: Modeling annotated data. In: Proceedings of the 26th Annual International ACM SIGIR Conference on Research and Development in Informaion Retrieval, pp. 127–134. ACM, New York (2003)
3. Blei, D.M., Ng, A.Y., Jordan, M.I.: Latent dirichlet allocation. J. Mach. Learn. Res. **3**, 993–1022 (2003)
4. Daud, A., Li, J., Zhou, L., Muhammad, F.: Knowledge discovery through directed probabilistic topic models: a survey. Front. Comput. Sci. China **4**(2), 280–301 (2010)
5. Hoffman, M.D., Blei, D.M., Bach, F.R.: Online learning for latent dirichlet allocation. In: NIPS, pp. 856–864. Curran Associates Inc. (2010)
6. Hofmann, T.: Probabilistic latent semantic indexing. In: Proceedings of the 22nd Annual International ACM SIGIR Conference on Research and Development in Information Retrieval, pp. 50–57. ACM, New York (1999)
7. Liu, Z., Zhang, Y., Chang, E.Y., Sun, M.: PLDA+: parallel latent Dirichlet allocation with data placement and pipeline processing. ACM Trans. Intell. Syst. Technol. **2**(3), 26:1–26:18 (2011)
8. Newman, D., Asuncion, A., Smyth, P., Welling, M.: Distributed algorithms for topic models. J. Mach. Learn. Res. **10**, 1801–1828 (2009)
9. Rubin, T.N., Chambers, A., Smyth, P., Steyvers, M.: Statistical topic models for multi-label document classification. Mach. Learn. **88**(1–2), 157–208 (2012)
10. Smola, A., Narayanamurthy, S.: An architecture for parallel topic models. Proc. VLDB Endow. **3**(1–2), 703–710 (2010)
11. Vorontsov, K.V.: Additive regularization for topic models of text collections. Dokl. Math. **89**(3), 301–304 (2014)
12. Vorontsov, K.V., Potapenko, A.A.: Additive regularization of topic models. Mach. Learn. **101**(1–3), 303–323 (2015)
13. Vorontsov, K., Potapenko, A.: Tutorial on probabilistic topic modeling: additive regularization for stochastic matrix factorization. In: Ignatov, D.I., Khachay, M.Y., Panchenko, A., Konstantinova, N., Yavorsky, R.E. (eds.) AIST 2014. CCIS, vol. 436, pp. 29–46. Springer, Heidelberg (2014)
14. Řehůřek, R., Sojka, P.: Software framework for topic modelling with large corpora. In: Proceedings of the LREC 2010 Workshop on New Challenges for NLP Frameworks, ELRA, Valletta, pp. 45–50, May 2010
15. Wang, Y., Bai, H., Stanton, M., Chen, W.-Y., Chang, E.Y.: PLDA: parallel latent dirichlet allocation for large-scale applications. In: Goldberg, A.V., Zhou, Y. (eds.) AAIM 2009. LNCS, vol. 5564, pp. 301–314. Springer, Heidelberg (2009)

Industry Talk

ATM Service Cost Optimization Using Predictive Encashment Strategy

Vladislav Grozin[1]([✉]), Alexey Natekin[2,3,4], and Alois Knoll[4]

[1] ITMO University, Saint-Petersburg, Russia
grozin@my.ifmo.ru
[2] Data Mining Labs, Saint-Petersburg, Russia
natekin@dmlabs.org
[3] Deloitte Analytics Institute, Moscow, Russia
[4] Technical University Munich, Garching Bei München, Germany
knoll@in.tum.de

Abstract. ATM cash flow management is a challenging task which involves both machine learning predictions and encashment planning. Banks employ these systems to optimize their costs and improve the overall device availability via reducing the number of device failures. Although cash flow prediction is a common task, complete design of the cost optimization system is a complex design problem. In this article we present our complete encashment strategy methodology. We evaluate the proposed system design on real world data from one of the Russian banks. We show that one can effectively achieve 18 % cost reduction by employing such strategy.

Keywords: ATM · Cash flow management · Cost optimization · Machine learning · Regression

1 Introduction

Cost saving has always been the cornerstone of business operations. There is no other industry but banking, where data-driven solutions to cost optimization have been so widely adopted. Some applications like risk management and fraud monitoring systems based on machine learning models, have been successfully used for several decades, whereas solutions like proactive customer retention modeling are only finding their way to bank application portfolios.

Nowadays banks pay a lot of attention to efficiency of cash management in their ATM networks [1]. ATM, or automated teller machines, are electronic telecommunication devices that enable the customers to perform financial transactions without the need for a human cashier, clerk or bank teller. It was estimated that improved inventory policies and cash transportation decisions for an ATM network can result in up to 28 % cost reductions [2].

In order to employ ATM cash management, a complete system with both predictive analytics and business logic has to be designed. In this article we present

© Springer International Publishing Switzerland 2015
M.Y. Khachay et al. (Eds.): AIST 2015, CCIS 542, pp. 385–396, 2015.
DOI: 10.1007/978-3-319-26123-2_37

our cost optimization methodology of an ATM network. This includes proactive maintenance with proper predictions of the amount of money to replenish. We take into account both the fact that encashment costs money, and that additional costs are imposed due to excessive unused money in the devices. We evaluate our proposed system on real world data from one of the Russian banks and analyze the results with respect to the work of actual human experts.

The paper proceeds as follows. Section 2 provides a literature review with the related work. Section 3 discusses the problem statement and the corresponding mathematical formulation. Section 4 describes our complete encashment strategy methodology. Section 5 outlines the architecture and technical details of the system. Section 6 reports the obtained results of our system on the real bank data. Section 7 concludes with a discussion about the potential technical and analytical improvements to the system design.

2   Related Work

Cash management systems of ATM networks consist of two primary components: cash flow prediction part and a part with the set of rules for cash replenishment. The former part is typically based on machine learning approach, while the latter focuses on optimization and encashment plan design.

Most of the published articles focus primarily on the prediction part of the problem. A popular way to approach this problem, followed by some authors [3,4], is to apply time series-based methods, which provide decent accuracy on their tasks. Unfortunately these methods do not allow one to use additional non-temporal factors like ATM placement (e.g. shopping mall or bank).

The second most popular way to predict cash flow is to build, or learn a functional relationship between the future cash flow and a set of independent factors which are likely to affect cash flow patterns. A common form of this relationship is a neural network [5–10] that also typically exploits time-series data of cash flows. With this approach one commonly uses factors like ATM operations in previous days, device placement, pay days, weekends etc. These methods are capable to capture complex relationships and take into account multiple factors from very different sources.

Despite the fact that it is the replenishment part that actually provides the capability to reduce costs, there is significantly less material on this part of the overall cash management problem. Cash prediction alone does not solve cost minimization, thus additional calculations have to be made in order to create an optimal encashment plan. One approach to encashment plan design is based on fuzzy expert systems [11]. These systems imitate human reasoning and the development of such systems involves human experts. However a common problem of such method is the limitation of experts' knowledge expressiveness in the form of fuzzy rule sets. Moreover, human input can be inconsistent and lead to unstable results.

Another opportunity is to apply stochastic programming for cash management [12]. Author deals with discretized money chunks in order to exploit the stochastic approach for short and mid-term optimization. Limitation to this approach is that it is hard to reason on the properties of the solution to be globally optimal or suboptimal. Integer programming can also can be applied to solve ATM-optimization problem [3]. In this case authors combines travelling salesman problem solution via integer programming and time-series based approach for predictions.

In our study we follow the machine learning approach of direct modeling an arbitrary functional relationship to get the future cash flow predictions. However for encashment planning we developed a greedy strategy based on cost balancing, which is different from the earlier discussed ones.

3   Business Problem Identification

We consider two types of ATM devices: cash-ins and dispensers. Cash-in devices support only debit operations and collect cash. Dispensers, on the contrary, support only credit operations and provide cash to the customers. These devices are placed across different locations of the city and suburbs. It is common that ATM devices come in couples, sometimes with more than one device of each type in one location.

Data on the total amount of cash in each device is collected daily. Typical cash flow patterns for each device type, drawn from two real devices, are shown on Fig. 1. Dotted lines correspond to encashment events, where money as either collected from the cash-in device, or was replenished in the dispenser.

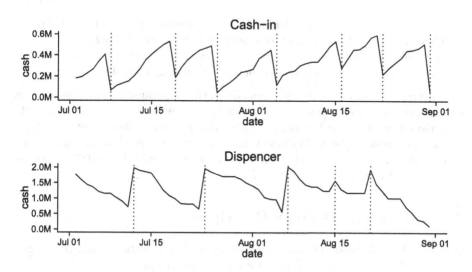

Fig. 1. Cash flow dynamics examples for two device types.

Our key objective is to provide ATM encashment plans that will minimize annual costs incurred by management of the ATM network. Given that ATM network consists of K devices, we can write these costs as follows:

$$AnnualCosts(D) = \sum_{i=1}^{K} \sum_{t=1}^{365} Cost_i(t)$$

K number of devices in network
$Cost_i(t)$ i-th device's expense at moment t

It is also important to keep the number of device failures minimal. By device failures we mean situations when the device is no longer operable. Dispenser can run out of money whereas cash-in can theoretically exceed its maximal volume. Considering the provided data, there were no registered cash-in failures, but dispensers indeed sometimes ran out of money. Unfortunately it is impossible to estimate the money equivalent of device failures, hence we will stick to the original objective of pure annual overall cost reduction.

The business problem can be decomposed further if we describe the $Cost_i(t)$ term in more detail. We have previously noted that incurred ATM costs consist of inventory costs of inefficiently used cash in the machines, and direct encashment costs. The former term equals to the amount of money that bank could have spent if kept it in turnover rather in devices while the latter is just the direct expense. The formula for costs can therefore be written as follows:

$$Cost_i(t) = r \cdot CashAmount_i(t) + EncashmentCost_i(t) \cdot IsEncashed_i(t),$$

r bank lending rate, converted into percents per day
$CashAmount_i(t)$ amount of money in i-th device at day t
$EncashmentCost_i(t)$ encashment price of i-th device at day t
$IsEncashment_i(t)$ 1 if i-th is encashed at moment t, 0 otherwise

In general, $EncashmentCost_i$ depends on time t because it is possible to save costs on efficient route planning. And routes are generated with different sets of devices daily thus for the same device i this cost can differ in between days. In our case most of the ATM devices were placed in the bank offices and didn't involve logistical planning. This allows us to proceed with fixed daily costs for each device $EncashmentCost_i(t) = FixedCost_i$, $i \in \overline{1, K}$.

4   Encashment Strategy Description

After we have defined the cost components that are to be optimized we can formulate the encashment strategy for cost optimization.

Our strategy is based on one general principle: balance between inventory and fixed costs in between two encashments. If for a particular device $i \in \overline{1, K}$ our cumulative inventory costs that start from previous encashment exceed $FixedCost_i$, we can cut off this excess by ordering a new encashment. This

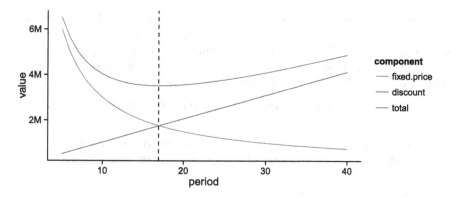

Fig. 2. Cost components (Color figure online).

idea of balancing costs components is illustrated on Fig. 2, calculated for an artificially simulated cash-in device.

Following this principle we seek a specific moment in time $t > t_0$ that starts from previous encashment time with $t_0 = 0$. This moment, defined as $\tilde{t}$ should be closest to the balance point when absolute difference between cumulative inventory cost and encashment cost is minimal:

$$InventoryCost_i(t) = r \cdot \sum_{j=1}^{t} CashAmount_i(j)$$

$$\tilde{t}_i = \operatorname*{argmin}_{t} |InventoryCost_i(t) - FixedCost_i|$$

After we obtain $\tilde{t}_i$ we could set it as the desired encashment time and forward it to the bank operations office for execution. For strategy design we will assume that $CashAmount_i(t)$ are known $\forall t \; \forall i$. Despite being unrealistic in practice we deal with their predictions and estimations, thus letting us to substitute these predictions into the derived encashment rules.

4.1   Technical Requirements

In the ideal world one could execute cash replenishment instantly. Unfortunately cash management suffers from a *lag* between the encashment decision of a particular device and the actual encashment execution. In our scenario with $lag = 2$ days, such decision of replenishment necessity of device i at time t could be executed no sooner than $\tilde{t}_i = t + 2$.

Despite the *lag*, in the ideal setting one can still make encashment decisions right away. However there is no encashment capability on weekends and holidays which can render optimal $\tilde{t}_i$ evaluation impossible for device i. To deal with this issue we generate a set of candidates $[t]_w = \overline{t, t + w}$ for further selection based on both cost balance and availability.

First a set of encashment candidates $[t]$ starting with $t + 2$ is defined: $[t] = \overline{t+2, t+7}$. Then $\forall t \in [t]$ we evaluate encashment capability of device i based on holiday calendar C_i and match dates. Thus we get $[t]^i_{real} = [t] \bigcap C_i$.

At time t we only need to make the encashment decision about the very first element of $[t]^i_{real}$. We can then formulate encashment rule as follows:

$$\tilde{t}_i = \underset{t \in [t]^i_{real}}{\operatorname{argmin}} |InventoryCost_i(t) - FixedCost_i|$$

IF $\tilde{t}_i = [t]^i_{1real}$ then order encashment for device i at time $\tilde{t}_i$,
ELSE   do nothing with device i and wait until tomorrow.

4.2   Device Thresholds

Physical constraints on ATM device storage capacity have to be mitigated as an additional constraint on encashment planning. Both cash-in and dispenser devices have an upper limit on the amount of cash they store. Moreover, dispenser can run out of cash. Situations when these capacity requirements are not met are considered as device failures and should be evaded.

Let us modify our strategy to the case of cash-in first. Without loss of generality put 0 as the moment of last encashment, t for current moment and lag as requirement input. Figure 3 illustrates the encashment process with complete money withdrawal at time $t + lag$. The line chart represents $CashAmount_i(t)$ while the filled red area corresponds to $InventoryCost_i(t)$ w.r.t. given r.

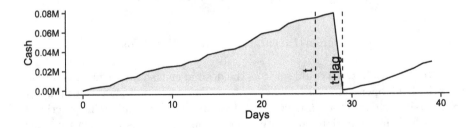

Fig. 3. Demonstration of cash-in withdrawal procedure.

Although there were no registered breaches of cash-in storage limit in our data we will still define the threshold rule for them. The only time when threshold can impact our decision is when we chose to omit encashment execution but should instead have provided cash withdrawal. This applies to either very first candidate $[t]^i_{[1]real}$ when we are already obliged to provide the encashment, or to the second one $[t]^i_{[2]real}$ when we see that it is the last moment before device failure. Thus given the $UpperThreshold$ our encashment rule modifies to:

IF $\tilde{t}_i = [t]^i_{1real}$ OR
$CashAmount_i([t]^i_{[1,2]real}) \geq$ THEN order encashment for device i at time $\tilde{t}_i$,
$UpperThreshold$
ELSE   do nothing with device i and wait until tomorrow.

The procedure of choosing $\tilde{t}_i$ for dispenser devices has only one difference – a reversed threshold rule. We have to define the *LowerThreshold* which serves as the indicator for necessary encashment:

IF $\tilde{t}_i = [t]_{1real}^i$ OR
$CashAmount_i([t]_{[1,2]real}^i) \leq$ THEN order encashment for device i at time $\tilde{t}_i$,
LowerThreshold
ELSE   do nothing with device i and wait until tomorrow.

Figure 4 demonstrates the dispenser replenishment rule with threshold in consideration. It is worth noting that the slope of this chart is different from that of a cash-in because the sign of $CashAmount_i(t)$ is never positive. *LowerThreshold* serves the purpose of device failure risk mitigation.

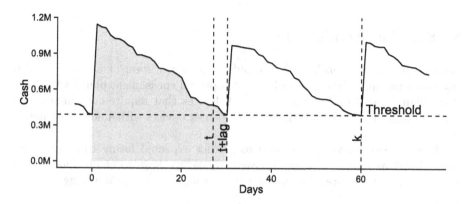

Fig. 4. Demonstration of dispenser replenishment procedure.

4.3   Encashment Amount Determination

The last component of our strategy is the amount of cash we have to leave in the device after the encashment execution. Cash-ins have a trivial answer to this problem given that we can always withdraw all cash from the device.

For the case of dispensers we can do the following. Consider we have identified that we want to execute encashment at time $\tilde{t}_i$. First, we have to predict one whole next month of $CashFlow_i(t) = \Delta CashAmount_i(t) \; \forall t \in \overline{\tilde{t}_i, \tilde{t}_i + 30}$. This is done in order to define $InventoryCostDispenser_i(t)$ as:

$$CashAmountDispenser_i(t) = LowerThreshold + \sum_{j=1}^{t} CashFlow_i(t)$$

$$InventoryCostDispenser_i(t) = r \cdot \sum_{j=1}^{t} CashAmountDispenser_i(j)$$

Now we can simulate the expected ATM behavior for the whole next month, taking into consideration a sequence of $InventoryCostDispenser_i(t)$, $t \in \overline{\tilde{t}_i, \tilde{t}_i + 30}$. Due to the fact that the second condition of crossing the threshold is fulfilled by design, We can then find the point of expected $t_{expected}$ next encashment as simply the most optimal one in terms of cost balance:

$$t_{expected} = \underset{t \in \overline{\tilde{t}_i, \tilde{t}_i + 30}}{\operatorname{argmin}} |InventoryCostDispenser_i(t) - FixedCost_i|$$

Finally, we can calculate this $AmountDispenser_i$ by taking into account that we will already have some cash remaining from the :

$$AmountDispenser_i = CashAmountDispenser_i(\tilde{t}_i + t_{expected}) -$$
$$- CashAmountDispenser_i(\tilde{t}_i)$$

5   Solution Architecture

Solution of the cost optimization problem takes the form of a system which takes as input historical data and outputs optimal encashment plan for the next date. Plan comes in the form of a list of devices that require encashment and supplemented with dates and in case of dispenser devices with amounts of money to deposit.

To make system viable we need to predict expected future cash flows and make decisions based on these forecasts. Thus, we have to create model that gives insight into future. Overall the system comprises of the following steps:

1. Data processing and feature extraction.
2. Predictive model evaluation.
3. Encashment plan provision.

5.1   Data Processing and Feature Extraction

In our study we were provided with historical about 115 devices from 2009–2013 years with the following information about each entry:

– ATM device ID.
– Date.
– Daily cash flow.
– Daily encashment (zero, if it didn't take place).

Additional data about each device included it's type (cash-in or dispenser), address and fixed encashment price. Due to the non-disclosure agreement we are unable to provide the raw data except for demo samples shown in figures.

There is considerable freedom in the choice of features to use in our models. Those can be separated into three different groups: point-wise dealing only with current date info; point-wise which aggregate data across some time window;

serialized representation of the time series. In terms of data domains we additionally extracted information about dates like in [13], holidays and weather. And besides initial raw time-series we also looked upon series of the closely correlated devices. For these series we extracted various aggregated summary statistics like mean, sd and quantile values. A complete map of all extracted features is presented in Fig. 5.

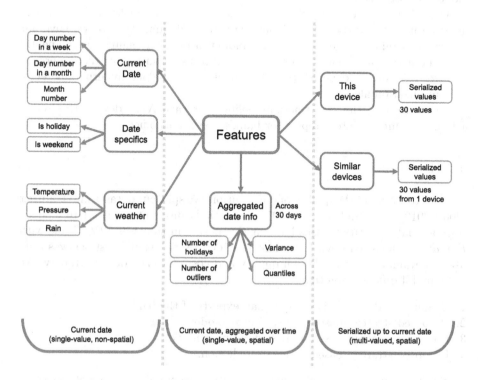

Fig. 5. Map of extracted features.

5.2 Predictive Model Evaluation

In order to predict device cash flows we considered two families of models: linear ones with linear regression and lasso, ridge and elastic net regularizations; and nonlinear machine learning models like gradient boosted decision trees, random forests and svms.

Each model was built separately to each device types and had it's own built-in parameter selection procedures. We employed 10-fold cross-validation for hyperparameter selection in each of these models. Finally we compared the resulting models in order to get the best linear and best nonlinear model representatives.

In terms of model goodness of fit criteria we chose the R^2 metric. Model with higher R^2 value doesn't necessarily provide better cost decrease, however one might reason that with the increase in mode accuracy one could achieve better cost reduction result.

5.3   Encashment Plan Provision

We follow our proposed strategy of keeping balance between $InventoryCost_i$ and $FixedCost_i$ for each device. Besides dispensers require exact amounts of cash to replenish if it reaches some critical threshold. The strategy is based on the predictions obtained from the considered linear and nonlinear machine learning models in discussion.

Choosing dispenser threshold is trade-off between safety and additional incurred inventory losses. The higher the threshold value, the less we have our failure probability at the cost of some amount of money remaining permanently idle. In our system we chosen threshold equal to two standard prediction errors, however one might consider optimizing it as a stand-alone hyperparameter of the optimization algorithm.

After we have the set of actions defined for each ATM device, this set of actions is compiled into a report for bank's operations office.

6   Results

In order to estimate the quality of our solution We split our dataset into train set (2009–2012 years) and test set (2013 year). Daily discount r was set to 0.00035 since annual loan rate for this bank at this period in time was 12 %. Table with $FixedCost_i$ for each device was also provided by the bank. Test set was only used to evaluate both prediction quality and cost efficiency of our strategy. We compared 4 different solutions:

1. Original predictions given by human experts of the bank.
2. Proposed strategy based on best linear model.
3. Proposed strategy based on best non-linear model.
4. Proposed strategy based on ideal prediction.

Solutions were compared based on 4 criteria resembling system objectives:

1. R^2 for cash-in devices.
2. R^2 for dispenser devices.
3. Number of device failures.
4. Cost reduction.

To assure stable and sound results, bootstrap simulations were applied to objective criteria estimates. Each model was fit on a resampled with replacement train dataset. Predictions were evaluated on the complete test set. The boostrap was repeated for 20 times and the resulting average and standard deviations of resampled statistics were collected. The resulting criteria for complete systems are presented in Table 1.

We can see that our strategy grants a reasonable 18 % cost reduction compared to the schedules prepared by human experts. Moreover if we consider ideally optimal predictions we see that the maximum cost reduction on this

Table 1. System comparison based on objective criteria.

Model name	R^2 cashin	R^2 dispenser	Failures	Cost decrease
Human experts	-	-	172	0.00 %
Best linear (LM)	0.59 ± 0.001	0.64 ± 0.002	104 ± 15.4	17.94 ± 0.11 %
Best nonlinear (GBM)	0.59 ± 0.028	0.64 ± 0.005	98 ± 9.2	17.89 ± 0.14 %
Ideal	1.00	1.00	0	27.12 %

data is around 27 %. This means that not only we achieved considerable results, but also the physical limit of productivity in this application is not far away.

It is worth noting that the resulting systems of both linear and nonlinear models behave nearly the same. Although nonlinear models provide a slightly lesser amount of failures, neither their costs decrease nor R^2 are significantly different from those of the linear model. It may be caused by the fact that the most important features of the resulting best nonlinear model (GBM) were very ARIMA-like (cash flow from previous 7,14,21 days) thus considering that the strong nonlinear model converged to a linear one.

7   Discussion

As noted in related works, there are many of articles on ATM cash flow prediction, but very few about actual optimization of ATM network costs. Authors usually deal with periodic data that is easy to predict. We have proposed a strategy that can decrease total expenses on ATM network management by 18 % and can be adopted to bank operations and other industry-related tasks dealing with supply chain management.

Our strategy however has considerable space for improvement. For example we can consider adaptive threshold selection for both types of devices and add additional hyperparameters on encashment decision support. It is the strategy that provides the cost reduction and potentially this would be the most fruitfull direction for future research of this problem.

Predicting the amount of cash to replenish is a more challenging task than simple cash flow predictions. Many different approaches can be taken to solve this problem. For example one could apply quantile regression for predictive risk estimation. It is even remove the threshold necessity if the system was completely rewritten in terms of predictive risks.

References

1. Westland, C.J.: Preference ordering cash, near cash and electronic cash. J. Organ. Comput. Electron. Commer. **12**(3), 223–242 (2012)
2. Wagner, M.: The optimal cash deployment strategy – modeling a network of ATMs. Master thesis, Swedish School of Economics and Business Administration, Finland (2007)

3. Gubar, E., Zubareva, M., Merzljakova, J.: Cash flow optimization in ATM network model. Contrib. Game Theory Manag. **4**, 213–222 (2011)
4. Wagner, M.: Forecasting daily demand in cash supply chains. Am. J. Econ. Bus. Adm. **2**(4), 377–383 (2010)
5. PremChand Kumar, P., Walia, E.: Cash forecasting: an application of artificial neural networks in finance. Int. J. Comput. Sci. Appl. **3**(1), 61–77 (2006)
6. Simutis, R., Dilijonas, D., Bastina, L.: Cash demand forecasting for ATM using neural networks and support vector regression algorithms. In: 20th EURO Mini Conference "Continuous Optimization and Knowledge-Based Technologies" (EurOPT- 2008), pp. 416–421 (2008)
7. Simutis, R., Dilijonas, D., Bastina, L., Friman, J.: A flexible neural network for ATM cash demand forecasting. In: 6th WSEAS International Conference on Computational Intelligence, Man-Machine Systems and Cybernetics, pp. 162–167 (2007)
8. Simutis, R., Dilijonas, D., Bastina, L., Friman, J., Drobinov, P: Optimization of cash management for ATM network. Inf. Technol. Control **36**(1A), 117–212 (2007)
9. Kamini, V., Ravi, V., Prinzie, A., Van den Poel, D.: Cash demand forecasting in ATMs by clustering and neural networks. Eur. J. Oper. Res. **232**(2), 383–392 (2013)
10. Zapranis, A., Alexandridis, A.: Forecasting cash money withdrawals using wavelet analysis and wavelet neural networks. Int. J. Financ. Econ. Econometrics (2009) (to appear)
11. Darwish, S.M.: A methodology to improve cash demand forecasting for ATM network. Int. J. Comput. Electr. Eng. **5**(4), 405–409 (2013)
12. Castro, J.: A stochastic programming approach to cash management in banking. Eur. J. Oper. Res. **192**(3), 963–974 (2009)
13. Esteves, P.S., Rodriguesm, P.M.M.: Calendar effects in daily ATM withdrawals. Econ. Bull. **30**(4), 2587–2597 (2010)

Industry Papers

Comparison of Deep Learning Libraries on the Problem of Handwritten Digit Classification

Dmitry Kruchinin[1], Evgeny Dolotov[1], Kirill Kornyakov[1,2],
Valentina Kustikova[1(✉)], and Pavel Druzhkov[1]

[1] Computational Mathematics and Cybernetics Department, Lobachevsky State
University of Nizhny Novgorod, Nizhny Novgorod, Russian Federation
{kruch.dmitriy,dolotov.evgeniy}@gmail.com
kustikova@vmk.unn.ru, pavel.druzhkov@itseez.com
[2] Itseez, Nizhny Novgorod, Russian Federation
kirill.kornyakov@itseez.com

Abstract. This paper presents a comparative analysis of several popular and freely available deep learning frameworks. We compare functionality and usability of the frameworks trying to solve popular computer vision problems like hand-written digit recognition. Four libraries have been chosen for the detailed study: Caffe, Pylearn2, Torch, and Theano. We give a brief description of these libraries, consider key features and capabilities, and provide case studies. We also investigate the performance of the libraries. This study allows making a decision which deep learning framework suites us best and will be used for our future research.

Keywords: Neural network · Deep learning · Classification · Caffe · Pylearn2 · Torch · Theano

1 Introduction

Nowadays deep neural networks are producing state-of-the-art results in a great number of complex problems, such as forecast, pattern recognition, data compression, etc. Thereby, many programming tools allowing the implementation of various deep learning methods have been developed [1]. The existing software possesses different functionality and requires different level of knowledge and skills for the detailed study. Therefore, a proper tool selection is an important factor to achieve the necessary result at the lowest time cost and effort.

The goal of the present work is a comparative analysis of several deep learning tools for future applications in the solving of the important problems of computer vision such as face, pedestrian and vehicle detection. As long as the detection problem can be reduced to the classification using "sliding" window approach, the analysis has been carried out for the handwritten digit classification problem. The main reader of this paper is a researcher who wants to make a quick decision about deep learning framework without any additional experiments.

M.Y. Khachay et al. (Eds.): AIST 2015, CCIS 542, pp. 399–411, 2015.
DOI: 10.1007/978-3-319-26123-2_38

A brief overview of the programming tools for the development and training of the neural models is represented. The main scope is concerned about the four libraries: Caffe [2], Pylearn2 [3], Torch [4], and Theano [5]. The basic capabilities of these libraries are considered, the examples of their applications are presented. An attempt to evaluate and compare their usability and performance has been undertaken.

2   The Problem of Deep Learning Libraries Comparison

The purpose of this work is to carry out the comparative analysis of the deep learning programming tools applied to the handwritten digit classification problem. This goal implies solving a number of problems:

1. An overview and comparison of the capabilities of the deep learning tools.
2. The selection of several tools for more detailed investigation.
3. The construction, training and testing of some neural networks.
4. A comparative analysis of the software performance in application to the handwritten digit classification problem.
5. The comparison of the software tools from the viewpoint of the offered functionality, usability, and performance in training of the typical neural nets.

3   Deep Learning Libraries, Capabilities and Features

3.1   Deep Learning Frameworks

A detailed comparison of the functionality of the well-known deep learning tools is presented in [1]. Below the tools, which might be the most interesting for the researchers are listed (Table 1). The first six libraries offer the widest functionality. They support fully connected neural networks (FC NNs) [6], convolutional neural networks (CNNs) [7], autoencoders (AEs) [8], and restricted Boltzmann machines (RBMs) [9]. The functionality of the rest is more limited, however, in some cases these ones overcome the software of the first group from the viewpoint of the performance, for example. As a consequence, one cannot ignore them.

Based on [1] as well as on the additional information and on the recommendations of the experts, let us select the following four libraries for further investigation: Theano [5], Pylearn2 [3] as one of the most functionally complete libraries, Torch [4], and Caffe [2] as a used widely by the community.

3.2   Caffe Library

The development of Caffe library [2] has been being carried out since September, 2013. The library is currently supported by Berkeley Vision and Learning Center (BVLC) as well as by GitHub Developers Community [18]. Caffe is implemented in C++ programming language, the Python and MATLAB wrappers are available. Linux and OS X operating systems are supported officially, an unofficial version for Windows is available too [19]. Caffe uses BLAS library for the vector

Table 1. The capabilities of some software tools for deep learning [1]

Name	Language	OS	FC NN	CNN	AE	RBM
DeepLearnToolbox [10]	Matlab	Win, Lin	+	+	+	+
Theano [5]	Python	Win, Lin, Mac OS	+	+	+	+
Pylearn2 [3]	Python	Lin, Vagrant	+	+	+	+
Deepnet [11]	Python	Lin	+	+	+	+
Deepmat [12]	Matlab	?	+	+	+	+
Torch [4]	Lua, C	Lin, Mac OS X, iOS, Android	+	+	+	+
Darch [13]	R	Win, Lin	+	–	+	+
Caffe [2]	C++, Python, Matlab	Lin, OS X, Win	+	+	–	–
nnForge [14]	C++	Lin	+	+	–	–
CXXNET [15]	C++	Lin	+	+	–	–
Cuda-convnet [16]	C++	Win, Lin	+	+	–	–
Cuda CNN [17]	Matlab	Win, Lin	+	+	–	–

and matrix computations. For the speed up of the computations, the support of the basic capabilities of CUDA technology and of cuDNN library for deep learning is available [20].

Caffe offers the capabilities of configuring, training, and testing of the FC NNs and CNNs. The input data and the transformations are described by the concept of *layer*. Subject to the storage format, the following types of input data layers may be used: DATA – defines a data layer in leveldb and lmdb formats; HDF5_DATA – a data layer in hdf5 format; IMAGE_DATA – a simple format, which implies a file to contain an image list with class labels, etc.

The transformations are also defined as the layers: the fully connected (INNER_PRODUCT) and the convolutional layers (CONVOLUTION); pooling layers (POOLING), and Local Response Normalization layers (LRN). When forming the transformations, various activation functions may be used: the sigmoidal one (SIGMOID), hyperbolic tangent (TANH), the absolute value (ABSVAL), etc. The last layer of a neural network should contain the loss function. Among the supported ones are: Euclidean loss, logistic loss, softmax loss, etc.

Along with various neural network configurations Caffe also provides different optimization methods for their training, such as Stochastic Gradient Descent (SGD) [21], Gradient descent with adaptive learning rate (AdaGrad) [22] and Nesterov's Accelerated Gradient Descent (NAG) [23].

The neural network structure, the initial data, and the training parameters are defined by the configuration files in the 'prototxt' format. Let us consider the stages of construction of such files using an example of a logistic regression network (Fig. 1). MNIST dataset [24] stored in lmdb format is used as a training set. The first layer named 'mnist' is of DATA type and extracts the network data (the images and the class labels, which these ones belong to). The images are transferred to the fully connected layer 'ip' (INNER_PRODUCT) for further processing. Afterwards, the output data from 'ip' layer and the class number are transferred to the 'loss' layer (SOFTMAX_LOSS) to compute the error.

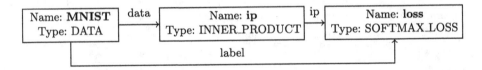

Fig. 1. The structure of the neural network

Actually, each layer is described by name, type, and by names of the input ('bottom') and output ('top') layers, as well as by a set of parameters. An example of a description of a fully connected layer is presented below.

```
layers {
  name:"ip"
  type: INNER_PRODUCT
  bottom: "data"
  top: "ip"
  inner_product_param {
    num_output: 10
  }
}
```

Then, it is necessary to define the parameters of the training procedure such as the path to the network configuration file, the periodicity of testing during training, the SGD parameters, the maximum number of iterations, the solver architecture, which the computations will be carried out on (CPU or GPU), and the path to save the trained network. Each row of the configuration file contains a corresponding 'name–value' pair.

The training is carried out using the main utility of the library. At that, certain set of options, in particular, the name of the file containing the description of the training procedure parameters is transferred.

After training, the obtained model can be used for the image classification, for example, using Python wrappers:

1. Import Caffe library. Set up the test mode and specify the solver architecture to execute the computations (CPU or GPU).

   ```
   import caffe
   caffe.set_phase_test()
   caffe.set_mode_cpu()
   ```

2. Load the trained model and the image to classify.

```
net = caffe.Classifier(MODEL_FILE, PRETRAINED, IMAGE_MEAN,
        channel_swap=(2,1,0), raw_scale=255, image_dims=(28, 28))
input_image = caffe.io.load_image(IMAGE_FILE)
```

3. Get the response of the network for the selected image.

```
prediction = net.predict([input_image])
print 'predicted class:', prediction[0].argmax()
```

Thus, by means of simple procedures one can obtain the first experimental results on the deep neural network models.

3.3 Pylearn2 Library

Pylearn2 [3] is a library having been being developed in LISA laboratory at Montreal University since February, 2011. There are about 100 developers at GitHub [25]. The library is implemented in Python, and actually is based on Theano. Currently Linux OS is supported; also running under any operating system using a virtual machine is possible, as the developers offer a configured virtual environment shell on Vagrant platform. In order to speed up the computations, Cuda-convnet library is used, which is written in C++/CUDA [16].

The capabilities to configure the FC NNs and CNNs, various types of the AEs (contractive and denoising) and of the RBMs are implemented in Pylearn2. Several kinds of the loss functions, particularly, the cross-entropy functions and log-likelihood measure are provided. Stochastic Gradient Descent (SGD), Batch Gradient Descent (BGD) and Nonlinear Conjugate Gradient Descent (NCG) are available for training.

In Pylearn2 library, the neural networks are defined by the description of these ones in a configuration file in YAML format. Since the library has been developed using the object-oriented approach, YAML is a fast method to serialize the configurations into the internal library objects. Let us consider the procedure of the construction of YAML files using an example of a logistic regression.

1. Define a training set. There is a ready-to-use class for processing MNIST database in the library. The first 50,000 images are employed in training.

```
!obj:pylearn2.train.Train {
dataset: &train !obj:pylearn2.datasets.mnist.MNIST {
which_set: 'train',
...
start: 0,
stop: 50000
},
```

2. Describe the network structure. For this, use the class implementing the logistic regression. It is enough to specify the necessary parameters only. The number of the input neurons in the fully connected layer is 784 (according to the number of pixels in the image), the number of the output ones is 10 (according to the number of the object classes), and the initial weights are defined to be zeroes (as a simplest way).

```
model: !obj:pylearn2.models.softmax_regression.SoftmaxRegression {
n_classes: 10,
irange: 0.,
nvis: 784,
},
```

3. Specify the algorithm for the training of the neural network and its parameters. Let us specify BGD method for training. At that, we will trace the results of classification by the training, validation, and test sets. The termination criterion is the maximal number of the optimization iterations.

```
algorithm: !obj:pylearn2.training_algorithms.bgd.BGD {
...
monitoring_dataset: {
'train' : *train,
'valid' : !obj:pylearn2.datasets.mnist.MNIST { ... },
'test'  : !obj:pylearn2.datasets.mnist.MNIST { ... }
},
},
```

4. For further application of the trained model, it is necessary to save the obtained result. Note that the model is saved in the 'pkl' format.

```
extensions: [
!obj:pylearn2.train_extensions.best_params.MonitorBasedSaveBest {
channel_name: 'valid_y_misclass',
save_path: "\%(save_path)s/softmax_regression_best.pkl"
},
]
```

Thus, the network configuration has been prepared and the necessary infrastructure for training and classification, which are performed by the calling of the corresponding Python script, has been defined. It's easy to see that the structure of configuration file is not obvious in the absence of class diagram.

3.4   Torch Library

Torch [4] is a library for the scientific computations with an extended support of the machine learning algorithms. It has been developed by Idiap Research Institute, New York University and by NEC Laboratories America since 2000.

The library is implemented in Lua and C. A fast script language Lua combined with SSE, OpenMP, and CUDA technologies allows Torch to demonstrate good performance. Linux, MacOS X, iOS, and Android operating systems are supported. The main modules can be executed under Windows also.

The library consists of a set of modules, each one being responsible for a particular stage of work with the neural networks. For instance, 'nn' module provides the configuring of a neural network, 'optim' module contains the implementation of optimization methods, used for training, and 'gnuplot' allows visualizing the data (graph plotting, image presentation, etc.). The installation of the additional modules allows extending the functionality of the library [26].

Torch provides the capability to create the complex neural networks by means of the container mechanism. *Container* is a class combining the declared components of a neural network into one common configuration, which can be transferred further to the training procedure. Torch allows using the following layers:

– Fully connected layers.
– Convolutional layers: convolution, subsampling, max-pooling, average-pooling, LP-pooling, subtractive normalization.
– Loss functions: mean square error (MSE), cross-entropy, etc.

Stochastic Gradient Descent (SGD), Averaged SGD, Broyden-Fletcher-Goldfarb-Shanno (L-BFGS) and Conjugate Gradient (CG) optimization methods can be used in training.

Let us consider the procedure of the configuring of a neural network using Torch. Actually, it includes the creation of a container and the successive adding of the network layers. The order of the layer adding is essential since the output of the preceding layer is an input of the next one. For example,

```
regression = nn.Sequential()
regression:add(nn.Linear(784,10))
regression:add(nn.SoftMax())
loss = nn.ClassNLLCriterion()
```

The use and training of a neural network consist of several stages:

1. The loading of the input data X. Function 'torch.load(path_to_ready_dset)' allows loading a dataset in the text format or in the binary one. As a rule, it is a Lua table consisting of the three fields: the size, the data, and the labels.
2. Definition of the response of the network on the input data X:

```
Y = regression:forward(X)
```

3. The computation of the loss function $E = loss(Y, T)$:

```
E = loss:forward(Y,T)
```

4. Update of the weights according to the back-propagation algorithm:

```
dE_dY = loss:updateGradInput(Y,T)
dE_dX = regression:updateGradInput(X,dE_dY)
regression:accGradParameters(X,dE_dY)
```

Thus, the declaration procedure as well as the training one takes less than 10 code lines that evidences the ease of use of the library. At that, the library allows working with the neural networks at low level and construct difficult structures such as multi-column deep neural networks [28].

In order to save and load a trained network, one can use the dedicated functions:

```
torch.save(path, regression)
net = torch.load(path)
```

After the loading one can use the network for the classification (see an example below) or for an additional training.

```
result = net:forward(sample)
```

Other neural network models can be found in the tutorials [4].

3.5   Theano Library

Theano [7] is an extension of Python language allowing computing the mathematical expressions involving the multidimensional arrays efficiently. Theano has been developed in LISA laboratory to support the fast development of machine learning algorithms. The library is supported for Windows, Linux, and Mac OS operating systems. Theano includes a compiler, which translates the mathematical expressions written in Python into an efficient C or CUDA code.

Theano offers a basic toolkit for the configuration and training of neural networks. The implementation of multilayered FC NNs (Multi-Layer Perceptrons, MLPs), CNNs, recurrent neural networks (RNNs), AEs, and RBMs are possible. Also various activation functions, particularly, sigmoidal, softmax, and cross-entropy one are provided. During the training Batch SGD is used.

The configuring of a neural network in Theano [5] is the implementing of a class with particular methods. For continence, implementation of the 'LogisticRegression' class should be provided, this class contains the fields for the training parameters W and b as well as the methods for the working with these ones, namely the computation of the network response ($y = softmax(Wx + b)$) and the error functions. In order to implement training method, it is necessary to describe the loss function, define the methods for the computation of gradients and for the modification of the neural network weights, and specify the dataset (the images themselves and the class labels). After the definition of all parameters, the function is compiled and transferred into the training loop. For fast saving and loading of the neural network parameters one can use the functions from cPickle package.

The process of the creation of a model and of the definition of its parameters requires the writing of a huge and noisy code [27]. The library is a low level one. Nevertheless, it is worth noting its flexibility as well as the possibility to implement and use of user-defined components, to construct multi-column DNNs [28]. Besides, there is a large number of tutorials [5].

4   Comparison of Libraries by the Example of Handwritten Digit Classification

4.1   The Handwritten Digit Classification Problem

Let us formulate the classification problem. A finite set of the grayscale images of digits $X = \{x_{i,j} | i \in \{0, ..., h-1\}, j \in \{0, ..., w-1\}, x_{i,j} \in \{0, ..., 255\}\}$ is given.

There is a subset X', for each image of the subset the class $y \in Y = \{0, ..., 9\}$, which it belongs to is known. Such subset is called *a training set*. Basing on the training set, it is necessary to construct an algorithm $\phi : X \rightarrow Y$, which can determine the digit number from Y for any element of the set X.

4.2   Infrastructure

All of the experiments on the evaluation of libraries performance have been conducted on a machine with Intel(R) Core i5-2430M @ 2.4 GHz CPU, NVIDIA GeForce GT 540M GPU and running Ubuntu 14.04. GCC 4.8 and nvcc 6.5 compilers were used to build C++/CUDA code.

4.3   Learning Parameters

The computation experiments have been carried out with the fully connected and convolutional neural networks:

1. Three-layered fully connected network (MLP):
   - 1st layer – FC (in: 784, out: 392, activation: tanh).
   - 2^{d} layer – FC (in: 392, out: 196, activation: tanh).
   - 3^{d} layer – FC (in: 196, out: 10, activation: softmax).
2. Convolutional neural network (CNN):
   - 1st layer – convolution (in_filters: 1, out_filters: 28, size: 5×5, stride: 1×1).
   - 2^{d} layer – max-pooling (size: 3×3, stride: 3×3).
   - 3^{d} layer – convolution (in_filters: 28, out_filters: 56, size: 5×5, stride 1×1).
   - 4th layer – max-pooling (size: 2×2, stride: 2×2).
   - 5th layer – FC (in: 224, out: 200, activation: tanh).
   - 6th layer – FC (in: 200, out: 10, activation: softmax).

All weights are initialized uniformly random in the range $(-\sqrt{6/(n\_in + n\_out)}, \sqrt{6/(n\_in + n\_out)})$, where n_in, n_out – number of input and output neurons. For training Stochastic Gradient Descent (SGD) method with a learning rate of 0.01, momentum of 0.9, weight decay of 5e-4, a batch size of 128, and maximum epochs number of 150 have been used. Epoch is the full pass through the data, and in current settings approximately equals to 390 iterations of SGD. Sotfmax loss function have been chosen for optimization.

4.4   Dataset

In the present work, MNIST [24] has been selected as a handwritten digits image database. The images were stored in a grayscale format with 28×28 pixels resolution. The digits were centered in the images. The training set consisted of 50,000 images, the test set – of 10,000 ones.

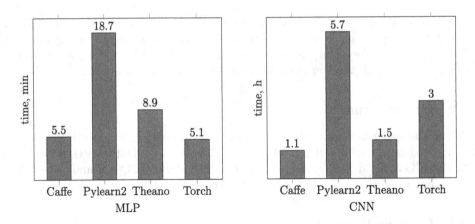

Fig. 2. The training time of MLP and CNN neural networks

4.5 Performance and Accuracy Analysis

The training time of neural networks described earlier is represented below (Fig. 2). Pylearn2 demonstrates the worst performance. As for the other libraries the training time depends on the structure of the neural network. Nevertheless, Caffe looks like more preferable from the viewpoint of both tests.

At the same time classification accuracy of MLP is above 97.4% for all libraries, of CNN — above 99% for all ones (Table 2). In some cases accuracy is a bit lower than the best known results achieved for these classifiers [24]. It is explained by slight differences of the network parameters. The variance of a classification accuracy for several experimental runs is small enough. Moreover, the variance for Pylearn2 and Theano equals to zero because of these ones perform stabilizing optimizations.

Table 2. The average classification accuracy (%) and the variance for 5 experiments

	Caffe		Pylearn2		Theano		Torch	
	Accuracy	Variance	Accuracy	Variance	Accuracy	Variance	Accuracy	Variance
MLP	98.26	0.0039	98.1	0	97.42	0.0023	98.19	0
CNN	99.1	0.0038	99.3	0	99.16	0.0132	99.4	0

4.6 Comparison of Chosen Libraries

Based on the performed investigations of the libraries functionality as well as of the performance analysis on the test problem, the places from 1 to 3 have been given by authors for each library according to the following criteria:

Table 3. The comparison results of the libraries in the scale from 1 to 3

	Performance	Usability	Flexibility	Functionality	Documentation	Total
Caffe	1	1	3	3	2	10
Pylearn2	3	2	3	1	3	12
Torch	2	2	2	2	1	9
Theano	2	3	1	2	2	10

- *The performance* reflects the learning time for the neural networks considered during carrying out the experiments as well as the classification time for the test sample.
- *The usability* evaluates the time required to study the library.
- *The flexibility* of the neural network configuration set up, the selection of the learning method options as well as the availability of various data processing methods.
- *The functionality* or the availability of typical deep learning methods (FC NNs, CNNs, AEs, RBMs, various optimization methods and loss functions).
- The availability and usability of the *documentation* and tutorials.

Let us consider the evaluations according to each of these criteria (Table 3). According to the computational experiments, Caffe library was the best from the viewpoint of the computation speed. At that, Caffe library appeared to be the most convenient in use. It is worth noting that Theano library has demonstrated the best results from the viewpoint of the flexibility. Pylearn2 has the largest functional capabilities. In the use of the documentation and tutorials, Torch developers were found to supply the most detailed and easy to understand educational materials from the viewpoint of the authors.

5   Conclusion

In the present work, an overview of deep learning tools is represented, the functional capabilities of these ones and the accessibility to use in solving the applied problems have been considered. The computational experiments have been carried out to evaluate the performance of a number of libraries (Caffe, Pylearn2, Theano, Torch) by the example of the handwritten digit classification problem. The comparison of the libraries in the practical applications has been performed.

The goal of the future work is to attempt the deep learning methods in solving the applied problems of face, pedestrian and vehicle detection but not to develop new methods or to modify the existing ones. The first stage is to estimate its applicability and to experiment with different structures of neural nets, and the next stage is to achieve good performance. At this moment, Caffe and Torch libraries are considered to be applied for solving these problems as the most convenient among described frameworks.

Acknowledgments. The work has been performed in Information Technologies Laboratory at Computational Mathematics and Cybernetics Department, Lobachevsky State University of Nizhni Novgorod under support by Itseez Co. and Argus Center for Computer Vision Co. Ltd.

References

1. Kustikova, V.D., Druzhkov, P.N.: A survey of deep learning methods and software for image classification and object detection. In: Proceedings of the 9th Open German-Russian Workshop on Pattern Recognition and Image Understanding (2014)
2. Caffe. http://caffe.berkeleyvision.org
3. Pylearn2. http://deeplearning.net/software/pylearn2
4. Torch. http://www.torch.ch
5. Theano. http://deeplearning.net/software/theano
6. Hinton, G.E.: Learning multiple layers of representation. Trends Cogn. Sci. **11**, 428–434 (2007)
7. LeCun, Y., Kavukcuoglu, K., Farabet, C.: Convolutional networks and applications in vision. In: Proceedings of the IEEE International Symposium on Circuits and Systems (ISCAS), pp. 253–256 (2010)
8. Hayat, M., Bennamoun, M., An, S.: Learning non-linear reconstruction models for image set classification. In: Proceedings of the IEEE Conference on CVPR (2014)
9. Restricted Boltzmann Machines (RBMs). http://www.deeplearning.net/tutorial/rbm.html
10. DeapLearnToolbox. https://github.com/rasmusbergpalm/DeepLearnToolbox
11. Deepnet Library. https://github.com/nitishsrivastava/deepnet
12. DeepMat Library. https://github.com/kyunghyuncho/deepmat
13. Package Darch. http://cran.r-project.org/web/packages/darch/index.html
14. nnForge Library. http://milakov.github.io/nnForge
15. CXXNET. https://github.com/antinucleon/cxxnet
16. Cuda-convnet - high-performance C++/CUDA implementation of convolutional neural networks. http://code.google.com/p/cuda-convnet
17. Cuda CNN Library. http://www.mathworks.com/matlabcentral/fileexchange/24291-cnn-convolutional-neural-network-class
18. Caffes repository. https://github.com/BVLC/caffe
19. Unofficial version of library Caffe to the Windows. https://github.com/niuzhiheng/caffe
20. NVIDIA(R) cuDNN - GPU Accelerated Machine Learning. https://developer.nvidia.com/cuDNN
21. Bottou, L.: Stochastic gradient descent tricks. In: Montavon, G., Orr, G.B., Müller, K.-R. (eds.) NN: Tricks of the Trade. LNCS, vol. 7700, 2nd edn, pp. 421–436. Springer, Heidelberg (2012). http://research.microsoft.com/pubs/192769/tricks-2012.pdf
22. Duchi, J., Hazan, E., Singer, Y.: Adaptive subgradient methods for online learning and stochastic optimization. J. ML Res. **12**, 2121–2159 (2011)
23. Sutskever, I., Martens, J., Dahl, G., Hinton, G.: On the importance of initialization and momentum in deep learning. In: Proceedings of the 30th International Conference on ML (2013)
24. The MNIST database of handwritten digits. http://yann.lecun.com/exdb/mnist

25. Pylearn2s repository. https://github.com/lisa-lab/pylearn2
26. Torch cheatsheet. https://github.com/torch/torch7/wiki/Cheatsheet
27. Example of logistic regression (Theano). https://github.com/ITLab-Vision/ DNN-develop/blob/master/theano/src/mnist/logistic_sgd.py
28. Ciresan, D., Meier, U., Schmidhuber J.: Multi-column Deep Neural Networks for Image Classification. http://arxiv.org/pdf/1202.2745v1.pdf

Methods of Localization of Some Anthropometric Features of Face

Svetlana Volkova[(✉)]

Vologda State University, Vologda, Russia
malysheva.svetlana.s@gmail.com

Abstract. In this paper a modified algorithm of localization of a face features based on the Viola-Jones method which is characterized by several classification stages is considered. Experiments show the improvement of the method performance that provides 98 % of correct localizations.

Keywords: Image classification · Viola-Jones method · A cascade of classifiers · Haar features · Facial points

1 Introduction

The task of localization of a human face features in a digital image is one of essential steps in biometric identification systems, in access control systems, and in pedestrian monitoring systems.

Algorithm of anthropometric features localization, as a rule, consists of several main stages: (1) frame capture; (2) preliminary processing; (3) detection of potential areas; (4) decision making on an accurate localization of the features.

At the first stage the division of video sequence into separate frames of the scene takes place. At the stage of preliminary processing filtering of images, application of defects elimination methods, luminance correction and detection of face are accomplished. At the third stage analyzing of the detected face regions with the purpose of searching for potential regions to find face features and identification of their coordinates are performed. At the last stage it is necessary to select that region out of the found ones which most accurately characterizes the target object. The paper describes third and fourth stage. So our system is designed for detecting facial points within a face image.

A human face is presented in an image by a region containing fixed features: outer and inner corners of eyes, tip of the nose, corners of lips, etc.

Necessary conditions for face features localization systems are the high speed and the immunity to conditions, because the image quality, scene luminance and the face orientation influences accuracy of the localization greatly.

Thus, the task of creating a quick and robust algorithm of anthropometric features localization tolerant to changing of shooting conditions is useful and timely.

In this paper we describe a hybrid algorithm of localization of some features of a face based on the Viola-Jones [1] algorithm and that performs a robust, accurate and speed-effective detection.

© Springer International Publishing Switzerland 2015
M.Y. Khachay et al. (Eds.): AIST 2015, CCIS 542, pp. 412–420, 2015.
DOI: 10.1007/978-3-319-26123-2_39

2   Solving the Task of Localization of the Potential Regions

There are 2 approaches to solving the task of localization of the potential regions - parametric and statistical.

In the parametric methods the characteristics of the facial area are measured according to the intuitive understanding of form, color, brightness. Many characteristics like the analysis of the absolute value of the brightness gradient to isolate the nose wings [2], the analysis of the color characteristics for searching mouth [3] have been studied and can be used in a practice. However, in the case of uncontrolled light, these methods don't provide consistent results.

The statistical approach involves constructing a statistical model using the features, which extracted from the part of the region [4, 5].

Popular feature sets include frequency information like Haar wavelets, Gabor features and features based on the gradient of the function, such as LBP [6], SIFT [7], GLOH [8].

Search for potential detection regions is based on the Viola-Jones [1] algorithm which utilizes the Haar-like features. The advantage of these simple features utilization is that they can be computed quickly enough with the use of the integral image. The integral image II for image I is defined as:

$$II(x,y) = \sum_{x' \leq x, y' \leq y} I(x',y') \tag{1}$$

The object being detected is represented as a combination of Haar-like features, an example of which is shown in Fig. 1. The computed value of the feature will be the difference of the sum of the pixels' luminance values overlayed by a light part of the feature and the sum of the pixels luminance values overlayed by the dark part of the feature.

The Viola-Jones algorithm reduces the task of detection to the task of classifying each region of the image: for each of the image sub-windows taken with all possible shifts, orientations and scales a hypothesis is being checked by a preliminary trained classifier. The detector is the array of strong classifiers with increasing complexity. A strong classifier is constructed by AdaBoost [9, 10] machine learning method as a linear combination of weak classifiers each of which is based on Haar-like feature described above.

Fig. 1. An example of Haar-like primitive features

3 Training of a Classifier for Face Feature Localization

For the description of the classifier training stage let us select a single feature of a face which corresponds to the tip of nose. To train the classifier two sets of samples are required: negative and positive. Negative samples (Fig. 2) contain images without nasal tip image; to improve differentiation of classes the negative samples contain not only arbitrary images, but also images of face regions which do not contain eyes. Positive images (Fig. 3) — images of nasal tip region, retrieved from the frames, captured by a camera, the size of which was formed with reference to the distance between the corners of eyes of a human being in dependence on the scale and image size. The size of the positive region cut was altered; however the sides' ratio remained unchanged. While training about 12 000 positive and 12 000 negative samples were used.

Fig. 2. Examples of negative images used for training the classifier focused on detection of the nasal tip

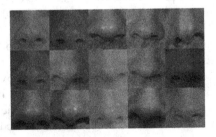

Fig. 3. Examples of positive images used for training the classifier focused on detection of the nasal tip

4 Finding the Resultant Square Out of the Array of Potential Regions

Basing on the fact that the sought-for region (the nose and eyes) within the human face region is presented in a single copy, out of several candidate regions defined by the sliding window method as two classes, it is necessary to accurately select the region characterizing the object.

The result of the classification algorithm operation is the list of squares detected as features of a face. Each of the obtained squares we characterize by a three dimensional vector containing the coordinate values of the centre (x and y) and the side's width (w). Then the normalization procedure of input vectors of the squares is accomplished in reference to the face square. Every parameter is expressed as a percentage of the face sizes. For this we use formulae (2):

$$X' = \left[\frac{x - x_0}{x_1 - x_0} \times 100 \right]$$

$$Y' = \left[\frac{y - y_0}{y_1 - y_0} \times 100 \right],$$

$$W' = \left[\frac{w}{w_0} \times 100 \right]$$

$$(2)$$

where X', Y', W' are normalized coordinates, (x_0, y_0), (x_1, y_1) are coordinates of the left top and the right bottom corners of the squares of the face detected, w_0 is the size of the face square.

Thus, the potential candidate-regions for the current image represent an array of three-dimensional vectors with normalized coordinates.

The rule for making a decision on this array is based on the model of naive Bayesian classifier [11], for which the statistical data were acquired using 50 000 images, for which as the resultant the tip of nose (or outer corner of an eye) was selected detected by a cascade of classifiers, the nearest to the accurate and deviated from the accurate position by not more than 10 % from the distance between the outer corners of the eyes. The accurate setting refers to location of the nasal tip (or outer corners of the eyes) indicated by an expert person. We consider as a hypothesis an event which corresponds to acceptance of the very square as a result. The total number of hypotheses equals the number of the potential regions detected +1.

We shall call "a zero hypothesis" the one which fits the case of rejecting all other hypotheses concerning the feature location. For the possibility of its detection let us define the limiting closed surface. To do this let us consider the distribution of points of resultants of the nasal tip (or outer corners of the eyes), participating in the forming of statistical data. In Fig. 4 the location of resulting corners of the tip of nose is shown.

The criterion for selection of the limiting close surface is the simplicity of implementation and the speed of utilization in the detecting system. So, there are two suitable variants for the task under consideration: a parallelepiped and a sphere. Analysis of the projections of the point cloud shows, that the sphere is more suitable for selection of the limiting close surface. Coordinates of the curve centers are computed as weighted-average for all the values of the respective coordinate out of the statistical sampling. The radius value was defined empirically based on the outcome of the algorithm operation in the whole with several test databases.

To check the validity of the zero hypothesis the face feature, assumed as the most fitting, undergoes the stage of additional analysis of introduction into the binding range in accordance with the classifier described above. In case of a negative answer the hypothesis of absence of the sought-for square within the range is accepted.

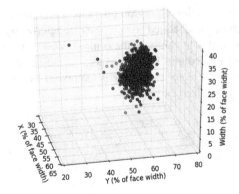

Fig. 4. Distribution of clusters of points resulting the tips of nose.

In case equal maximum probabilities are obtained for two or more hypotheses, the introduction of the bounding sphere is necessary. In this case measurement of a distance from the points of the hypothesis analyzed to the center of the bound array takes place. As the resultant will be assumed the one, the points of which are closer to the center, and, consequently, to the cloud of experimental data values. In case of equiprobable hypotheses with the same distance to the center of the bounding array any of them may be assumed as the resultant.

5   Modification of Localization Algorithm by Adding More Classifiers

To improve the results of classification the following modifications with the use of multiple pass algorithm were proposed: a parallel processing of classification results and a serial processing of classification results (Figs. 5 and 6).

In the serial processing two cascades of the classifiers trained with different samples and with different parameters are involved. For their optimal convergence the first to be used is the stricter one which allows to more accurately detect the location of the face feature, but which, however, does not have the ability to detect all possible detection regions. The second classifier is weaker, but less sensitive to changes. The second classifier allows detecting less accurately but with a greater probability compared to the first one.

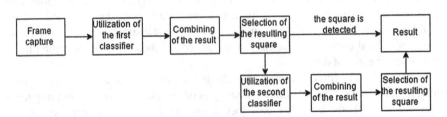

Fig. 5. Modification of the face feature localization algorithm with serial operation of classifiers

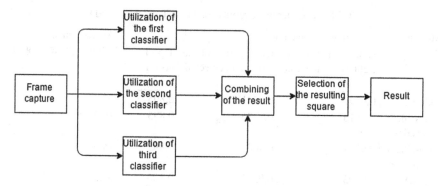

Fig. 6. Modification of the face feature localization algorithm with parallel operation of classifiers

In the parallel processing three classifiers are involved. They are trained with the parameters allowing getting an accurate enough result but are not capable of detecting all possible regions. By training parameters all the three classifiers are similar to the first one used in the case of serial processing. As opposed to the serial processing where the second classifier is added in case of the first classifier failure, in the parallel processing they are operating jointly. And only after all the potential candidate regions have been detected by all the classifiers the combining of results for filtering out of unit squares takes place, and then a decision is made as per the statistical method described above.

This approach allows increasing the detection accuracy, and the frame processing time remains consistent.

6 Results of the Experiment

The proposed algorithm was tested experimentally with the use of several test samplings:

Images from the FERET [12] database. It contains 1187 images of faces. Image size is 256 × 384, the average distance between the outer corners of eyes is 90 pixels;

Database consisting of 3000 images acquired from Internet. Image size is 768 × 1024, average distance between the outer corners of eyes is 200 pixels;

Database consisting of 3002 images obtained from an IP-camera. Image size is 2560 × 1920, average distance between the outer corners of eyes is 150 pixels.

A PC with the following specifications was used in the experiment: CPU is Core (TM) i3 CPU 530 @ 2.93 GHz, graphics card is NVidia GeForce GTX 580 3 Gb, RAM 2 Gb.

The proposed algorithm was implemented as a software program in the object-oriented programming language C++ (Visual Studio 2008).

In Tables 1, 2, and 3 the results of the test for samplings are presented. The feature of face is considered to be detected correctly if the deviation from the one detected by an expert does not exceed 10 % of the distance between the outer corners of eyes. Also

Table 1. Experiment results for FERET database (№1)

Algorithm	Accuracy of detection			Average time of 1 frame processing, ms
	Outer corner of the right eye is detected	Outer corner of the left eye is detected	Tip of nose is detected	
1 classifier	0,9591	0,9721	0,895	13,3
2 classifiers in series	0,9616	0,9736	0,895	16,3
3 classifiers in parallel	0,9764	0,9686	0,895	14,9
OpenCV	0,9393	0,9553	0,806	34,3

Table 2. Experiment results for faces database from Internet (№2)

Algorithm	Accuracy of detection			Average time of 1 frame processing, ms
	Outer corner of the right eye is detected	Outer corner of the left eye is detected	Tip of nose is detected	
1 classifier	0,8959	0,903	0,5383	19,2
2 classifiers in series	0,8844	0,8914	0,5146	20,2
3 classifiers in parallel	0,9566	0,9553	0,6	19
OpenCV	0,638	0,695	0,51	316

Table 3. Experiment results for database of images from IP-camera (№3)

Algorithm	Accuracy of detection			Average time of 1 frame processing, ms
	Outer corner of the right eye is detected	Outer corner of the left eye is detected	Tip of nose is detected	
1 classifier	0,7324	0,7349	0,7616	20,8
2 classifiers in series	0,6003	0,8955	0,7349	24,1
3 classifiers in parallel	0,8994	0,9853	0,865	18,7
OpenCV	0,159	0,143	0,08	420

in Tables 1, 2, and 3 result of free eye and nose detectors that come with OpenCV is presented.

As the result of modeling it was ascertained, that the use of 2 classifiers in series or of 3 classifiers in parallel increases the percentage of correct localization for all the three databases, however the use of 3 classifiers in series allows achieving 98 % accuracy of detection. Due to the utilization of the graphics card the time of operation practically does not change for the modification of the algorithm.

In the process of comparing the described algorithm to the existing methods of localization of the features of the face, while collating the tables of results, the following peculiarities have been identified: average time of a frame processing allows utilization of the algorithm in the real time mode. The index of correctly detected face features coordinates comes up to 98 %, the use of several classifiers trained with a small value of the parameter of false detected, allows improving the result of localization. As drawbacks of the algorithm one can consider the requirement to big volumes of training samplings and a long time of training.

7   Conclusion

In this paper a modified algorithm for localization of some anthropometric features of a face in the video stream on the basis of the Viola-Jones method and the Haar-like features is proposed. A model of the localization system based upon utilization of multi-pass scheme on classifiers trained with different parameters and on different samplings. The novelty of the method is in utilization of the multi-pass serial and parallel scheme and in the method of selection of the final answer according to statistical data on the distribution of the features of a face.

It is experimentally displayed that utilization of this approach allows achieving the accuracy of localization of the outer corners of eyes and the tip of nose up to 98 %; the time of one frame processing does not exceed 25 ms, which allows for utilizing the algorithm in real time systems of biometrical identification.

References

1. Viola P., Jones, M.: Rapid object detection using a boosted cascade of simple features. In: Proceedings of IEEE Computer Society Conference on Computer Vision and Pattern Recognition, vol. 1, pp. 1063–6919, December 2001
2. Yin, L., Dasu, A.: Nose shape estimation and tracking for model-based coding. In: Proceedings of IEEE International Conference on Acoustics, Speech, Signal Processing, pp. 1477–1480 (2001)
3. Vezhnevets, V.: Face and facial feature tracking for natural human-computer interface. In: Proceeding of conference GrafiCon 2002 (2002)
4. Liao, C.T., Wu, Y.K., Lai, S.H.: Locating facial feature points using support vector machines. In: Proceedings of the 9th International Workshop on Cellular Neural Networks and Their Applications, pp. 296–299 (2005)
5. Rowley, H.A., Baluja, S., Kanade, T.: Neural network-based face detection. IEEE Trans. Pattern Anal. Mach. Intell. **20**(1), 23–38 (1998)
6. Ahonen, T., Hadid, A., Pietikainen, M.: Face description with local binary patterns: application to face recognition. IEEE Trans. Pattern Anal. Mach. Intell. **28**(12), 2037–2041 (2006)
7. Lowe, D.G.: Distinctive image features from scale-invariant keypoints. Int. J. Comput. Vis. **60**(2), 91–110 (2004)
8. Mikolajczyk, K., Schmid, C.: A performance evaluation of local descriptors. IEEE Trans. Pattern Anal. Mach. Intell. **27**(10), 1615–1630 (2005)

9. Schapire, R.E.: The Strength of Weak learnability. Mach. Learn. **5**, 197–227 (1990)
10. Freund, Y., Schapire, R.E.: A decision theoretic generalization of on-line learning and an application to boosting. J. Comput. Syst. Sci. **55**(1), 119–139 (1997)
11. Vapnik, V.N.: Vosstanovlenie zavisimostei po empiricheskim dannym, pp. 71–83, Moscow, Nauka (1979) (in Russian)
12. Phillips, J., Moon, H., Rizvi, S.A., Rauss, P.J.: The FERET evaluation methodology for face-recognition algorithms. IEEE Trans. Pattern Anal. Mach. Intell. **22**(10), 1090–1104 (2000)

Ontological Representation of Networks for IDS in Cyber-Physical Systems

Vasily A. Sartakov[✉]

Ksys Labs, Moscow, Russia
sartakov@ksyslabs.org

Abstract. Cyber-Physical System (CPSs) combine information and communication technologies and means controlling physical objects. Modern infrastructure objects such as electrical grids, smart-cities, etc. represent complex CPSs consisting of multiple interconnected software and hardware complexes. The software contained in them requires development, support, and in case of updates termination can be the target for malicious attacks. To prevent intrusion into networks of cyber-physical objects one can use Intrusion-Detection System (IDS) that are widely used in existing noncyber-physical networks. CPSs are characterized by formalization and determinacy and it allows to apply a specification-based approach for IDS development.

This paper is devoted to IDS development using the ontology-based representation of networks. This representation allows to implement both at the software level – by means of comparing movement of network traffic with its model, and at the physical level – by means of controlling connections of network devices. Ontological representation provides a model of network which is used for creation specifications for IDS.

Keywords: Networks · Intrusion detection systems · Ontology

1 Introduction

Modern approaches to the development of critical infrastructure elements are characterized by a high degree of penetration of information and communication technologies. For example, an up-to-date object of energy infrastructure is a substation and it represents a cyber-physical object which main elements are data transmission facilities and control elements. In other words, an infrastructure object represents a network of interacting intellectual components, each element of which can be implemented as a separate hardware and software complex having its own software and implementing a specific function.

Individually, hardware and software complexes include application and system software that shall be updated with the course of time (for some software solutions) for the purposes of security improvement. In case software is not updated, CPS elements may be exposed to invasions and used by malicious users to affect other components of the network because of vulnerabilities.

M.Y. Khachay et al. (Eds.): AIST 2015, CCIS 542, pp. 421–430, 2015.
DOI: 10.1007/978-3-319-26123-2_40

This work is devoted to the development of a network instruction detection system. Among many types of attacks aimed to cause abnormal operation of critical infrastructure elements we are focused on network intrusions. Our model of an intruder is based on the assumption that intrusion is carried out inside the network by means of abnormal affection of some elements of the critical infrastructure object on the others. As a consequence, the applied protection method utilizes the formalized model of the system describing all kinds of interactions of network elements and detects deviations from this model in network traffic.

2   Intrusion Detection Systems

IDSs are a widely used tool of information security provision. Coupled with active network security tools such as firewalls, these technologies allow to protect and control interaction of network components, detect malicious traffic and intrusions. There are a lot of developers and producers of such equipment on the market, new systems are being actively developed both in industry and in the academic community.

Conceptually, all kinds of IDSs can be referred to two opposite approaches [1]. The first approach is based on the analysis of network traffic and detection of specific data sequences in it. Such traffic analysis is called *signature analysis* and, consequently, the IDS is also called signature IDS. If the attack signature is correctly described, such method of instruction detection is distinguished by a high rate of actuations and a low rate of type II errors. This method requires continuous support of signature database without which it is impossible to counter instructions of known types as well as new ones and it is a disadvantage of this method.

Another approach to development IDS is based on detection of *anomalies*. Unlike the signature-based approach where specific intrusion signatures are detected in traffic, a deviation in the behavior of the controlled system is detected when dealing with anomalies. This approach applies statistical profiling, machine learning methods, etc. to detect "normal" behavior and deviations. The complexity of development of IDS of this type is a correct description of the model of "normal" behavior. An incorrectly constructed model of behavior can result in false actuations and malfunctioning. At the same time, a correctly described system can detect both known attacks and, in some cases, new ones.

Besides detection of anomalies and attack signatures there are a few intermediate approaches that include some elements of the above mentioned ones. For example, a probabilistic approach, sometimes called statistical or Bayesian [2]. Within the framework of this approach a description of the system, its states and probabilities of transition of one state to another is generated. As for the other approach based on the specification (a specification-based approach) a model of a working network is created, i.e. it is detected what interactions are allowed in the network, then the model is expanded by means of a set of security policies, and further on, the IDS responds to deviations in the operation of the observed system from its model [3].

As a general approach for the development of IDS in critical infrastructures we use the approach based on the specification. First, as compared with other methods, this method is distinguished by the lowest rate of type II errors. Moreover, unlike the signature approach, it does not require continuous support and development of signature databases and it allows to detect unknown types of attacks and intrusions [4]. Third, despite the high complexity of network specification development, the field of application allows to decrease the complexity of this work. Since we consider IDS in the context of CPSs and critical infrastructure, we assume that the structure of this infrastructure will vary slightly in course of time and thus a model of the network can be developed once and it will not require any further changes.

3   Architecture

As it was mentioned earlier, modern objects of critical infrastructure actively use information and communication technologies for components communication and remote control. Since infrastructure objects fulfill such functions as control, monitoring, impact on the physical environment, being almost fully autonomous in the process of taking decisions, we call them CPSs. Examples of such systems: power plants, smart buildings, etc.

A few peculiarities are typical for CPSs. Besides intensive communication, it is extremely important that these systems are specialized, i.e. developed specifically to be used under certain conditions and it makes the architecture of these

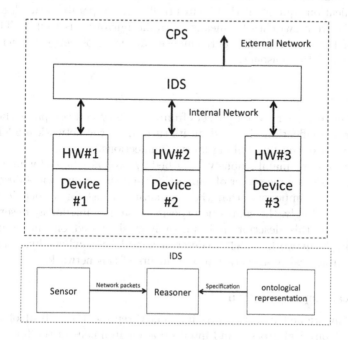

Fig. 1. IDS for CPS

systems to be known and determinated. In other words, all interactions of system components can be formalized and their changes are negligible with course of time. This peculiarity allows us to focus on the development of IDS based on specifications.

Another peculiarity of CPSs is working with the physical environment, and in the process of CPS development models are created, both models of the physical environment and models of the entire system. System engineering methodologies actively use digital representations of physical components in the development of industrial complexes. These models include components of different in nature, for example, people, their actions, physical devices and paper documents, but they are combined in the process of development.

These two peculiarities determine the architecture of developed IDS. An IDS is a separate component of critical infrastructure; It detects intrusions and malfunctioning of cyber-physical devices of one network.

A diagram of a network of cyber-physical devices is shown in Fig. 1. As for the logistics an IDS consists of three components: a sensor, a reasoner and a network model. The sensor captures traffic. The reasoner compares network traffic with the specification obtained on the basis of the network model and performs logging of intrusions. The network model is the third component; it describes the logical and physical model of the network.

Sensor

The sensor shall access to all network traffic. Physically the sensor may be a part of communication equipment that routes traffic, or a separate network device to which traffic from switches by means of traffic mirroring is routed. The main function of the sensor is high-performance capturing of network packets and their transfer to the reasoner.

Reasoner

The reasoner receives network packets from the sensor and compares them with the network specification obtained, in its turn, from the network model. Specification is a set of rules describing known connections.

Connections are unambiguously described by two pairs *ip:port* where the first pair describes the source server of connection, and the second one – destination. In addition, it is important that the rule actuates just for a specific network protocol. Accordingly, the connection specification unambiguously describes the connection and this description is unambiguously based on the connections described in the network model. To describe a model of a cyber-physical network we apply ontological representation of elements of this network.

Ontological Representation

As it was mentioned before, CPSs inherently represent a hybrid of a device controlling a physical process and intellectual environment in the form of hardware and software. CPSs differ from embedded ones by an available model of a physical process and interaction with the physical environment. Interaction with the physical environment and formalization of this interaction forces to use the ontological representation of this interaction, because physical and logical components of the model contact in the process of this interaction.

IDS for CPSs is a part of a critical infrastructure object. It is involved in controlling system operation and takes decisions on the separation of components of the network (in the case of operation in the mode of intrusion prevention) or informs network operators of any changes. The importance of physical representation of network components is the same as the importance of logical representation - presence of a network connection and a formal description of the connection protocol are important, but also representation of the physical location of devices is important. Physical location of devices is necessary, for example, to remove devices from the network quickly, or monitor the integrity of the physical location of the network components. As a result, we use the ontological representation of the network that allows to combine logical and physical elements of network.

Ontology

The necessity to combine logical and physical objects in the reasoner forces to use the ontological representation of these entities. Ontology is a formal, clear, precise definition of conceptualization. Conceptualization is an abstract, simplified view of the world formed for specific purposes [5]. In other words, it is possible to create a description of entities through conceptualization, and we need it to describe a network of CPS. Formalization of the same description will allow to use machine learning tools in future for semi-automatic changing of models of these networks.

To ensure the basic functionality of IDS it is necessary to describe the following entities ontologically: a server (physical, logical), a service, a network port, a protocol, a network interface, a network device, and a switch port. Moreover, it is necessary to describe relationships of these entities.

In the physical representation a server describes the location of physical communication equipment of a hardware and software complex included into the CPS. A logical server describes the entity inside which software is located. A physical server is characterized by a model of a hardware platform, its physical characteristics (for example, capacity of power supply, location), a logical server is characterized by the version of software (OS) and a set of specific services.

The service is the basic unit of communication as he fulfills network functions. The logical connection of services describes network interaction out of which may a specification for the reasoner can be eventually extracted. The service is characterized by a network port on which traffic is received and a protocol. In case the program performs client function, the logical connection of the "client" program with the "server" program shall be specified. The logical connection is required to visualize the network configuration for the network operator.

A network port, a network interface and a network device are physical and logical entities allowing to turn the logical connection of the services in the specification. A logical network interface allows to extract network address of the program, the network port - the port to which the program associated and the network device – the hardware address. Extraction of these entities allows to attach the logical service to a specific network device. The logical connection of physical network devices describes the physical connection of devices in the switch that allows to control the integrity of the physical connection of devices.

4   Evaluation

A prototype of an IDS built using the ontological representation of the network is presented in this section. To try the concept of the ontological representation we developed a sensor, a reasoner, software allowing to create a network specification, as well as network model in accordance with which specifications are created.

Sensor

A development board based on system-on-chip PPC QorIQ P2040 was used as the hardware platform. This system-on-chip allows to receive network packets directly avoiding device drivers; it helped to save on data transfer between the kernel and the user space. QorIQ system-on-chip is equipped with 4 cores; it allowed to extract network packets out of the network device in the multithreading mode and send them to the reasoner. Empirically, it was found that the maximum performance is achieved when 3 kernels are used to capture traffic (one capture thread per kernel) and one kernel - for reasoner threads. Data between the sensors and the reasoner were transferred by means of shared memory using signals to inform the reasoner of packets extraction from the network device.

Reasoner

A modified version of a multithreaded instruction detection system called Suricata[1] was used as the reasoner. Suricata is an IDS focused on the analysis of the contents of packets – searching in the contents of packets and logging. Suricata uses rules in which a pattern (a line, a data set) is described; availability of these rules is checked in all packets as well as a response to rule activation (logging, packet removal).

We do not use packet analysis as the developed IDS is built on specifications, so the basic functionality of Suricata associated with the analysis of the contents of packets was removed. At the same time we used a system of rules selection and logging used in Suricata. We used syntax to describe rules (specifications) identical to syntax of Suricata as well as the format of events representation in order not to develop separate event processing tools for the network operator.

---

[1] http://suricata-ids.org.

Rules

Suricata uses the rules format developed for IDS Snort [6]. The rules are as follows:

```
action proto src_ip src_port -> dst_ip dst_port
                        (msg:"Message";content:"body")
```

Where *action* means action, for example, alert, *proto* is a protocol type, for example, IP, *src_ip:src_port* is a pair describing the source address of connection, *dst_ip:dst_port* is, respectively, the destination address of connection, *msg* is a message (in the case of alert) that will be logged when the rule is actuated, body is a data sequence which availability will be checked by Suricata in traffic. This syntax unambiguously describes the actuation conditions and is actively used in IDS. But as for IDS working with specifications, rules shall fulfill the opposite function - describe how it is allowed to interconnect but not vice versa.

We modified the rules processing subsystem and it allowed to describe allowed connections and not to violate the accepted rules syntax. And two rules are created for each connection two rules – on the part of the client machine and on the part of the server.

Moreover, we added binding of the machine address to IP addresses by means of inclusion of an additional keyword:

```
alert ip [3.1.1.1] any -> any any
        (msg:"Wrong hw addr";eth_src:00-00-03-01-01-01;).
```

Ontology

To describe the network model we used language of ontologies description Resource Description Framework (RDF) [7] in notation n3 [8]. RDF describes interactions

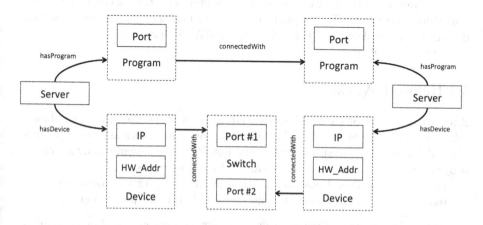

Fig. 2. Connection

of entities in format "subject-predicate-object". It allows to connect both physical and logical objects. For example, below is a description of the server identified through #S2 bound by means of predicates hasDevice, model and replica with other entities, some of which describe physical devices (for example, network device S2_eth0) and others - logical devices (for example, program s2_sps_1414).

```
@forSome <#S2>.
<#S2>          a              <Server>.
<#S2>          <name>         "S2".
<#S2>          <hasDevice>    <#S2_eth0>.
<#S2>          <model>        <#simple1U>.
<#S2>          <hasProgram>   <#s2_sps_1414>.
```

The network description in n3 format presented above is processed by a script on Python that creates a complete set of rules based on the model. Output data of this script are a set of rules as well as a visual representation of the network and physical location of devices.

A simplified example of connection of two programs[2] located in different hardware complexes of one CPS is shown in Fig. 2. The client is on the left of the diagram and the server is on the right. A network connection at the software level is presented at the top and a network connection at the hardware level is presented at the bottom. The essence of the server is connected with the essence of the program through hasProgram predicate. The program, in its turn, can be connected with another program through conectedWith entity. Both programs contain ports, but as for the client program this port is not required unlike the program serving as the server program. At the logical level the difference of the client program from the server one is detected thanks to connectedWidth predicate.

A similar situation is at the physical level. The server is connected with the essence of describing the network device through hasDevice predicate. The description of the network device contains the network address and the hardware address. Moreover, the network device is connected with the concentrator Switch port through connectedWidth predicate. The mechanism of hardware connection description is implemented through connectedWidth predicate for the purpose of its further control.

5   Related Works

A lot of different IDSs have been developed in recent years. We chose several most relevant related projects out of the wide range of technologies and architectures. Since IDSs are based on specifications and ontological representation, the related projects cover the area of ontologies and IDS architectures.

The Reaction after Detection Project (RED) proposes an ontology-based approach to instantiate security policies and determine policy violations [9].

---

[2] A detailed description of the ontology, examples, network graphical representation, etc. is presented on project website http://github.com/Ksys-labs/fdnet.

The technology developed in the RED project provides a way to map alerts into the attack context. This context is used to identify the policies that should be applied in the network to solve the treat. The ontological representation of policies and alerts is used for this purpose. In contrast with this project we use ontological representation to describe possible interactions inside networks and we do not describe intrusions.

Another conception of ontology-based IDS was proposed by Udercoffer et al. [10]. The idea of this project is to develop an ontology that describes intrusion in the form of anomaly. The ontology describes the attack, its consequences and the means of the attack. As another anomaly-based IDS project, the prototype demonstrates ability to detect complex attacks, but all components of the network should be installed into the IDS. In contrast with this work, our ontology is not developed to detect anomalies. We detect deviation in communications inside a network. Moreover, our prototype does not require installation of additional exemplars of IDS: we use just one IDS in each network.

The ontological representation of the networks and attacks is also provided by Frye et al. [11]. The main goal of the paper is to present experimental project TRIDSO (Traffic-based Reasoning Intrusion Detection System using Ontology) which detects complex attacks. Complex attacks are presented by a formal method that basically provides a formal description of generic attacks and identification of new attacks. From the point of view of ontological representation our work is closely related to this work. Firstly, TRIDSO uses snort rules for traffic detection, and we use the same mechanism for the same purpose. Secondly, TRIDSO uses the ontological representation of attacks and network traffic, but we use network representation of CPS, where a network sub-system is just one component of the system. Our ontological representation does not provide ontology of an attack because we represent intrusion as deviation from the ontological model.

Ontological and semantic representations are widely used approaches for networks description. For example, Neuhaus et al. describes [12] the ontological representation of sensors networks, while Ustalov demonstrates [13] a semantic approach to describes the configuration of a cloud environment.

6 Conclusion

This paper is dedicated to the ontological representation of the network to create specification-based IDS. Within the framework of this project we developed and created an intrusion detection system to be applied in networks of CPSs. The ontological representation of the network allowed to create formal models of CPSs that combine both physical and logical entities. IDS in current implementation require a network model based on which a network specification is created. This process requires involvement of a human. We see development of this project in automation and it means that we would like to minimize the presence of a human in the system. For this purpose means of semi-automatic creation of the ontological representation will be developed. As the ontological representation uses a machine-interpreted language, we assume that most of the work related to detection of connections and their description can be done automatically, without a human.

References

1. Zhu, B., Sastry, S.: Scada-specific intrusion detection/prevention systems: a survey and taxonomy. In: Proceedings of the 1st Workshop on Secure Control Systems (SCS) (2010)
2. Kruegel, C., Mutz, D., Robertson, W., Valeur, F.: Bayesian event classification for intrusion detection. In: Computer Security Applications Conference, Proceedings. 19th Annual, pp. 14–23. IEEE (2003)
3. Ko, C.C.W.: Execution Monitoring of Security-Critical Programs in a Distributed System: A Specification-Based Approach. Ph.D. thesis, UNIVERSITY OF CALIFORNIA DAVIS (1996)
4. Balepin, I., Maltsev, S., Rowe, J., Levitt, K.: Using specification-based intrusion detection for automated response. In: Vigna, G., Kruegel, C., Jonsson, E. (eds.) RAID 2003. LNCS, vol. 2820, pp. 136–154. Springer, Heidelberg (2003)
5. Gruber, T.R.: Toward principles for the design of ontologies used for knowledge sharing? Int. J. Hum. Comput. Stud. **43**, 907–928 (1995)
6. Roesch, M., et al.: Snort: lightweight intrusion detection for networks. In: LISA, vol. 99, pp. 229–238 (1999)
7. Klyne, G., Carroll, J.J.: Resource description framework (rdf): concepts and abstract syntax (2006)
8. Berners-Lee, T., Connolly, D.: Notation3 (n3): a readable rdf syntax. W3C Team Submission, January 2008. http://www.w3.org/TeamSubmission, (3) (1998)
9. Cuppens-Boulahia, N., Cuppens, F., Autrel, F., Debar, H.: An ontology-based approach to react to network attacks. Int. J. Inf. Comput. Secur. **3**, 280–305 (2009)
10. Undercoffer, J., Joshi, A., Pinkston, J.: Modeling computer attacks: an ontology for intrusion detection. In: Vigna, G., Kruegel, C., Jonsson, E. (eds.) RAID 2003. LNCS, vol. 2820, pp. 113–135. Springer, Heidelberg (2003)
11. Frye, L., Cheng, L., Heflin, J.: An ontology-based system to identify complex network attacks. In: 2012 IEEE International Conference on Communications (ICC), pp. 6683–6688. IEEE (2012)
12. Neuhaus, H., Compton, M.: The semantic sensor network ontology. In: AGILE Workshop on Challenges in Geospatial Data Harmonisation, Hannover, Germany, pp. 1–33 (2009)
13. Ustalov, D.A.: A semantic approach for representing the cloud computing environment configuration. In: Proceedings of the 14th International Supercomputing Conference "Scientific Service on the Internet", Moscow, MSU, pp. 706–710 (2012)

Determination of the Relative Position of Space Vehicles by Detection and Tracking of Natural Visual Features with the Existing TV-Cameras

Dmitrii Stepanov$^{(\boxtimes)}$, Aleksandr Bakhshiev, Dmitrii Gromoshinskii,
Nikolai Kirpan, and Filipp Gundelakh

Russian State Scientific Center for Robotics and Technical Cybernetics,
Saint Petersburg, Russia
{dnstepanov,alexab,d.gromoshinskii,
n.kirpan,f.gundelakh}@rtc.ru

Abstract. During spacecrafts maneuvers, especially at the rendezvous and docking stages, one of the most important tasks is to determine the relative positions of the vehicles. The current Russian "Course" and recently proposed ATV/HTV docking systems are complex and require mounting of specific cumbersome equipment on the outer sides of both vehicles. The proposed TV-based docking control system uses the existing cameras, "natural" visible features of the ISS and an ISS laptop to determine all six relative coordinates of the vehicles. At the training stage the ISS 3D-model and video recordings are used. This paper describes the algorithm flow and the problems of the passive, TV-only approach. The system efficiency is tested against models, mockups and the recordings of previous rendezvous of "Progress", "Soyuz" and ATV spacecrafts. The nearest goal of the system is to become an independent docking control system helping the ground docking control team and the cosmonauts.

Keywords: Rendezvous · Docking control · Spacecraft · ISS · TV · Television · Features · Machine learning

1 Introduction

In the field of space exploration one of the most important tasks is to determine parameters of relative motion of spacecrafts, especially for the rendezvous and docking stages [1]. All modern cargo and passenger spacecrafts are already equipped with television systems allowing observation of each other and the International Space Station (ISS) during rendezvous and docking. The article proposes a passive, TV-only approach for the docking control and describes the corresponding algorithms, software implementation and the evaluation results.

The proposed software system runs on an ISS laptop (or a PC at the Mission Control Center), receives a video signal from a spacecraft's camera, simultaneously detects and tracks natural visible features of the ISS and calculates the relative 3D-position of the spacecraft. The features are selected at the learning stage using previous docking recordings and a detailed ISS 3D model. The 3D-position is acquired

© Springer International Publishing Switzerland 2015
M.Y. Khachay et al. (Eds.): AIST 2015, CCIS 542, pp. 431–442, 2015.
DOI: 10.1007/978-3-319-26123-2_41

by a PnP solution from the correspondence of the image coordinates and the a priori known 3D-coordinates of the features.

2 Statement of the Problem

The goal is to determine the spatial position of an observed spacecraft (or any object with a known geometry) relative to the camera installed on another spacecraft. The observed object is usually called passive one and the spacecraft with the camera is called the active one. The problem can be solved using the following information:

1. 3-dimensional spatial coordinates of N > 3 points of the passive object relative to its own coordinate system (i.e. the geometry of the object);
2. 2-dimensional image coordinates of the same N points;
3. The camera model (internal camera calibration parameters).

This set of data allows solving of the PnP-problem [2, 3] relative to the camera coordinate system. If the camera position is known relative to the active spacecraft then the spacecraft's position (and a position of any of its point) can be determined too.

Figure 1 introduces the camera and the target coordinate systems used in the article. One origin of the camera coordinate system is located at the camera principal point. The corresponding Z-axis is directed along the optical axis of the camera, and X- and Y-axes are parallel to the image horizontal and vertical axes. This coordinate system is "natural" for the camera (with identity external calibration matrix) and a PnP solution is usually presented in this system.

The second origin is located at the docking target origin. The Z-axis is directed along the axis of the remote element (cross) so that in the docked state the direction is same as of the camera Z-axis.

Figure 2 demonstrates the ZY-plane (a projection of the real 3D case) of the docking coordinate system. The line of sight connects the principal point of the camera

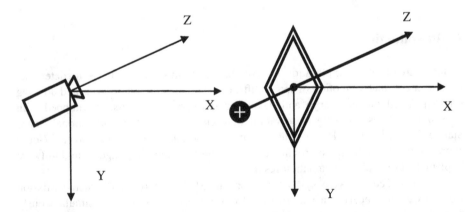

Fig. 1. Camera and target coordinate systems

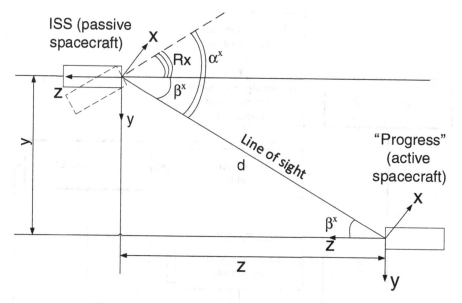

Fig. 2. Docking coordinate system (YZ-plane projection)

with the docking target origin. In the camera coordinate system there are x, y, z – linear displacements between the origins, Rx, Ry, Rz – rotation angles along the X-, Y- and Z-axis.

The two angles β^x, β^y of the camera Z-axis to the line of sight, the two angles α^x, α^y of the target Z-axis to the line of sight, the distance d between the origins (along the line of sight), and the angle θ of mutual roll are the six coordinates often used for the docking control.

The distance d can be calculated from the x, y and z using the Eq. (1), the active angles β are calculated according (2), and the passive angles α are calculated according (3).

$$d = \sqrt{x^2 + y^2 + z^2} \tag{1}$$

$$\begin{cases} \beta^x = atg\dfrac{x}{z} \\ \beta^y = atg\dfrac{y}{z} \end{cases} \tag{2}$$

$$\begin{cases} \alpha^x = \beta^x + Ry \\ \alpha^y = \beta^y + Rx. \end{cases} \tag{3}$$

3   Algorithm Description

The Fig. 3 shows the proposed algorithm of relative 3D pose estimation. The blocks are described below.

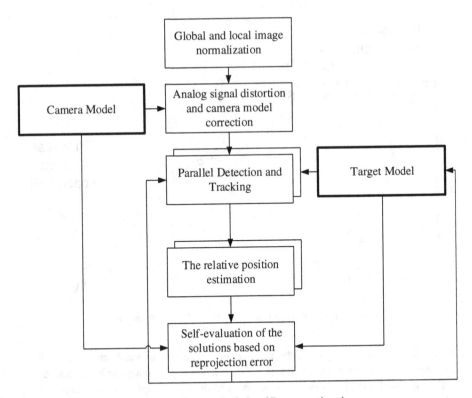

Fig. 3. Algorithm of relative 3D pose estimation

3.1 Camera Model

The camera model describes the relationship between the 3D-coordinates of scene points and the 2D-coordinates of the corresponding image points. Parameters of the model (internal camera calibration parameters) include focal lengths, principal point and distortion coefficients. They can be estimated either offline using a calibration pattern (while the camera is on Earth) using methods like [4, 5], either online using multiple frames containing objects of known geometry. For example, the ISS itself (its feature points) can be used for the camera calibration if at least 8 points with known 3D- and image coordinates are presented to an algorithm like [6, 7].

3.2 Target Model

The target (ISS) model keeps the information of the target feature points: the 3D-coordinates of the feature points relative to the target coordinate system and their descriptors used for detection. The points are combined to point sets according to the distance between the spacecrafts, lighting conditions, active docking module etc. The official detailed ISS 3D-model is used as a source for the precise points coordinates.

3.3   Analog Signal Distortion and Camera Model Correction

The problem with the current docking television system is that it is completely analog, and the signal suffers from the corresponding distortions. The Fig. 4 shows the different images from the same rendezvous recording.

The most often distortion leads to a shift of the camera image frame (the center of the image and its borders) inside of the captured frame as shown on the top right picture. To cope with this problem a method is proposed for detection of the vertical and horizontal strokes, which are included in the image right in the camera and so distorted together with the camera image. The detected strokes coordinates are then used to correct the camera model (the principal point and the field of view) for each new frame.

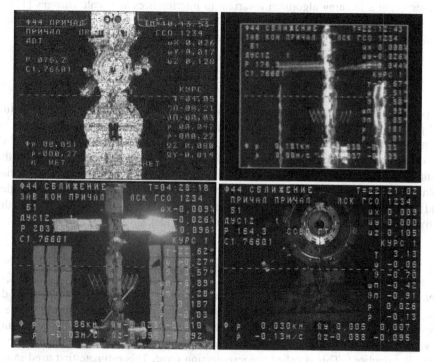

Fig. 4. Analog signal distortions: noise, overscan, blur and normal underexposed picture

3.4   Global and Local Image Normalization

The illumination of the observed objects changes drastically at the rendezvous stage mostly because of changes of the positions of the spacecrafts relative to the Sun. The global preprocessing normalizes the images so that the influence of illumination change is minimized.

3.5   Parallel Detection and Tracking

First, a corresponding set of feature points is selected from the target model according to the observed situation (that could include the target docking module, the orientation of the spacecrafts, Earth and Sun visibility, current distance etc.). The selection can be done using the priori information for the distance (for example it may be given by the operator) or using our own distance and situation estimate.

Then, the detection algorithm searches the features by applying a sliding window to the image, extracting Statistical Effective Multi-scale Block Local Binary Patterns (SEMB-LBP) [8] as descriptors and using a pre-learned cascaded classifier to get a class of each position in the image [9–11]. The descriptors are mostly chosen because of their calculation speed which is much faster than for more robust descriptors like SIFT.

At the same time, the points from the previous frame are tracked across the new frames using a learning algorithm invariant to interferences and scale changes [12].

The newly detected points have priority over the tracked ones because the tracking tends to slowly drift (accumulate error) as a result of its online learning process.

3.6   The Relative Pose Estimation

Depending on the number of points found by the detection and tracking, the full 3D-position (PnP-solution) or a partial solution can be estimated. If the number of points is greater than three, a combination of PnP-solvers [2, 3] is used to calculate all six coordinates of the pose. The combination includes a RANSAC-based solver for the initial pose estimation (also filtering outliers) and an iterative solution for a precise solution. If the number of points equals three or two, only the distance, active angles and the roll are estimated (supposing passive angles equal zero). For one point, only the active angles can be calculated using the camera model.

3.7   Self-Evaluation of the Solutions Based on Reprojection Error

The reprojection module calculates the reprojection error for each new frame. To do this the position estimation together with the camera model are used to project the a priori 3D-coordinates of the features to the image coordinates. Then the maximum deviation of the real (tracked or detected) image point from the corresponding projected point is calculated. This is called the reprojection error. This estimate first used in the RANSAC-based PnP solver to select the best group of points and finally to check the overall solution. If the reprojection error is too big, the whole solution for the current frame may be declared wrong.

3.8   Distance-Based Sets of Points Switching

Figure 5 demonstrates images from the "Progress" docking camera received at different distances to the ISS. As can be seen, there is no single set of points that can be used for the whole docking process. This means that the algorithm should be able to switch

Fig. 5. Images of ISS from different distances: 200, 60, 20 and 10 meters

between different sets of points while the distance reduces (or changes in general case) seamlessly. This is done by assigning of different point sets to the different distance ranges and selecting of the set according to the estimated distance. At the docking of "Progress" to ISS the typical ranges are:

1. From 200 to 100 m – the overall ISS body is used;
2. From 100 to 35 m – the active docking module is used together with its surroundings;
3. From 35 to 10 m – the docking hatch is used as one point together with the surroundings. The docking target can also be used as a single point;
4. From 10 to contact – the docking target is used as five points.

4   System Implementation

The structure of the software implementation mostly repeats the algorithm and adds the following software-specific modules:

1. The frames capture module receives the frames from the "Progress" television system. It can also receive frames from the Mission Control Center (MCC) television system (the "Progress" signal transmitted to MCC via ISS in near-real time)

or from the video files containing recordings of the previous rendezvous. This is very helpful for the evaluation purposes.

2. The camera internal calibration module keeps the pre-calibrated camera parameters.
3. The global preprocessing module normalizes the images, detects and corrects the analog video signal distortions by correction of the camera internal calibration parameters.
4. The target model keeps the information of the target feature points: the coordinates of the points and their descriptions used for detection.
5. The detection and tracking module implements the most important methods of the concurrent detection and tracking of the target visible features. It returns the current image coordinates of the feature points.
6. The relative position calculation module solves the final problem of estimation of the coordinates using the image and model coordinates of the feature points. The completeness of the solution depends on the number of given feature points. The position is estimated using several methods, and the best solution is selected using the reprojection error measure.
7. The "Course" coordinates calculation module converts the docking coordinates to the coordinate systems of antennas of the Russian docking control system "Course". This is useful for the evaluation process in case the corresponding real "Course" measurements are provided.
8. The user (cosmonaut) interface module displays:
   (a) the realtime image from the TV-system (or from video file);
   (b) the sets of feature points and the active set;
   (c) the calculation results including the estimated coordinates and the accuracy of the estimate (measure based on the reprojection error);
   (d) information about selected and manually reference points;
   (e) selection of the feature point sets;
   (f) control buttons and other elements.
9. The software also allows for manual points referencing (point and/or enter data). If the model (3D-) coordinates are given then these points can be used for position estimation in the same way as the automatically detected. For example, two manually specified points allows estimation of the distance, roll and active angles of the target. For each manually specified point, its active angles can always be estimated. The manually and automatically specified points can be mixed to get a complete or partial solution. The manually specified points are tracked continuously across the new frames using the tracker module.

5 Evaluation

The system efficiency have been tested against models, mockups, recordings of previous rendezvous of "Progress", "Soyuz" and ATV spacecrafts and finally at the Mission Control Center. First, the main modules of the algorithm have been evaluated separately. The corresponding detailed results are presented in [13].

5.1 Modules Evaluation

The detection and tracking module has been tested on 3D-models of ISS, on mockups and finally on the video recordings. It shown the possibility of pixel coordinates estimation with the mean square error of about 2 pixels in most cases. The error is calculated as a difference between the estimated pixel coordinates and the ground truth, which is known from the models or given by an expert in case of a real camera evaluation. The points with bigger displacements are mostly rejected by the self-evaluation algorithms described in 3.7. The method demonstrated high detection rate and low errors even with the presence of the analog distortions and the textual information (Fig. 4).

The 3D-pose estimation module has been evaluated on the same data. Table 1 provides an example of the coordinates estimation error at different mean pixel errors. The experimental point set shown on the Fig. 6 has a small distance between the points. This resulted in large relative errors in passive and roll angles.

The passive angles appear to be very sensitive to the pixel error (which is not surprising). Therefore, a real TV-system will often fail to determine them precisely. It is one of the most important drawbacks of the proposed single camera system. The drawback may be eliminated by using so called "mutual measurements" with cameras installed on both docking spacecrafts and pointed at each other.

Table 1. Dependency of the estimation error (Δ) on the mean pixel error

Pixel error, pix.	$\Delta\beta^x,^\circ$	$\Delta\beta^y,^\circ$	$\Delta d,\%$	$\Delta\alpha^x,^\circ$	$\Delta\alpha^y,^\circ$	$\Delta\theta,^\circ$
0	0,00	0,00	0,00	0,00	0,00	0,00
1	0,06	0,05	2,29	1,65	1,73	1,29
2	**0,10**	**0,09**	**4,23**	**2,90**	**2,60**	**2,43**
3	0,15	0,21	6,20	4,81	4,24	3,60

Fig. 6. An experimental set of model points (image fragment, distance 200 meters)

5.2 Evaluation Using Recordings of Rendezvous and the "Course" Telemetry

The complete system performance has been checked against the real rendezvous video recordings from the "Progress", "Soyuz" and ATV docking control cameras. Examples are shown on Figs. 4 and 5.

The recordings are split into training and testing subsets. The training set is used for the detector learning and the feature points selection. The testing videos imitate the real rendezvous.

First, the detection and tracking stability has been evaluated. At this stage, many improvements have been proposed and implemented to cope the analog distortions and

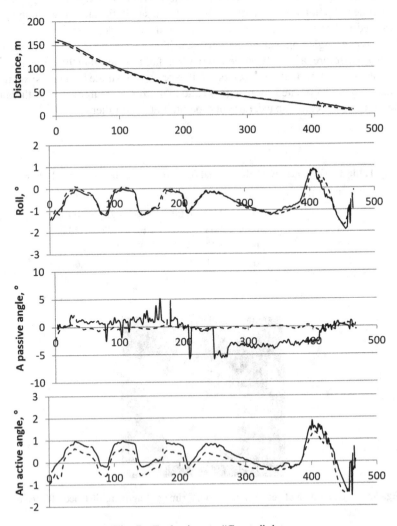

Fig. 7. Evaluation on "Course" data

lighting changes. They allowed passing most of the testing videos in automatic or semi-automatic mode.

The "Course" module compares the TV-based estimation of coordinates to the real telemetry. An example of such comparison is shown on Fig. 7. The values are provided without any filtering so the outliers are visible at some figures.

The "Course" evaluation confirmed that the passive angles are estimated with error up to several degrees. This error significantly reduces when the image coordinates of the points of the current point set are having big variance or when the set contains points with different Z (depth). The docking target allows precise passive angles calculation at the final docking stage (from about 10 m to contact). The distance, roll and the active angles are estimated with an appropriate precision in the whole rendezvous range.

The noticeable constant gap between the "Course" and the "TV" lines in some of the figures is because of an error recently found in the "Course" system thank to our system independent docking control.

5.3   Real-Time Experiments at the Mission Control Center

The evaluation at the Mission Control Center (MCC) has been provided using a copy of an ISS laptop (Lenovo T61) and a video signal translated from the "Progress M-26 M" ship in real-time during its docking to Zvezda module on 17 February 2015. The analogue (PAL) signal from the ship's KL-154 camera is translated to ISS where it's encoded to MPEG-2 stream and then forwarded to Earth (MCC).

The single-threaded prototype has shown the processing speed of about 1 frame per second. It was enough for stable features tracking and the position estimation on distances from 140 to 3.5 m (rendezvous, docking and coupling).

6   Conclusion

The proposed approach to the determination of the relative position of spacecrafts is based on detection and tracking of "natural" visual features. It can be applied to the most of the modern spacecrafts because they are already equipped with the television systems, and the 3D-models of the objects they dock (mostly ISS) are well known.

The system has been evaluated against models, makeups and recordings of previous rendezvous of "Progress", "Soyuz" and ATV spacecrafts (with the corresponding reference data received from the "Course" rendezvous system) demonstrating promising results.

The most important problems of the proposed approach are dependency on the visual conditions and the analog distortions and relatively high error of the passive angles estimation.

The first problem can only be solved using new digital television equipment and fusion of several systems based on different physical principles (for example by adding a LIDAR or structured lighting).

The second problem may be solved by using two cameras installed on both docking spacecrafts (i.e. the "Progress" camera and the ISS docking camera). In such case, the passive angles can be measured from the second side as active angles, i.e. with high precision. Both cameras exist already, and the so-called "mutual measurements" are to be tested at the flight experiments stage. One other way of increasing of the passive angles measurement is using feature points located along the ISS (with very different Z coordinates).

At the final stage (10 to 0 m), the docking target can be successfully used to calculate all the coordinates because it has a remote element (cross).

The nearest goal of the system is to become an independent docking control system helping the ground docking control team and the cosmonauts on the ISS. To reach this, the system has to pass real-time experiments in the Mission Control Center and then flight tests at the ISS using the existing equipment (cameras and laptops).

References

1. Bakhshiev, A.V., Stepanov, D.N., et al.: Proposals development for the design and technical implementation of the determining parameters relative motion system based by video processing. Technical report. Saint-Petersburg, Russia, RTC, 143 (2010)
2. Moreno-Noguer, F., Lepetit, Y., Fua, P.: Accurate noniterative o(n) solution to the PnP problem. In: de Janeiro, R. (ed.) Proceeding of the IEEE International Conference on Computer Vision, Brazil, October 2007
3. Gao, X.S., Hou, X.R.: Complete solution classification for the perspective-three-point problem. IEEE Trans. Pattern Anal. Mach. Intell. 25, 930–934 (2003)
4. Zhang, Z.: A flexible new technique for camera calibration. IEEE Trans. Pattern Anal. Mach. Intell. 22(11), 1330–1334 (2000)
5. Bouguet, J.Y.: MATLAB calibration tool. http://www.vision.caltech.edu/bouguetj/calib_doc
6. Tsai, R.Y.: A versatile camera calibration technique for high accuracy 3D machine vision metrology using off-the-shelf TV cameras and lenses. IEEE J. Robot. Autom. RA–3(4), 323–344 (1987)
7. Horn, B.K.P.: Tsai's Camera Calibration method Revisited. MIT Press, McGraw-Hill, Cambridge, New York (2000)
8. Liao, S., Zhu, X., Lei, Z., Zhang, L., Li, S.Z.: Learning multi-scale block local binary patterns for face recognition. In: Lee, S.-W., Li, S.Z. (eds.) ICB 2007. LNCS, vol. 4642, pp. 828–837. Springer, Heidelberg (2007)
9. Viola, P., Jones, M.J.: Rapid object detection using a boosted cascade of simple features. Comput. Vis. Pattern Recogn. 1, 511–518 (2001)
10. Freund, Y., Schapire, R.E.: A short introduction to boosting. In: International Joint Conference on Artificial Intelligence, vol. 2, pp. 1401–1406 (1999)
11. Viola, P., Jones, M.J.: Robust real-time face detection. Int. J. Comput. Vis. 57, 137–154 (2004)
12. Kalal, Z., Mikolajczyk, K., Matas, J.: Tracking-learning-detection. IEEE Trans. Pattern Anal. Mach. Intell. 34, 1409–1422 (2012)
13. Bakhshiev, A.V., Korban, P.A., Kirpan, N.A.: Software package for determining the spatial orientation of objects by tv picture in the problem space docking. Robot. Tech. Cybern., Saint-Petersburg, Russ., RTC 1, 71–75 (2013)

Implementation of Agile Concepts in Recommender Systems for Data Processing and Analyses

Alexander Vodyaho and Nataly Zhukova[✉]

Department of Information Systems,
Saint-Petersburg State Electrotechnical University, Saint Petersburg, Russia
nazhukova@mail.ru

Abstract. Recommender systems have recently become an essential part of the majority of modern information systems. In the paper recommender systems oriented on supporting data and information processing and analyses are considered. The systems are aimed to give recommendations to end users on selection and usage of data processing methods and algorithms. Both commonly used and newly developed algorithms are taken into account by the systems. The algorithms have diverse technological and program implementation. Agile features of the recommendation systems allow continuously modify and enlarge the set of the used methods and algorithms. The key advantage of the systems is possibility to test new algorithms on real data processing tasks. An example of recommender system for binary data streams processing is described.

Keywords: Data processing and analyses · Agile concept · Recommender systems

1 Introduction

Nowadays Recommender Systems (RS) become an effective tool for solving the information overload problem [1–5]. Personalized recommendations become an important part of many on-line e-commerce applications like Amazon.com, Netflix and Pandora. Modern RS are usually personalized, different users or user groups receive diverse suggestions. They try to predict the most suitable products or services based on the user's preferences and constraints. RS collect preferences from users, which are either explicitly expressed or are inferred by interpreting user actions [4]. Now RS are treated as efficient tools for business for increasing revenue and profitability of the product portfolio.

Data processing and analyses is a subject domain that now strongly requires techniques and tools for making recommendations, because a considerable part of solved applied problems is based on gathering and processing data from technical and natural objects or complexes of objects [6, 7]. Very often amount of gathered data is huge, data are heterogeneous and can have bad quality. At the same time period of data deterioration is very short. Commonly used algorithms cannot deal with continuously incoming streams of operational data and provide solutions that meet requirements of

© Springer International Publishing Switzerland 2015
M.Y. Khachay et al. (Eds.): AIST 2015, CCIS 542, pp. 443–457, 2015.
DOI: 10.1007/978-3-319-26123-2_42

end users to precision of formed results or to computational time [8]. It forces the scientific community to develop hundreds of different methods and algorithms for data processing and analyses. The algorithms are highly specialized, sensitive to input data and have restrictions on the fields of their application. They require defining up to dozens of configuration parameters to adapt the algorithms to the contexts of their application. Notion of data processing and analyses context is considered in [9]. As a rule operators that solve applied problems in different subject domains are not specialists in the area of data processing and analyses and they are not able to select appropriate algorithms and configure them. In order to investigate new algorithms using real data for various subject domains, implement them in data processing and analyses systems (DPAS) and train operators require time and financial resources. RS for data processing and analyses (DPA RS) are expected to make suggestions for selecting and configuring algorithms using available information and knowledge about the subject domain of data processing and applied subject domains.

The solutions developed for commerce oriented systems cannot be directly used for data and information processing because they are based on information about users, items and transactions. In order to process data it is necessary to consider information that defines context of data processing, in particular, information about solved problems, characteristics of analyzed objects, features of input data, and parameters of the data processing systems.

For building effective data processing and analyses RS it is proposed to use agile concepts at the majority of the stages of systems life cycle, in particular, the stages of systems design, development and support. The created systems assume to be knowledge based systems with agile architecture. Support of agile features in RS allows add, use and estimate new methods and algorithms with different implementations without modifying program code. The RS are developed as separate IS that are integrated into DPAS. DPAS provide real contexts for algorithms application and means and tools for making decisions.

2 Agile Concepts for DPAS and RS

Agile concept was introduced in 2001 for software development in the form of a manifesto [10]. The concept is operating on simple notion to provide prototype environment for end users interested in learning, experimenting and adopting new solutions. Environment gives opportunity to apply working solutions to solve real problems. It can be configured and modified for exact needs of a user in short time and requires few resources for development and support.

Modern information systems (IS) for data processing and analyses widely use the agile concept, for example, agile approach is applied for IS design, knowledge based architectural descriptions defines the architecture of the systems, agile features of the systems allow build, execute and modify business processes, configure methods and algorithms in dynamics [11, 12].

DPAS mostly refer to industrial highly reliable systems. As an example of highly reliable DPAS for processing telemetric information from space objects can be considered [13]. Telemetry is data transmitted at regular intervals from sensors [14]. In the systems it is admissible to use only well designed and well implemented program components. New algorithms that are developed by the scientific community and have prototype implementation cannot be integrated into DPAS.

RS developed as separate systems don't influence reliability of DPAS and can contain methods and algorithms implemented using different technologies and tools. The algorithms are supposed to be used in test purposes. A selected subset of the efficient algorithms can be implemented according to industrial standards and included in DPAS.

In Fig. 1 support of agile features at different stages of the IS life cycles is shown. The stages at which the agile concept is implemented are dashed. The arrows in Fig. 1 show providers and consumers of the solution that allow support agile features of the systems.

Agile features are supported at four levels. The levels reflect needs of different classes of IS, in particular corporative IS, specialized industrial IS and program packages applicable only for research purposes. IS use and provide methodological solutions for systems agile development produced in the sphere of information technologies (IT). Prototype systems created by a scientific community implement the principles of agile concept at all stages of their life cycle. For DPAS specialized solutions for agile features support have been developed for the stages of systems design, execution and support [9, 11]. Implemented solutions are based on using IT agile methodologies and suggestions for new algorithms made by DPA RS. At the stage of program components and modules implementation in DPAS agility is not supported. To support agility of DPA RS technological solutions provided by DPAS and implementation of methods and algorithms developed for scientific prototype systems are used.

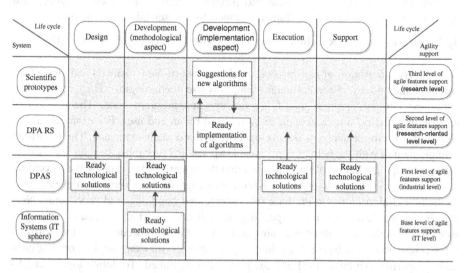

Life cycle / System	Design	Development (methodological aspect)	Development (implementation aspect)	Execution	Support	Life cycle / Agility support
Scientific prototypes			Suggestions for new algorithms			Third level of agile features support (research level)
DPA RS			Ready implementation of algorithms			Second level of agile features support (research-oriented level level)
DPAS	Ready technological solutions	Ready technological solutions		Ready technological solutions	Ready technological solutions	First level of agile features support (industrial level)
Information Systems (IT sphere)		Ready methodological solutions				Base level of agile features support (IT level)

Fig. 1. Support of agile features at the stages of IS life cycles

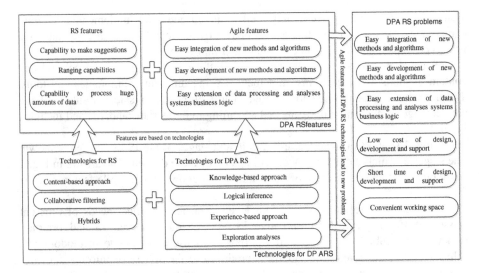

Fig. 2. Features, technologies and problems of DPA RS

3   Common and Agile Features of DPA RS

RS for data processing and analyses are expected to support a set of agile features that allow end users test, evaluate and adapt new algorithms for their need in different subject domains. The set of the features is limited by the restriction on the cost of their implementation. In order to minimize expenses they have to be implemented and supported using ready technological and program solutions provided by DPAS and open source methods and algorithms developed by the scientific community.

Agile features of DPA RS. Below the list of agile features of RS is given (Fig. 2).

(a) *Easy integration of new methods and algorithms.* New methods and algorithms are implemented by the researchers using diverse technologies. They have to be integrated into RS without modifying the systems program code. The process of integration must be simple enough to be realized by an end user. For example, a new algorithm for cluster analyses is developed by applied mathematicians. The algorithm is implemented with the help of R-package [15]. The RS user must have an opportunity to import implementation of the algorithm into the system and start using the algorithm.

(b) *Easy development of new methods and algorithms.* Users can develop new methods and algorithms on the base of implemented computational procedures. New algorithms are developed using special graphical editors which represent all implemented methods, algorithms and simple computational procedures in the form of nodes. New nodes can be created using script languages, for example, Python. For each node properties can be defined and the nodes can be linked. To define logical conditions and loops supplementary nodes are provided. Debugging of algorithms and modifying of earlier developed algorithms is also supported.

(c) *Easy extension of data processing and analyses business logic.* Separate algorithms integrated into RS are executed in the context of business processes defined in DPAS. New approaches to data processing and analyses can be created by a user on the base of combination of existing and new methods and algorithms. Implementation of the approaches assumes definition and execution of new business processes that extend the logic of DPA RS. For example, a new approach for situations assessment proposed in [16] is based on using data mining methods and algorithms adapted for objects parameters measurements processing and analyses.

Technological Solutions for DPA RS. RSs commonly implement content based approach or collaborative filtering algorithms for making suggestions [17]. The set of the technological solutions used in RS for data processing and analyses is extended with a number of additional technologies supported in DPAS.

(a) *Support of agile business logic.* Agile logic is used to select the algorithms for solving the tasks of data processing and analyses using knowledge based descriptions of the algorithms and tasks. The descriptions allow estimate applicability of the algorithms in the defined contexts. Estimations are made according to the defined field of application, restrictions and specialized features of the algorithms. Additional knowledge about the algorithms, for example, its complexity or approaches for a priori estimation of the formed results are also considered.

(b) *Logical inference.* Recommendations of experts are represented in the form of logical rules interpretable by an inference machine. The recommendations usually refer to the conditions of algorithm application, in particular, features of input data, and to approaches to results evaluation and interpretation.

(c) *Usage of experience of the algorithms application.* Both positive and negative cases of algorithm application are formalized and used as precedents. Positive cases can be considered separately or generalized and represented in a form of best practices. Information about history of the algorithms application by all users is analyzed, formally described and integrated. On the base of gained experience logical rules are defined.

(d) *Usage of the exploration approach.* The approach assumes that a RS attempts to apply various algorithms to solve the task in view. The results of the successful attempts are estimated, compared and ranged. The systems make suggestions to users according to the list of the ranged algorithms.

Key Problems of Agile Features Support in DPA RS. To support agile features in a RS it is necessary to solve a set of problems.

(a) *The algorithms must have knowledge based formalized descriptions.* For making recommendations RS need knowledge about the available methods and algorithms. End users have to provide desired knowledge about all methods and algorithms. To describe the algorithms a specialized tool can be used. The tool is required to support simple human readable descriptions of algorithms and allow generate formal descriptions interpretable by the systems. One can use ontological models for data, information and knowledge representation adapted to the subject domain of data processing and analyses and ontology editors for the models creating and manipulating.

(b) *RS have to inherit business logic and program components of DPAS.* Integration with RS assumes that the system use program components provided by DPAS and support all business processes defined in data processing and analyses system. RS use information model, inference machine, mathematical and modeling libraries, business process engine, graphical tools for describing ontologies and business processes. RS have to be designed and implemented as separate systems that don't influence the state of DPAS.

(c) *RS must have flexible structure.* RS must satisfy actual needs of end users at all steps of data processing and analyses. It requires continuous modification of the recommender systems. Flexile features of RS can be provided by designing systems with agile architecture.

(d) *RS must meet common requirements of research-oriented systems.* To be demanded by the consumers RS must be created in short time, with low financial expenses and require few resources for support. The systems must have user friendly graphical interface and be convenient for end users work.

The problems of the agile features support in RS can be overcome through defining the information model for RS and integrating RS with DPAS.

4 DPA RS Information Model

Information model in DPA RS is aimed to store and provide data, information and knowledge (DIK) required for using and evaluating new methods and algorithms for testing purposes. The model is defined on the base of the information model developed for DPAS and inherits its structure and its features. The key features of RS information model are model relevance due to continuous updating, compliance to standards of DIK representation, human readable form, interpretability by a virtual machine.

The DPAS model consists of static (SIM) and dynamic (DIM) information models. The SIM provides DIK that are rarely changed, for example the descriptions of the objects constant characteristics and structure. Actual data about the entities described in SIM is stored in dynamic information model (Fig. 3), for example, the objects current positions, speeds and accelerations. DIM also contains operational data and information about the environment. In the Fig. 3 models that contain actual data are marked with dashed lines.

DPAS information model contains two types of DIK. To the first type of DIK refers data, information and knowledge about the subject domain of data processing and analyses. It includes various classifiers of initial and processed DIK, descriptions of technologies, processes, methods, algorithms and procedures for data processing and analyses. DIK of the second type include data, information and knowledge about the applied subject domains, including DIK about the initial data, results of data processing, solved tasks, end users, environment, analyzed objects, possible situations and their features and etc. Environmental data is provided to the systems by data centers in the form of initial measurements, for example, time series' of values, or results of measurements processing, for example, regular grids [12]. Detailed description of the DPAS information model can be found in [18].

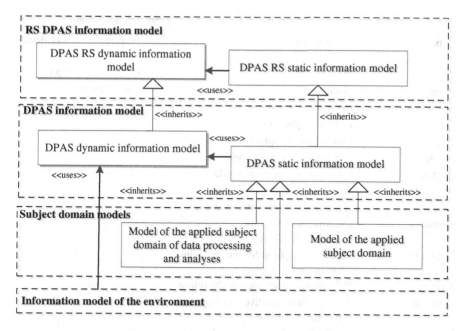

Fig. 3. DPA RS and DPAS systems models

The information model of DPA RS extends the dynamic and static models of DPAS. The static model of DPA RS contains additional information about new methods and algorithms. It also stores the results of DIK processing formed in RS. In separate cases the model contains additional information about applied subject domains and/or environment required in RS. For describing DIK similar formats are used in DPAS and DPA RS. For example, the methods and algorithms are described by the set of properties presented given in Table 1 [11]. The properties can be defined manually (M) and/or automatically (A).

The links between the entity "algorithm" and other entities of the subject domains are described with a set of relations. The relations are defined in the format:

<relation id> [<relation description>] (object of the relation, subject of the relation).

An example of relations:

usage [business processes use algorithms] (processes, algorithm);

tasks [tasks can be solved using] (tasks, algorithm);

domain [algorithm is applicable in subject domains] (algorithm, subject domain);

type [algorithm refer to the type according to the algorithms types classifier] (algorithm, algorithms types [18]).

DIK in DPAS and DPA RS are represented in a form of a system of interconnected OWL ontologies, sets of data bases and knowledge bases. The ontologies of the DPAS and DPA RS are two separate ontologies with similar structure and partly different content. The DPA RS ontology is formed as a clone of the DPAS ontology and extended with additional DIK. Existing DPA RS ontologies can be updated by merging with the DPAS ontologies. Merge operations require few resources due to similar

Table 1. Format of the methods and algorithms properties description

Property id	Property name	Property type	Defined M/A
Id	Identifier	OWL description	M
Name	Name		
Description	General description		
Application	Field of application		
Conditions	Conditions of application		
Restrictions	Restrictions on application		
Input data	Requirements to the input data		
Output data	Description of the output data		
Settings	Description of the setting parameters		
Estimation	Description of the procedures for the results estimation		
Complexity	Resources required for the algorithm execution		
Analogs	Information about similar algorithms		
Experience	Description of positive and negative precedents and best practices	Links	A
Implementation	Available implementations of the algorithm		
Recommendations	Recommendations for the algorithm application	Logic rules	M/A

structure of the ontologies. Conflicts of ontologies merging are solved by the user manually or DPAS ontologies are considered as master one. The settings of the merging operations allow define the period of ontologies update. Diverse settings can be defined for different ontologies. Commonly ontologies that contain operational DIK are updated continuously and ontologies with static DIK are updated by users' requests. Data bases store operational and historical data referred to the applied subject domains. DPA RS are not allowed to modify data used by DPAS. Additional data required by DPA RS is stored in separate tables or separate databases. Knowledge bases contain knowledge represented in the form of logic rules. The rules formed and used by DPA RS are considered as supplementary and are not used by DPAS. The rules can be placed in a separate knowledge base or stored in the knowledge base of DPAS with specialized tags.

5   DPA RS Architecture

The structure of a DPAS and DPA RS is given in Fig. 4. The DPA RS architecture is defined by the architecture of DPAS [19] and includes components provided by DPAS and contains also a number of specialized components.

Typical DPAS is service oriented knowledge based system which includes 3 layers: DIK layer, front end layer and back end layer.

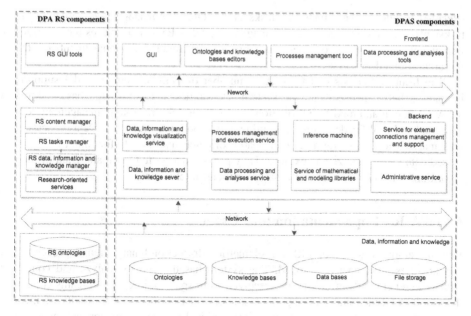

Fig. 4. DPA RS and DPAS architecture

DIK layer contains ontologies, knowledge bases, data bases and file storage which contains historical data. DPA RS systems at this layer contain RS ontologies and RS knowledge bases with information and knowledge of the recommender systems.

The back end provides a number of infrastructure services and services specialized for data processing and analyses. Processes management and execution service, inference machine and administrative service refer to infrastructure services. Processes management and execution service provide access to process manager of DPAS. Data, information and knowledge server managers, stores and provides access to DIM. The server implements the information model of DPAS. Service for external connections management and support is aimed to establish connections and exchange data, information and knowledge with data centers and external systems. Data processing and analyses service and service of mathematical and modeling libraries allow solve end users applied problems. At the back end content manager, tasks manager, DIM manager and set of research-oriented services refer to DPA RS components. Content manager is responsible for adding, modifying and managing the set of the methods and algorithms for data processing and analyses. Methods and algorithms are integrated into mathematical and modeling libraries. Data, information and knowledge manager solves the tasks of DIM management, modification and support. It verifies consistency of DIM, solves the tasks of RS DPA ontologies clone, merge and update. The operations are executed by the DIK server. Task manager provides access to specialized research-oriented services of DPA RS and DPAS services and graphical tools. The specialized services are new experimental services for data processing and analyses and services for mining the historical data of the algorithms usage. The task manager is responsible for solving non-functional tasks that guarantee the security of the processes executed in DRAS. The functional tasks of the DPA RS manager are similar to the tasks solved by the DPAS process manager. The DPA RS manager uses the solutions implemented in DPAS or their modifications adapted for management of the research-oriented processes.

The front end consists of GUI and a number of tools and editors. GUI supports interaction between users and the system. Ontologies and knowledge bases editors allow view, edit and modify information model of DPAS. Processing management tool is a graphical tool for designing, executing and debugging processes. Data processing and analyses tools include a set of research oriented instruments for initial and processed data view, analyze and explore. They provide access to OLAP and Data Mining algorithms and other means and tools of artificial intelligence.

6 DPA RS Implementation and Usage

Recently among others [20] a RS for telemetric information binary streams processing and analyses have been developed. The system has the aim to analyze and control structure and contents of the streams received from space objects. The streams contain the results of functional and signal parameters measurements. The results of measurements are represented in the form of time series. Different parameters are measured with different frequency depending on their physical nature. Frequency can vary from several Hz up to hundreds Hz. For example, frequency of measurements of temperature parameters is low and frequency of vibration measurements is high. All measurements are packed into binary steams and send to data centers. The streams consist of a set of separated channels. In order to avoid gaps in streams, measurements of several parameters can be placed in one channel (subcommutation of the parameters, Sub) or several channels can be used to locate measurements of one parameter (supercommutation of the parameters, Sup). Structure of the streams is defined by the number of channels (the length of card, L_f), size of the channels (the length of the words, L_w) and the scheme of the parameters subcommutation and supercommutation. Words contain information about the results of the parameters measurements are represented as sequences of bits. Cards are separated sequences of words. The example of the binary stream is given in Fig. 5. The lines and the columns of the presented matrix correspondingly reflect the cards and the channels of the stream. The bytes that equal 'zero' are show as black points and the bytes that equal 'one' – with white points.

There are cases when information about structure of the streams is not available or lost. In order to restore structure 3 groups of methods and algorithms has been developed. To the first group refer the methods and the algorithms that allow represent structure of the streams in a graphical form. The methods and the algorithms of the first group are developed within the field of cognitive graphics. The second group contains methods that have been commonly used for streams processing, in particular, methods of correlation analyses, application of differential operators. The third group is a group of intelligent methods and algorithms. They are strongly oriented on using knowledge and means and tools of artificial intelligence. The third group includes methods based on calculation of frequency rank distribution, methods for building and comparing

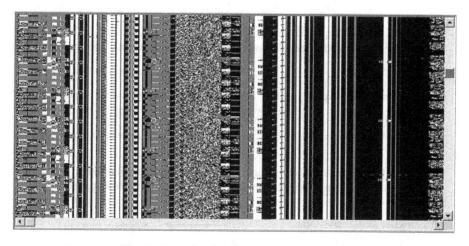

Fig. 5. Example of the binary steams structure

graphs. The ontology of methods for processing structures of binary streams is presented in [18].

Part of the methods and algorithms were implemented as separate tools that till now have been widely used by experts for streams structure analyses. They were developed using different programming languages, in particular, C ++, Delphi, Java, etc. New algorithms were organized in a form of a java library. Existing and new algorithms were integrated into the RS and tested by experts on eighty binary streams with different structure.

The procedure of streams structure analyses and evaluation consisted of a sequence of steps:

(i) structures of the binary streams were defined manually using graphical tools;

(ii) different algorithms implemented in RS were applied to define the structures;

(iii) results of the algorithms application were estimated by the experts, formally described and saved;

(iv) saved results were compared and ranged by the system. On the base of the results the general procedure for the algorithms application was defined (Fig. 6). In the similar way the steps of the presented procedure were described.

The algorithms for processing the streams and logic of the processing were implemented in several program complexes developed in [20]. In Fig. 7 several graphical forms of the complexes are given. The presented forms allow define structure of the streams (a), analyze information transferred using one channel (b), reveal structure of the stream (c), represent the structure of the stream in the form of a graph.

The program complexes were tested on sixty binary streams. Structure of the 57 % of the streams was restored using the algorithms defined by the system. 41 % of the structures required additional iteration for improving the results. The suggestions for the algorithms application at the second iteration were provided by the systems on the base of the estimations of the results of the first iteration. 2 % of the streams were not processed and required manual processing. The percentage of the correctly defined

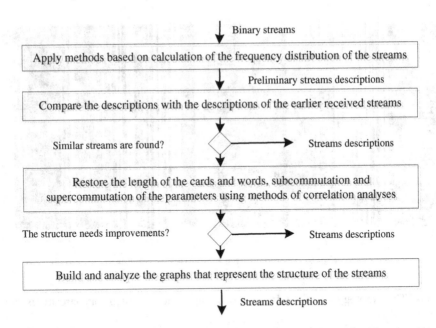

Fig. 6. General procedure for the binary streams processing and analyses

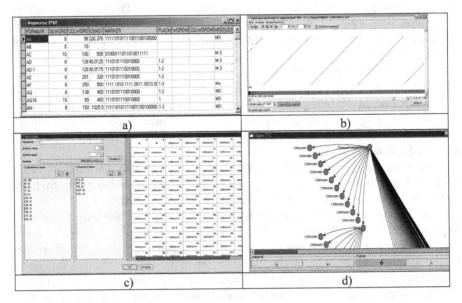

Fig. 7. GUI of program complexes for processing stream structure

elements of the streams structures is presented in Table 2. The presented results are average results of 60 streams processing.

Table 2. Format of the methods and algorithms properties description

Parameters of the stream	L_f	L_w	Sub	Sup
$|Sub| > 60\%$	98	80	91,3	90,1
$|Sup| > 60\%$	92	40	90,2	92,4
$DP = 5\%, SVP = 35\%, FVP = 40\%, FPP = 10\%, OP = 10\%$	98	78	93	89,5
$DP = 5\%, SVP = 35\%, FVP = 10\%, FPP = 40\%, OP = 10\%$	80	20	72,3	71,7

The set of the analyzed streams contained streams with different structures. Structure of the streams is characterized with the number of subcommutations and supercommutations, the number of the parameters of different types. Parameters of five types were considered: discrete parameters (DP), parameters with slowly changing values (SVP), parameters with fast changing values (FVP), parameters that have values represented as numbers with floating points (FPP) and other parameters (OP).

7 Conclusion

In the paper RS for solving data processing and analyses tasks are considered. The systems allow select and make suggestions for using new methods and algorithms developed by the scientific community for solving problems in the applied subject domains. The set of the selected algorithms are supposed to be integrated into DPASs. The algorithms are accomplished with extended semantic descriptions that allow apply the algorithms by the end users of DPAS that are not experts in the subject domain of data processing and analyses. The DPA RS are to be used by highly skilled experts that test and evaluate new algorithms.

The DPA RS are designed as agile knowledge based systems. Agile features of the systems allow integrate new methods and algorithms implemented using different technologies without modification of the program code, provide access to the extended set of means and tools for data processing and analyses, provide environment for new algorithms testing and evaluation. The algorithms can be used in processes created for solving applied problems of different subject domains. It allows test the algorithms on real data and in real conditions, significantly reduce amount of time and financial resources required for the RS design, development, implementation and support due to usage of commercial off-the-shelf (COTS) technological and program solutions.

The DPA RS is developed as a separate system that is integrated into DPAS. The information model of the RS is defined as a separate model that inherits the model of DPAS.

A RS for data processing and analyses was implemented for the subject domain of complex dynamic objects analyses. The system was used to select the algorithms for processing binary data streams of telemetric information received from space objects. Application of the system allowed reduce the number of the algorithms that have been earlier considered for solving the tasks of binary streams processing in two times.

References

1. Resnick, P., Varian, H.: Recommender systems. Commun. ACM **40**(3), 56–58 (1997)
2. Mahmood, T., Ricci, F.: Improving recommender systems with adaptive conversational strategies. In: Proceedings of the 20th ACM Conference on Hypertext and Hypermedia, pp. 73–82. ACM, New York (2009)
3. Burke, R.: Hybrid web recommender systems. In: Brusilovsky, P., Kobsa, A., Nejdl, W. (eds.) Adaptive Web 2007. LNCS, vol. 4321, pp. 377–408. Springer, Heidelberg (2007)
4. Ricci, F., Rokach, L., Shapira, B., Kantor, P. (eds.): Recommender Systems Handbook, 1st edn. Springer, Heidelberg (2011)
5. Costa, H., Macedo, L.: Emotion-based recommender system for overcoming the problem of information overload. In: Corchado, J.M., Bajo, J., et al. (eds.) PAAMS 2013. CCIS, vol. 365, pp. 178–189. Springer, Heidelberg (2013)
6. Pankin, A., Zhukova, N.: Data Representation Dynamic Model for Distributed Urban IGIS. In: Proceedings of 18th International Conference on Urban Planning and Regional Development in the Information Society, Italy, Rome, 20–23 May, 2013, pp. 1139–1145. CORP, Schwechat (2013)
7. Ignatov, D., Smirnova, O., Zhukova, N.: Dynamic information model for oceanographic data representation. In: Proceedings of the Workshop «Modeling States, Events, Processes and Scenarios» co-located with 20th International Conference on Conceptual Structures, India, Mumbai, 10–12 Jan 2013, pp. 82–97. Higher School of Economics Publishing House (2013)
8. Smirnova, O., Zhukova, N.: Smart navigation for modern cities. In: Proceedings of 19th International Conference on Urban Planning and Regional Development in the Information Society, Austria, Vienna, pp. 593–602. CORP, Vienna, 21–23 May, 2014
9. Vodyaho, A., Zhukova, N.: Implementation of JDL Model for multidimensional measurements processing in the environment of intelligent GIS. Int. J. Conceptual Struct. Smart Appl. **2**(1), 36–56 (2014)
10. Manifesto for Agile Software Development. http://agilemanifesto.org/
11. Vitol, A., Pankin, A., Zhukova, N.: Adaptive multidimensional measurements processing using IGIS technologies. In: Proceedings of the 6th International Workshop on Information Fusion and Geographic Information Systems: Environmental and Urban Challenges. SPb., 12–15 May, 2013. Springer, Heidelberg, pp. 179– 200 (2014)
12. Smirnova, O., Zhukova, N.: Atmosphere and ocean data processing in decision making support system for Arctic exploration. In: Proceedings of the 6th International Workshop on Information Fusion and Geographic Information Systems: Environmental and Urban Challenges, SPb., 12–15 May, 2013. Springer, Heidelberg, pp. 305–324 (2014)
13. Frank, C., Russell, P, Henry, R.: Telemetry Systems Engineering, p. 632. Artech House (2002)
14. Glossary of Software Engineering Laboratory terms. Software engineering laboratory series. SEL-82-105 (1995). http://ntrs.nasa.gov/archive/nasa/casi.ntrs.nasa.gov/19840015067.pdf
15. The R Project for Statistical Computing. http://www.r-project.org/
16. Pankin, A., Vodyaho, A., Zhukova, N.: Situation assessment using results of objects parameters measurements analyses in IGIS. In: Proceedings of the Workshop «Formal Concept Analysis Meets Information Retrieval» co-located with the 35th European Conference on Information Retrieval, Moscow, 24 March, 2013, pp. 134–148. National Research University Higher School of Economics (2013)
17. Segaran, T.: Programming Collective Intelligence. Building Smart Web 2.0 Applications, p. 362. O'Reilly Media, USA (2007)

18. Vodyaho, A., Zhukova, N.: System of ontologies for data processing applications based on implementation of data mining techniques. Supplementary Proceedings of the 3rd International Conference on Analysis of Images, Social Networks and Texts, Yekaterinburg, 10–12, April, 2014, pp. 107–120. Higher School of Economics Publishing House (2013)
19. Vodyaho, A., Zhukova, N.: Building smart applications for smart cities – IGIS -based architectural framework. In: Proceedings of 19th International Conference on Urban Planning and Regional Development in the Information Society, Austria, Vienna, 21–23 May, 2014, pp. 109–118. CORP, Vienna (2014)
20. St. Petersburg Electro-Technical University Research and Engineering Center. http://rec-etu.com

Author Index